# NATURAL
# WONDERS
## OF THE WORLD

# NATURAL
# WONDERS
## OF THE WORLD

FOREWORD BY CHRIS PACKHAM

# CONTENTS

## INTRODUCTION
## 08

10  EARTH'S STRUCTURE

12  PLATE TECTONICS

14  EARTH'S PAST

16  LIFE ON EARTH

## NORTH AMERICA
## 18

20  MOUNTAINS, LAKES, AND PRAIRIES

24  MOUNTAINS AND VOLCANOES

38  GLACIERS AND ICE-SHEETS

48  RIVERS AND LAKES

62  COASTS, ISLANDS, AND REEFS

66  FORESTS

72  GRASSLANDS AND TUNDRA

74  DESERTS

## CENTRAL & SOUTH AMERICA
## 82

84  ANDES TO AMAZON

88  MOUNTAINS AND VOLCANOES

100  GLACIERS AND ICE-SHEETS

104  RIVERS AND LAKES

114  COASTS, ISLANDS, AND REEFS

118  FORESTS

122  GRASSLANDS AND TUNDRA

124  DESERTS

## EUROPE
## 128

130  LAND OF RIVERS AND PLAINS

134  MOUNTAINS AND VOLCANOES

146  GLACIERS AND ICE-SHEETS

152  RIVERS AND LAKES

162  COASTS, ISLANDS, AND REEFS

172  FORESTS

175  GRASSLANDS AND TUNDRA

## AFRICA
## 176

178  LAND OF THE RIFT

182  MOUNTAINS AND VOLCANOES

190  RIVERS AND LAKES

198  COASTS, ISLANDS, AND REEFS

200  FORESTS

204  GRASSLANDS

208  DESERTS

**DK** | Penguin Random House

### DK LONDON

**Senior Editor** Peter Frances
**Project Editors** Gill Pitts, Miezan van Zyl
**Editors** Claire Gell, Frankie Piscitelli, Kaiya Shang
**Indexer** Elizabeth Wise
**Jacket Editor** Claire Gell
**Jacket Design Development Manager** Sophia MTT
**Managing Editor** Angeles Gavira Guerrero
**Associate Publisher** Liz Wheeler
**Publishing Director** Jonathan Metcalf

**Senior Art Editor** Ina Stradins
**Project Art Editors** Francis Wong, Steve Woosnam-Savage, Mik Gates
**Senior Graphics Co-ordinator** Sharon Spencer
**Picture Researcher** Liz Moore
**Senior Jacket Designer** Mark Cavanagh
**Pre-production Producer** Gillian Reid
**Senior Producer** Anna Vallarino
**Managing Art Editor** Michael Duffy
**Art Director** Karen Self
**Design Director** Phil Ormerod

**Illustrators** Adam Benton, Peter Bull, Dominic Clifford, Dynamo Ltd, Arran Lewis, Sofian Moumene, Michael Parkin
**Cartographer** Ed Merritt

### DK INDIA

**Senior Editor** Dharini Ganesh
**Editors** Riji Raju, Sonia Yooshing
**Assistant Editor** Priyanjali Narain
**Picture Researchers** Ashwin Raju Adimari, Surya Sankash Sarangi
**Jackets Editorial Coordinator** Priyanka Sharma
**Managing Editor** Rohan Sinha
**Managing Jackets Editor** Sreshtha Bhattacharya
**Pre-production Manager** Balwant Singh

**Senior Art Editor** Vaibhav Rastogi
**Art Editor** Debjyoti Mukherjee
**Assistant Art Editors** Yashashvi Choudhary, Simar Dhamija, Vaishali Kalra
**Managing Art Editor** Anjana Nair
**Picture Research Manager** Taiyaba Khatoon
**Jacket Designer** Dhirendra Singh
**Senior DTP Designer** Harish Aggarwal
**DTP Designers** Mohd Rizwan, Anita Yadav
**Production Manager** Pankaj Sharma

# ASIA
## 214

**216** TUNDRA TO TROPICS

**220** MOUNTAINS AND
VOLCANOES

**232** GLACIERS
AND ICE-SHEETS

**236** RIVERS AND LAKES

**250** COASTS, ISLANDS,
AND REEFS

**254** FORESTS

**260** GRASSLANDS
AND TUNDRA

**262** DESERTS

# AUSTRALIA & NEW ZEALAND
## 268

**270** AN ANCIENT LAND

**274** MOUNTAINS
AND VOLCANOES

**280** GLACIERS
AND ICE-SHEETS

**282** RIVERS AND LAKES

**284** COASTS, ISLANDS,
AND REEFS

**292** FORESTS

**294** GRASSLANDS

**295** DESERTS

# ANTARCTICA
## 300

**302** THE FROZEN
CONTINENT

**304** MOUNTAINS
AND VOLCANOES

**306** GLACIERS
AND ICE-SHEETS

**309** TUNDRA

# THE OCEANS
## 310

**312** THE MID-ATLANTIC
RIDGE

**314** THE SEYCHELLES

**315** THE GREAT
CHAGOS BANK

**315** THE MALDIVES

**316** THE HAWAIIAN
ISLANDS

**320** THE GALAPAGOS
ISLANDS

**322** THE EAST PACIFIC
RISE

**324** THE MARIANA
TRENCH

# EXTREME WEATHER
## 326

**328** CYCLONES

**330** THUNDERSTORMS

**332** TORNADOES

**334** SANDSTORMS AND
DUSTSTORMS

**336** ICE STORMS

**338** AURORAE


**342** DIRECTORY

**404** GLOSSARY

**424** INDEX

**438** ACKNOWLEDGMENTS

First published in Great Britain in 2017 by
Dorling Kindersley Limited, 80 Strand, London WC2R 0RL
Copyright © 2017 Dorling Kindersley Limited
A Penguin Random House Company

2 4 6 8 10 9 7 5 3 1

001-297088-October/2017

A CIP catalogue record for this book
is available from the British Library.
ISBN: 978-0-2412-7629-7
Printed in China

A WORLD OF IDEAS:
SEE ALL THERE IS TO KNOW
www.dk.com

# Contributors

**Jamie Ambrose** is an author, editor, and Fulbright scholar with a special interest in natural history. Her books
include DK's *Wildlife of the World*.

**Robert Dinwiddie** is a science writer who specializes in Earth science, the Universe, the history of science, and
general science. When not writing, he enjoys exploring the world's volcanoes, glaciers, coral reefs, and other
natural phenomena.

**John Farndon** has written extensively on Earth sciences and natural history. His books include *How the Earth
Works*, the acclaimed *The Oceans Atlas*, and *The Practical Encyclopaedia of Rocks and Minerals*. He has been
shortlisted a record six times for the Royal Society's Young People's Science Book Prize.

**Tim Harris** has been fascinated by all aspects of the natural world since a young age. He has studied glaciers in
Norway and has written books about natural history for children and adults, including the award-winning
*Migration Hotspots: The World's Best Bird Migration Sites*.

**David Summers** is a writer and editor with a training in natural history filmmaking. He has contributed to books
on a range of subjects including natural history, geography, and science.

**Half-title page** Tristan da Cunha Island, south Atlantic Ocean
**Title page** avalanche wall near Mount Everest, Nepal
**Foreword** Mount Wanaka Mountains, New Zealand

# FOREWORD

I dare you to turn a page and not want to be there. Standing aloft on a mountain peak, cresting a dune or peering through a mask into crystal-clear waters. Flying over forests, skimming swamps, or ducking vast waves, this is the ultimate destinations wish list for anyone who dreams of exploring planet Earth. But before you set off, you'd need to pack quite a kit, from snow boots to a snorkel, from crampons to a compass, because your tour would encompass every terrain, every altitude, every habitat on Earth. You'd be frozen, soaked, and sweating, testing your body to the limits.

Such is the stuff of fantasy, but this book reveals the staggering diversity and awe-inspiring beauty of our planet's natural resources. For me, it's the ultimate page-turner, providing not only remarkable photographs of each wonder but also concise and fascinating facts about the where, how and why. It's a fabulous mix of the sciences: geology, geography, zoology, botany ... even physics and chemistry, all carefully woven together to provide simple explanations of extraordinary phenomena. There are so many juicy nuggets of information here, so many "Ah, so that's how it was made, that's how it works, that's why or where or when". But this never intrudes on the sheer "wow" of the revelations, never dulls the essential romance of the natural world. When you pick this up, you can escape into a realm of magnificent beauty and be excited by learning more about it. It's certainly not a children's book, but every child should have it to either ignite or satisfy a curiosity about their world and its astonishing diversity.

In this age of screens and tablets, I still love books, particularly sumptuous tomes like this. There's something nice about having them lying on the kitchen, lounge, or bedside table, to pick up and browse over breakfast, lunch, or before you sleep. They feel good, smell real, they are tangible. I have favourite books – I don't have favourite downloads, because they don't really actually "exist". I have my old childhood books. Will your children have their old childhood apps or e-books? I don't expect so. And this book is certain to enter my charts because it will never be out of date, will always be an invaluable source of reference, and because my desire to know the natural world is fuelled by the wonders on its pages.

**CHRIS PACKHAM**
FLYING OVER LAKE MICHIGAN (DETAILS ON PAGES 50-51), 10 MAY 2017

**Lake in the rift**

Kenya's Lake Magadi is one of many lakes in East Africa's vast Great Rift Valley, formed as blocks of Earth's crust shift downwards into the underlying mantle. The unusual colours come from minerals washed into the lake.

# Introduction

# EARTH'S STRUCTURE

Our planet has a diverse surface, with a huge variety of landscapes and materials, ranging from water and gases, to living matter, soil, ice, and rock. However, its interior is much less varied, consisting mainly of just rock, metal, and some water.

## Internal structure

Much of what is known about Earth's structure has been established from the study of seismic waves created by earthquakes. Internally, our planet has three main layers: the core, mantle, and crust. The core is made mainly of iron, with some nickel, and is itself double-layered– a liquid outer core enclosing a solid inner core.

Surrounding the core is the mantle, which consists mainly of solid (but in parts, deformable) silicate rock. Its top part is called the upper mantle and has two layers. The uppermost of these, sometimes called lithospheric mantle, is solid and brittle and is fused to the crust. Below it is a more deformable layer, the asthenosphere. Below the upper mantle is a region called the transition zone, now known to contain significant amounts of water "locked in" to its rocks. Beneath this is the thickest of all of Earth's internal layers, the lower mantle.

The outermost layer is the crust. This is solid, and there are two different sorts. Thick continental crust, made up of many different rock types, foms the land surface, while thin, relatively dense oceanic crust – containing just a few rock types – lies under the oceans. Both types are fused to the underlying lithospheric mantle, forming a combined rigid shell, the lithosphere.

## ▽ EARTH'S LAYERED STRUCTURE

Each layer from the crust inwards has a temperature and density higher than the layer surrounding it. The temperature ranges between -50°C (-58°F) and 50°C (122°F) at the surface to more than 6,000°C (10,800°F) in the inner core.

Atmosphere
more than 1,000 km
(620 miles) thick

Crust

Lithospheric mantle
40-120 km (25-70 miles) thick

Asthenosphere
250-340 km (155-210 miles) thick

Transition zone
250 km (155 miles) thick

Lower mantle
2,230 km (1,385 miles) thick

Liquid outer core
2,200 km (1,370 miles) thick

Solid inner core
1,275 km (790 miles) radius

## Mantle convection

A slow, gradual, circulation of rock occurs in the mantle. This circulation is called mantle convection and is driven by heat flowing out of Earth's core. The mantle is also thought to contain plumes of rising, hotter semi-solid or liquid rock, called mantle plumes. Where they penetrate into the crust, these create surface hotspots (see p.13). Earth holds a colossal amount of internal heat energy, much of it generated during the planet's original formation and still trapped inside. This energy is continually added to by the radioactive decay of unstable isotopes of various chemical elements scattered throughout

### ▷ WHOLE-EARTH CONVECTION

The overall pattern of convection in Earth's mantle is not fully understood. One hypothesis, shown here, suggests that a series of convection cells slowly carry rock from the bottom of the mantle to the top and then back again.

hotspot, where extra-hot material reaches crust

mantle plume

convection cell

the interior. Earth's energy is constantly trying to escape, and mantle convection and mantle plumes, as well as phenomena such as earthquakes, are an expression of this. Although our planet's mantle and crust are predominantly solid, as a result of processes connected to mantle convection and plumes, both contain collections of hot melted rock and dissolved gases, called magma. The presence of magma in the crust is associated with volcanic and geothermal activity (such as hot springs and geysers) at Earth's surface.

## Rocks and the rock cycle

Earth's crust, the part of our planet that has been shaped into a myriad of landscapes and physical features, is made mostly of rocks. Rocks are assemblages of chemical substances called minerals, and are of three main types: igneous, sedimentary, and metamorphic. Igneous rocks are the result of magma from deep within the planet being erupted at the surface or injected into the crust and then cooling and solidifying. Sedimentary rocks are formed by the deposition and cementation of mineral or rock particles created by the weathering or breakdown of other rocks. Metamorphic rocks are formed by the alteration of other types of rock by high temperature, pressure, or a combination of the two. A variety of processes occurring within Earth's crust and surface continually change crustal rocks from one type into another, in a never-ending succession of events known as the rock cycle. The various components of the rock cycle, which include, for example, volcanism and numerous erosional and depositional processes, are extremely important in shaping Earth's surface landscapes.

### Earth's atmosphere

The atmosphere is part of our planet's overall structure and has several layers. Air circulates only in the lowest layer, which is called the troposphere, and it is only in this layer that weather occurs – taking the form of, for example, winds, precipitation (rain, hail, and snow), and changes in temperature, pressure, and humidity. Weather is driven largely by a combination of energy coming from the Sun and from Earth's rotation. The averaged-out weather in a region over a long period of time is termed that region's climate. Both weather and climate, and the connected processes of the water cycle (which include, for example, precipitation, the flow of water over Earth's land surface in streams and rivers, and the evaporation of water

# The mantle transition zone is thought to **contain more water** than **all the oceans**

from the sea) are other vital factors in shaping landscapes. Changes in Earth's climate, in particular a persistent warming caused by increased levels of carbon dioxide in the atmosphere, are having a widespread effect on surface landscapes. For example, a general retreat and shrinkage of glaciers is occurring worldwide.

◁ HURRICANE
One of the more extreme forms of weather, a hurricane originates from solar warming of the surface of a tropical ocean. This creates a region of low pressure, with dense clouds above it and winds circulating around it.

# PLATE TECTONICS

Numerous features and events at Earth's surface, from volcanic activity to earthquakes and mountain-chain formation, are explainable on the basis that our planet's outer shell is split into fragments called tectonic plates, which slowly move around on the surface.

### Earth's plates

Earth's crust is fused to the top layer of the underlying mantle (see p.10), forming a shell-like structure called the lithosphere. The lithosphere is split into several chunks, called tectonic plates, which slowly move around on Earth's surface, driven by convection in the mantle. There are seven large and numerous medium-sized and smaller plates. As they move, Earth's continents are slowly shuffled around. The rate of movement is small, but over tens or hundreds of millions of years, it can produce major rearrangements. During Earth's long history, continents carried by different plates have collided from time to time, pushing up mountains, or combined to form supercontinents. Conversely, large landmasses have sometimes split into smaller chunks in a process called rifting.

### Plate boundaries

The science of plate tectonics largely revolves around the study of what happens near the edges of plates as they move relative to each other. These plate boundaries

upwelling magma solidifies to add new material to plate edges

plate movement

**SPREADING RIDGE**

◁ **TECTONIC PLATES**
Earth's plates have irregular shapes and a wide range of sizes. They cover Earth's surface like a jigsaw. Here, several plates are visible (their boundaries highlighted in orange).

△ **DIVERGENT BOUNDARY**
At a divergent boundary, plates move apart (top). This type of boundary can be seen at the Silfra fissure in Iceland (bottom), where the Eurasian and North American plates are separating.

are dynamic places where important landscape-altering events take place, such as volcanic activity, mountain-building, island formation, rifting, and earthquakes.

Plate boundaries fall into three main classes. The first, called divergent boundaries, occur where two plates are moving apart. At this type of boundary, material continuously welling up from the mantle fills the gap and creates new plate. Divergent boundaries are found extensively on the ocean floors, in the form of mid-ocean spreading ridges. But they also occur, among other places, in Iceland and in East Africa, along what is called the Great Rift Valley (see pp.184–87). They are always associated with volcanic activity.

The second class of boundaries, convergent boundaries, are found where two plates move towards each other. At these boundaries, all or some of one plate moves down, or subducts, under the other and is destroyed. Where both plates carry continental crust, mountains form as chunks of crust are forced together. This process accounts for the origin of many mountain chains, such as the Himalayas. Other instances involve a plate carrying oceanic crust subducting the other plate. Boundaries of this type are characterised by features such as a deep-sea trench along the line of the boundary; a chain of volcanoes, always on the side of the plate that is not subducting; and the frequent occurrence of major earthquakes.

The third class of boundaries, called transform boundaries, occur where plate edges push past each other, without new plate being created or existing plate destroyed. These boundaries are also a site and source of earthquakes. Examples can be found in California (along the famous San Andreas Fault, see pp.30–31), in New Zealand's South Island, and elsewhere, including extensively on the ocean floors.

### Hotspot volcanism

Although two of the three main types of plate boundary are frequently linked to volcanism, not all volcanoes develop at plate boundaries. Some appear in the middle of plates. This type of volcanism is usually accounted for by the presence of mantle hotspots – locations at the top of Earth's mantle that appear to be the source of peculiarly large amounts of energy. The movement of a plate across a hotspot (whose position is fixed) can, over a long period of time, create a chain of volcanic features at the surface. This explains, for example, evidence of much ancient volcanic activity in a line to the southwest of Yellowstone National Park (due to movement of the North American Plate over this hotspot). It is also regarded as the best explanation for the formation of some chains of volcanic islands, including the Hawaiian Islands (see pp.316–19).

## Plates move at between 7 mm (¼ in) per year – one-fifth the rate fingernails grow – and 150 mm (6 in) per year – the rate human hair grows

SUBDUCTION ZONE

SLIDING PLATES

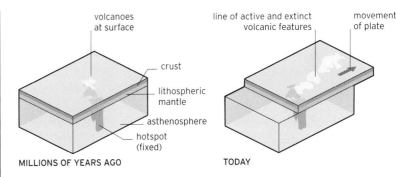

MILLIONS OF YEARS AGO      TODAY

△ **CONVERGENT BOUNDARY**
At this type of boundary, one plate moves under, or subducts, another and is gradually destroyed (top). If both plates carry continental crust, mountains are pushed up (bottom).

△ **TRANSFORM BOUNDARY**
At a transform boundary, two plates push past each other in opposite directions (top). This type of boundary may be visible as a linear fault running across the landscape (bottom).

△ **VOLCANIC CHAINS**
Where a plate moves over a mantle hotspot, it can lead to a chain of volcanic features at the surface. The Hawaiian Islands formed in this way through movement of the Pacific Plate over a hotspot currently located under the volcano Kilauea (bottom) on the Big Island of Hawaii.

# EARTH'S PAST

The landscape features we see on Earth today are the result of events and processes stretching back more than 4 billion years. It is only in the last few hundred years, however, that scientists have pieced together some of the main details of the story.

### Earth's age and origins

A precursor body of what is now planet Earth formed around 4.55 billion years ago, from collisions of smaller bodies in a disc of material spinning around the Sun. When it was perhaps 40 million years old, the proto-Earth is thought to have experienced one final collision, leading to the formation of what we now call Earth and its natural satellite, the Moon.

The early Earth was probably born in an extremely hot, molten state. Heavier materials, mainly iron, sank to the centre, with lighter materials forming layers around this. For around the first 150 million years, no solid crust formed as there were continuing impacts from comets and asteroids, while high volcanic activity continually reworked the surface. Around 4.37 billion years ago, the oceans had

◁ **ANCIENT COLLISION**
A collision between a proto-Earth and a Mars-sized body, some 4.51 billion years ago, is thought to have led to the formation of the Earth-Moon system.

begun to form, through condensation of water released into the atmosphere by ancient volcanoes. By 4 billion years ago, the first pieces of continental crust had also formed. Movements of early tectonic plates caused landmasses to collide and merge, gradually forming the ancient cores of today's continents.

△ **CRYSTAL EVIDENCE**
A zircon crystal, from Jack Hills, Australia, has been dated at 4.375 billion years old and is the oldest known material of terrestrial origin.

### The geological timescale

The fact that Earth is extremely ancient, and that the sedimentary rock layers (strata) of its continental crust were laid down in sequence over a vast period of time, first became clear to scientists between the 17th and 19th centuries. They noticed that many rock strata contain fossils – the remains of ancient, apparently extinct, animals and plants. They divided the fossil-bearing rock strata – and thus Earth's history as a life-bearing planet – into three eras: the Paleozoic (meaning ancient life), Mesozoic (middle life), and Cenozoic (recent life). Later, these eras were subdivided into geological periods. In rock strata deeper and older than the fossil-bearing rocks, there seemed at first to be no life. These rock strata were later subdivided into three extremely long time

## ▷ GEOLOGICAL TIME

Over the enormous span of Earth's history, times during which there was no life or only the simplest life forms have lasted longer than the eras where multicellular life-forms exist.

### KEY

| | |
|---|---|
| ▢ Cenozoic 66-0 MYA | ▢ Proterozoic 2,500-541 MYA |
| ▢ Mesozoic 252-66 MYA | ▢ Archean 4,000-2,500 MYA |
| ▢ Paleozoic 541-252 MYA | ▢ Hadean 4,550-4,000 MYA |

4,550 MYA
Earth's formation

land plants

animals

multicellular life

prokaryotes

eukaryotes

intervals called eons – the Hadean followed by the Archean and Proterozoic eons. It is now known that simple organisms called prokaryotes (single-celled bacteria-like organisms) first evolved during the Hadean. Some 2 billion years then passed before more complex single-celled organisms called eukaryotes evolved (see pp.16–17).

### Agents of change

Geologists have argued for centuries about the nature of the processes, operating through Earth's past, that have brought about the varied landscapes seen today. In the late 18th century, support grew for the argument that Earth's features were mostly formed by slow, gradual, and ongoing changes over time. It was clear, for example, that erosion had shaped the planet everywhere. This view contrasted with the opposing doctrine, which proposed that a series of cataclysms was responsible for most of the Earth's surface features. Today, it is agreed that most landforms on Earth now are the result of gradual processes. Nevertheless, there have been catastrophes: for example, an asteroid or comet impact about 67 million years ago (MYA) is thought to have caused fires, massive earthquakes, worldwide darkness, and possibly wiped out the dinosaurs.

## Changes in climate and sea level

Earth's climate has changed quite drastically over long periods of time. At one extreme, there have been times when there were no ice-caps and temperate deciduous forests grew at the poles. At the other, it seems probable that Earth was at least once completely frozen over. Similarly, sea level has fluctuated dramatically. For much of the past 540 million years, it has been higher than it is today. But, for the last 2.5 million years or so, sea level has mostly been lower, since during this time Earth has been in an ice age with much of its water locked up in polar ice-sheets. Over the past 17,000 years or so, Earth has been in a warming period within this ice age, called an interglacial, and sea level has risen again. Since the late 18th century, the planet has warmed by around 1°C (1.8°F), due to increased atmospheric carbon dioxide, and sea level has risen by about 30 cm (12 in).

◁ **SLOW BUILD-UP**
Deltas, such as the Yukon Delta seen here, are examples of a landform built up slowly over a long period of time.

▽ **ERODED LANDFORM**
Horseshoe Bend in Arizona exhibits the effects of 75 million years of erosion. The rock layers in its walls range from around 180 to 200 million years old.

# LIFE ON EARTH

Earth's land surface, oceans, and atmosphere teem with life, which has colonized every corner of our planet. This is remarkable given that when Earth formed, it was just a dead rock. Life has drastically altered the appearance of our world and the chemistry of the atmosphere.

## Early life on earth

Life on Earth probably originated around 4.25 billion years ago, from precursor chemicals. The oldest definite evidence of life is around 3.7 billion years old and suggests that simple single-celled organisms existed in oceans at that time. These microbes were already able to photosynthesize – that is, to use light energy to manufacture sugars from water and dissolved carbon dioxide, releasing oxygen in the process. By around 1.9 billion years ago, photosynthesizing organisms had released significant quantities of oxygen into the atmosphere. This was important to further evolution of life, as some of the oxygen (consisting of $O_2$ molecules) was converted into ozone ($O_3$ molecules), which formed a protective layer around the planet, filtering out harmful incoming ultraviolet rays. The presence of oxygen in the atmosphere, and dissolved in the oceans, also made animal life possible. By 1.2 billion years ago, multicellular organisms made from eukaryotic cells (more complex than the earlier prokaryotic cells of the simple microbes) had evolved.

◁ **OLDEST SIGNS OF LIFE**
The oldest fossils known are 3.7 billion-year-old traces of stromatolites – layered structures formed by microbes living on an ancient seafloor – within this rock found in Greenland.

## Evolution, spread, extinctions

Between around 525 and 500 million years ago (MYA), during a time in Earth's history called the Cambrian Period, a great proliferation and diversification of life with mineralized shells and skeletons occurred in the oceans. Today, this is known as the Cambrian explosion. By 490 MYA, the first plants had appeared on land, and by 420 MYA, there were also animals on land. From there onwards, life continued to spread and diversify rapidly, but there were also extinctions.

One major wave of extinctions occurred at the end of the Permian Period, around 252 MYA. The exact cause is not known, although a series of massive volcanic eruptions are one suspect. Another occurred at the end of the Cretaceous Period, 66 MYA, when more than 75 per cent of all species died out, including most dinosaurs. However, one dinosaur group survived and evolved into birds.

▷ **EXTINCT FORM**
*Tylocidaris* sea urchins, of which this is a fossil example, lived between about 140 million and 40 million years ago, but then died out.

## Kingdoms of life today

Life on Earth is extremely diverse. Estimates of the number of Earth's current species range from 2 million to 1 trillion, of which only about 1.6 million have been identified and described so far. Various schemes have been put forward for dividing known life-forms into kingdoms. A consensus scheme, proposed by scientists at the Smithsonian Institution in 2015, defines seven kingdoms. Two consist solely of

△ **ARCHAEA**
Many of these simple, single-celled lifeforms live in extreme environments, such as acid hot springs.

△ **PROTOZOA**
Some protozoa cause disease. This one, *Giardia lamblia*, causes a type of human intestinal infection.

△ **PLANTS**
There are over 300,000 known species of plants. They are the basis of most ecosystems, especially on land.

## Life has been found inside rocks 19 km (12 miles) below Earth's surface

prokaryote organisms – bacteria and archaea. The other five are based on eukaryotic cells: protozoa (single-celled life-forms such as amoebae); chromista (algae of various sorts, including diatoms and water moulds); and three more familiar groups: fungi, plants, and animals. The various life-forms interact with Earth, affecting landscapes in many ways. For example, plants cover a significant proportion of the land surface, and their roots help keep soil from being carried away by rainwater. Life has also changed the chemistry of Earth's atmosphere, making it uniquely oxygen-rich. Without life, Earth would have evolved and would appear and work quite differently.

▷ **RIVER CROSSING**
One of the most awe-inspiring natural sights on Earth is the annual migration of vast numbers of wildebeest, zebras, and gazelles – here about to cross a river – across the Serengeti ecosystem.

**Heating system**
Powered by a vast plume of hot, semi-solid rock, or magma,
many kilometres below the surface, Yellowstone National
Park is the site of spectacular geothermal features,
including geysers, hot springs, and pools of boiling mud.

# North America

# MOUNTAINS, LAKES, AND PRAIRIES

## North America

North America is the third largest continent. Geographically, it is also considered to include Greenland. Most of the continent lies on a single tectonic plate, with small parts of Mexico and California lying on the Pacific Plate to the west, which borders the North American Plate along the infamous San Andreas Fault.

Unusually, North America has a vast and ancient lowland heart, almost completely surrounded by younger belts of fold mountains thrown up by continental collisions in the past. This lowland heart, with its Great Plains and Great Lakes, is drained almost entirely by one of the world's great river systems, the Mississippi–Missouri, which channels water south into the Gulf of Mexico. The Appalachian range in the east

is relatively old and has been worn low over time. But the Western Cordillera, which includes the Alaska, Cascade, and Coast ranges and the inland Rocky Mountains, running down the western limb of the continent, is geologically young and still high – so high that these mountains block the flow of moist air from the Pacific. This gives much of the interior a markedly continental climate with bitter winters and hot summers punctuated by tornadoes, and in the south the lack of rain creates large deserts. Only to the north are there no mountains, and the expanse of the Canadian Shield stretches north into the bitter wastes of the Arctic.

### KEY DATA

▲ **Highest point** Denali, Alaska: 6,190 m (20,308 ft)

▼ **Lowest point** Death Valley, California: -86 m (-282 ft)

● **Hottest record** Death Valley, California: 57°C (134°F)

● **Coldest record** Northice, Greenland: -66°C (-87°F)

### CLIMATE

Most of North America lies in the temperate zone, with a subtropical south and an Arctic north. The coasts are moist, but the interior is drier.

AVERAGE TEMPERATURE

AVERAGE RAINFALL

ARCTIC OCEAN

Greenland

Northice
-66°C

Labrador Sea

Baffin Bay

Davis Strait

Hudson Strait

Labrador

Péninsule
d'Ungava

Ellesmere Island

Queen Elizabeth
Islands

Parry Islands

Baffin Island

Victoria
Island

Hudson Bay

Banks
Island

Lake
Athabasca

Great Slave
Lake

Reindeer

Beaufort
Sea

Great Bear
Lake

Mackenzie

Mackenzie Mountains

Rocky

Brooks Range

Casc

Bering Strait

Denali
6,190 m

Alaska Range

Yukon

Queen
Charlotte
Islands

Gulf of
Alaska

Kodiak
Island

PACIFIC
OCEAN

Bering
Sea

Aleutian Islands

ATLANTIC OCEAN

Nova Scotia
Cape Cod
Laurentian
St Lawrence
Canadian Shield
Lake
Ontario
Niagara
Falls
Lake
Erie
Ohio
Appalachian Mountains
Blue Ridge
Lake Superior
Lake
Huron
Great
Lakes
Tennessee
Alabama
Lake
Michigan
Illinois
Mississippi
Arkansas
Lake Okeechobee
The
Everglades
Straits of Florida
Mississippi
Delta

Gulf of Mexico

Caribbean Sea

Yucatán
Peninsula

Canadian Shield
Lake
Winnipeg
Lake
Manitoba
Saskatchewan
South Saskatchewan

NORTH AMERICA

Great Plains

Missouri
Platte
Kansas
Red River
Rio Grande
Sierra Madre Oriental
Lago de
Chapala
Sierra Madre del Sur

Yellowstone
Colorado
Plateau
Colorado
Sierra Madre Occidental

Mountains

Snake
Great
Salt
Lake
Great
Basin
Grand
Canyon
Mojave
Desert
Sonoran
Desert
Gulf of California
Baja California
Mount
St Helens
2,549 m
Mt Whitney
4,418m
Death Valley
-86m, 57°C
Sierra Nevada
Coast Ranges
Vancouver
Island

## ECOSYSTEMS

North America has five main ecosystems: temperate woods in the east, interior grasslands, northern boreal forest, deserts and scrub in the southwest, and temperate coniferous forests in the west.

**KEY**

- Tropical broadleaf forest
- Tropical dry broadleaf forest
- Tropical coniferous forest
- Temperate broadleaf forest
- Temperate coniferous forest
- Mediterranean woodland and scrub
- Tropical and subtropical grassland
- Temperate grassland
- Wetland
- Desert and scrub
- Boreal forest and taiga
- Tundra
- Ice

## GEOLOGY

North America is built around the huge Canadian shield - expanses of ancient metamorphic rock that stretch across Canada. Around this lie young sediments, often folded into mountains.

**KEY**

- Precambrian (pre-541 million years ago)
- Paleozoic (541-252 million years ago)
- Mesozoic (252-66 million years ago)
- Cenozoic (66 million years ago to present day)

**ELEVATION**

| M | FT |
|---|---|
| 3,000 | 10,000 |
| 1,500 | 4,900 |
| 0 | 0 |

Coast
Ranges
Sierra Nevada
Great
Salt Lake
southern tip of
Rocky Mountains
Great Plains
Appalachian
mountains

A
B
CROSS-SECTION

Km
0   250   500
Miles
0   250   500

△ VAST SLAB
Parts of the Canadian Shield – a huge slab of tough and very ancient volcanic and metamorphic rocks – are exposed by erosion in areas such as riverbanks.

◁ SCULPTED PEAKS
One of the world's greatest mountain ranges, the Rockies run all the way along the western edge of North America. They have been heavily sculpted by glaciers.

# THE SHAPING OF NORTH AMERICA

North America began life as part of the great, ancient landmass geologists call Laurentia, which journeyed across the world, colliding with and separating from other continents. The continent we know today came into being less than 200 million years ago.

## Ages of rock

Underlying much of North America is a large shield of ancient rock that once formed the core of Laurentia. In the US, it is covered by a blanket of more recent sedimentary rock. But in Canada, where the shield is exposed on the surface, geologists have found some of the world's oldest rocks – survivors from the dawn of Earth's story, up to 4.4 billion years old.

About 750 million years ago, this Laurentian heartland formed part of the supercontinent of Rodinia, but after Rodinia broke up Laurentia drifted almost to the South Pole before drifting north again. For a while, Laurentia was alone, but eventually it was drawn into the regathering of continents and became part of a new supercontinent – Pangaea.

About 480 million years ago, the plate Laurentia sat on crunched into neighbouring ocean plates, causing the Appalachian mountains to start rising. It took 250 million years for the mountains to reach their maximum height. At first, they were probably as tall and majestic as the Himalayas. In the southwest, rocks in Nevada and Utah were twisted by the same tectonic forces.

## America's backbone

About 200 million years ago, Laurentia split from Pangaea to drift northwest, opening up the Atlantic Ocean and creating North America. The new continent's east coast settled down as it moved away, and the Appalachians were worn gradually lower and lower by millions of years of erosion and weathering. This is why the Appalachians are smaller than the Rocky Mountains even though they are older.

The west, on the other hand, became a titanic battleground between opposing tectonic plates, as the North American Plate ploughed into and rode over the oceanic crust of the Pacific Plate, forcing it to subduct, or descend into the mantle. The Rocky

# The Appalachians are over 480 million years old

## KEY EVENTS

**480 million years ago**
The Appalachians start to rise. They continue rising for another 250 million years to Himalayan heights, but have since been lowered by erosion.

**80-55 million years ago**
The rising Rocky Mountains cause the surrounding land to uplift. As a result, the Western Interior Seaway, a large inland sea, dries out.

**18 million years ago** The Greenland ice-sheet forms around this time. However, it may be even older than this, and it is likely the ice-sheet has receded and advanced many times since then.

**11,000 years ago** The last of the ice-sheets covering much of North America start to recede, relieving most of the continent of its covering of ice.

**1.5-1 billion years ago** Most of the land comprising the crust of central and eastern North America today forms.

**200 million years ago**
Tectonic activity splits Laurentia apart from Europe, allowing the land that will become North America to drift northwest.

**26-22 million years ago** The Earth's climate becomes cooler, favouring the spread of grasslands. The Great Plains first appear, covering a swathe of flat land that used to be the Western Interior Seaway.

**15,000 years ago,**
During the last Ice Age, some glaciers start to melt, and water pools to form the Great Lakes.

**10,000 years ago** Ice melts and sea levels rise, causing the land bridge connecting Alaska and Russia to submerge.

△ DRY FALLS
The Palouse river flows through a giant gorge that was carved out by the Missoula floods.

▽ ISOSTATIC REBOUND
As North America's ice-sheets melted, the weighed-down land beneath slowly began to rise – a process called isostatic rebound. This effect can be seen, for example, in the exposure of old shorelines.

▽ CRACKS IN THE LAND
The San Andreas faultline visibly slices across the Mojave Desert, dividing the Pacific and North American tectonic plates.

Mountains rose as the submerged oceanic plate melted to create volcanoes. Islands slammed against the continent, and the collisions piled up layers of rock on the continent's coast. The Rocky Mountains are unusual in that they rose up further inland than coastal mountain ranges, such as the Andes in South America, normally do. The reason remains a mystery for geologists – it may be that the ocean plate descended at a much flatter angle and so melted in the mantle at a point further beyond the coast.

### ICE AGE
During the coldest period of the last ice age 24,500 years ago, a giant ice-sheet covered the northern half of North America, and the Arctic Ocean had permanent sea ice.

sea ice

ice-sheet

### American icebox
There have been five major ice ages in Earth's history. The most recent began 1.8 million years ago and lasted until just 11,700 years ago. During that time, much of Canada and the northern US were covered by a vast ice-sheet, and most of the Arctic Ocean had permanent sea ice. The ice-sheet's massive weight gouged deep scars into the land that would later become basins for the Great Lakes, and a huge lake called Lake Agassiz was ponded up against it. As the giant ice-sheet finally melted, it unleashed cataclysmic torrents called the Missoula floods – in Washington and Oregon – carving out giant canyons and vast waterfalls that are now left as huge, bare cliffs, such as those seen at Dry Falls, Washington.

### Active faultline
The small Farallon Plate was once sandwiched between the Pacific and North American plates. It was entirely swallowed up 30 million years ago as the two bigger plates collided and subducted. When the subduction stopped, the two gigantic plates began sliding past each other rather than smashing together. It is here that the world's most famous strike-slip fault formed – the San Andreas Fault. The fault is 1,300km (810 miles) long and stretches from the Mendocino coast south to the San Bernardino Mountains and the Salton Sea. Since it formed, the land on either side has moved at least 550km (340 miles). Every juddering movement unleashes devastating earthquakes.

## BUILDING THE NORTH AMERICAN CONTINENT

KEY ▬ Convergent boundary ▬ Divergent boundary

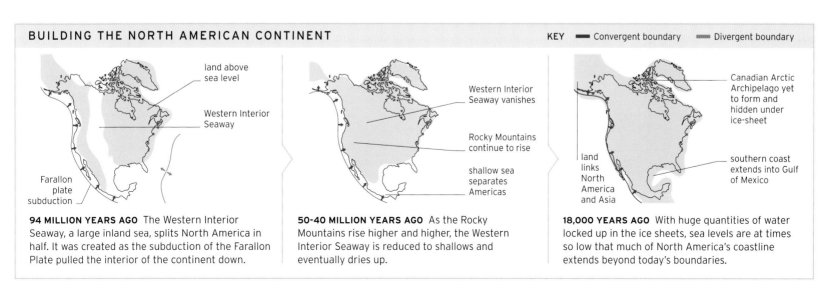

land above sea level

Western Interior Seaway

Farallon plate subduction

Western Interior Seaway vanishes

Rocky Mountains continue to rise

shallow sea separates Americas

Canadian Arctic Archipelago yet to form and hidden under ice-sheet

land links North America and Asia

southern coast extends into Gulf of Mexico

**94 MILLION YEARS AGO** The Western Interior Seaway, a large inland sea, splits North America in half. It was created as the subduction of the Farallon Plate pulled the interior of the continent down.

**50-40 MILLION YEARS AGO** As the Rocky Mountains rise higher and higher, the Western Interior Seaway is reduced to shallows and eventually dries up.

**18,000 YEARS AGO** With huge quantities of water locked up in the ice sheets, sea levels are at times so low that much of North America's coastline extends beyond today's boundaries.

NW. North America

# The Alaska Range

*The world's highest mountain range outside of Asia and the South American Andes*

Stretching for some 650 km (400 miles) like a crescent moon across south-central Alaska, this range appears as an awe-inspiring barrier of high, snow-capped peaks. The loftiest mountains are clustered around Denali, a name given by the local Native American Koyukon people to the highest summit in the Alaska Range. The mountains act as a formidable obstruction to the flow of moist air from the Gulf of Alaska, which lies further to the south, and – as a result – they experience very high levels of snowfall. Due to this, the valleys in the range are filled with some large glaciers, such as the Black Rapids Glacier (see p.43). These, along with the dramatic Arctic scenery, have made the Alaska Range a magnet for hikers.

### HIGHEST PEAKS IN THE ALASKA RANGE

1 Denali 6,190 m (20,308 ft)
2 Mt Foraker 5,304 m (17,402 ft)
3 Mt Hunter 4,442 m (14,574 ft)
4 Mt Hayes 4,216 m (13,832 ft)
5 Mt Silverthrone 4,029 m (13,219 ft)

▽ **HIGH POINT**
Snow-clad Denali, also known as Mount McKinley, is the highest mountain peak not just in the Alaska Range but in the whole of North America.

(see p.43)

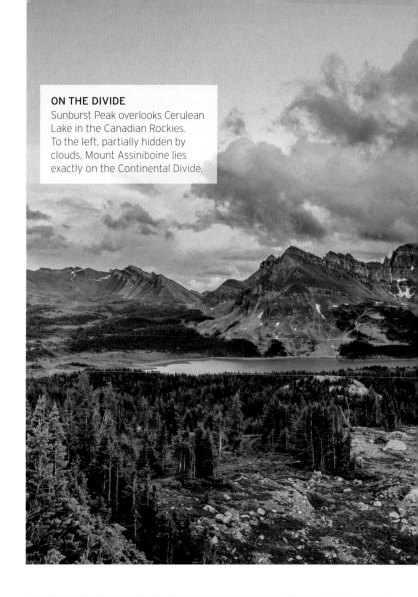

**ON THE DIVIDE**
Sunburst Peak overlooks Cerulean Lake in the Canadian Rockies. To the left, partially hidden by clouds, Mount Assiniboine lies exactly on the Continental Divide.

### THRUST FAULTING

A thrust fault is a break in Earth's crust along which deeper (usually older) layers of rock are pushed up and over shallower layers. An overthrust occurs when they are pushed hundreds of kilometres forward, too. After erosion, ancient rocks may sit on top of younger layers.

older rock layers
stress
thrust fault
younger rock layers
**THRUST FAULT DEVELOPS**
movement along fault

stress
**OVERTHRUST**
overthrust

**AFTER EXTENDED EROSION**
older rocks exposed by erosion of younger rocks

## The Rockies boast **more than 100 summits** that are **over 3,000 m (10,000 ft)** high

# The Rocky Mountains

*A region of sublime natural beauty comprising the central part of one of the greatest mountain belts on Earth, the Western Cordillera of North America*

W. North America

The Rocky Mountains, or Rockies, extend for almost 4,800 km (3,000 miles), running southwards from the northern part of British Columbia in Canada and passing through parts of six US states, from Montana to New Mexico. Located within the range is North America's Continental Divide: to the west of this divide, rivers flow into the Pacific, and to the east they flow into the Atlantic or Arctic oceans. The highest peak in the Rockies is Mount Elbert in Colorado at 4,401 m (14,440 ft).

## Shaping the Rockies

The Rockies mainly formed in stages over the last 80 million years and have a geologically complex history. The range contains many examples of folding and thrust faulting (see panel, left), for example, in the Canadian Rockies and Montana. The Rockies also include areas, such as Yellowstone National Park (see pp.30–33), that have been extensively affected by volcanic activity, as well as regions impacted by glaciation during various ice ages.

A number of the peaks are covered in coniferous forests, with alpine tundra above the treeline. Many animals inhabit the Rocky Mountains, including elk, moose, mountain goats and sheep, bears, wolverines, and bald eagles. With its stunning scenery, this range is a popular leisure-time destination, especially for activities such as camping, hiking, fishing, and mountain biking.

▽ **YOUNG AND RUGGED**
Pushed up in the past 13 million years, the Teton Range in Wyoming is one of the youngest parts of the Rockies. It features several peaks exceeding 3,700 m (12,100 ft) in height.

# El Capitan and Half Dome

*Massive rock outcrops in Yosemite National Park, California, made almost entirely of granite*

W. North America

## EL CAPITAN ROCK TYPES

El Capitan is composed mainly of a 100-million-year-old igneous rock known as El Capitan granite. Taft granite, a slightly younger rock, makes up much of the rest.

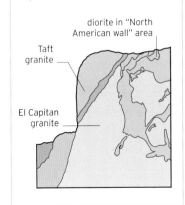

diorite in "North American wall" area

Taft granite

El Capitan granite

Tens of millions of years ago, numerous rounded masses of igneous rock, called plutons, formed underground in what is now central California as enormous globs of magma rose from deep in Earth's crust and then cooled. Following several periods of uplift and erosion, many of these plutons are now visible above the surface as massive rock cliffs and outcrops. Two of the most famous, both located in Yosemite National Park, are El Capitan and Half Dome.

### Imposing cliffs

El Capitan ("the Chief") is a granitic behemoth that is around 1.5 km (1 mile) wide and some 900 m (2,950 ft) from base to summit along its tallest face. It was given its name in 1851 by a roving Spanish militia force. Equally impressive, overlooking Yosemite Valley about 8 km (5 miles) away,

is Half Dome, named for its distinctive shape. One of its sides is a vertical face around 600 m (2,000 ft) high, while the other three sides are smooth and round, making it appear from some angles like a dome cut in half.

The two formations are famous for the challenge they pose to rock climbers. Once considered impossible to climb, El Capitan has now been conquered thousands of times over, and there are more than 70 different routes to the top, of varying difficulty. At Half Dome, a 13-km (8-mile) hiking route has been constructed to the top from the valley floor for non-rock climbers. This includes several hundred steps cut into the granite.

◁ COMEBACK RAPTOR
Peregrine falcons have been nesting on El Capitan and other sites in Yosemite since 2009, following a 16-year absence.

W. North
America

# The Cascade Range

*A stunning array of volcanoes and other peaks in the North American west*

The Cascade Range is a part of the "Ring of Fire" – a horseshoe-shaped band of volcanoes and associated mountains that encloses the Pacific Ocean. Extending for 1,100 km (680 miles), the Cascades formed over the last few million years as a result of the process of subduction as the Juan de Fuca oceanic plate slides underneath the North American Plate.

## Volcanic hotbed

The Cascade Range includes many large and magnificent-looking volcanoes, such as mounts Baker, Rainier, and St Helens in Washington; Mount Hood in Oregon; and Mount Shasta in California. Several Cascade volcanoes have erupted at least once in the last 200 years, and all volcanic eruptions during this period in the contiguous United States (the whole country other than Alaska and Hawaii) have been from Cascade volcanoes.

▽ **BROODING GIANT**
Glacier-shrouded Mount Rainier is considered to be one of the world's most dangerous volcanoes, due to its eruption potential and proximity to the city of Seattle.

▷ **MEGABLAST**
The 1980 eruption of St Helens is the deadliest to have occurred in the US in recorded history. Most of the ash emitted settled over vast areas of the northwestern US.

## THE ERUPTION OF MOUNT ST HELENS

In May 1980, a violent eruption tore apart the snow-capped peak of Mount St Helens, the most active volcano in the Cascade Range. Before the eruption, a bulge had been growing for several weeks on the volcano's flank, due to magma being forced up inside. At 8:32am on 18 May, an earthquake triggered a landslide and a sideways blast. This was soon followed by total disintegration of a part of the flank and the summit of the volcano.

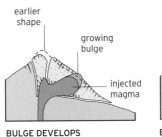

earlier
shape

growing
bulge

injected
magma

**BULGE DEVELOPS**

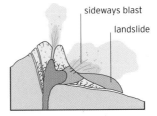

sideways blast

landslide

**EARTHQUAKE-TRIGGERED BLAST**

cataclysmic blast of hot gases, ash,
and pulverized rock

**FLANK DISINTEGRATION**

The eruption of **Mount St Helens** in 1980 produced an **ash plume** that **rose 24 km (15 miles)** into the air

◁ **DACITE CLIFFS**
The cliffs shown here, which rise vertically nearly 550 m (1,800 ft) above the lake and centre around an edifice called Llao Rock, are made from a type of solidified lava called dacite.

**Caldera rim**
The rim is blanketed in a thick layer of pumice and ash from the eruption that formed the caldera.

**Submerged platform**
This broad mound is made of solidified lava that erupted onto the caldera floor soon after Mount Mazama's collapsed.

**Underwater cinder cone**
The summit of Merriam Cone is 154 m (505 ft) below the lake surface.

▶ **CRATER LAKE CALDERA**
This cross-section of Crater Lake cuts through Wizard Island. The caldera is about 9 km (6 miles) across, and the lake that fills it is famous for its exceptionally high water clarity and deep-blue colour, as well as for being one of the deepest lakes in the world at almost 600 m (2,000 ft) deep.

**Hillman peak**
Rising 606 m (1,988 ft) above the lake surface, this is the highest point on the caldera rim.

**Basalt layer**
This is the uppermost of the rock layers on which Mount Mazama grew. It is solidified lava from an earlier volcano.

**Collapsed block**
Part of the floor under Mount Mazama fractured into blocks, which dropped downwards.

**Magma conduit**
Rising magma produced volcanic features on the caldera floor. The conduit is now blocked off with solidified lava.

**Breccia**
The remains of the top part of Mount Mazama formed a thick layer of cemented sedimentary rock fragments called breccia.

sediments are a few hundred metres thick

**Magma chamber**
Hydrothermal activity on the lake floor suggests a magma chamber still exists under the caldera.

**Ring fracture**
This developed during the eruption that caused Mount Mazama's collapse.

◁ **WIZARD ISLAND**
This volcanic cinder cone grew up within a few hundred years of the collapse of Mount Mazama and currently reaches 230 m (750 ft) above the lake surface. It has not erupted in thousands of years.

**△ VOLCANIC REMNANT**
This oddly shaped island, protruding 50 m (165 ft) above the lake surface, is known as Phantom Ship. Made primarily from a rock called andesite, it is a remnant of a volcano that existed long before Mazama.

water level is fairly constant at 1,883 m (6,178 ft) above sea level

**Mazama rock layers**
Buried around the sides of the lake are the undisturbed remains of Mount Mazama, consisting of lava and ash.

W. North America

# Crater Lake

*The collapsed crater of an ancient volcano in the Cascade Range now filled by a deep lake*

Crater Lake, in south central Oregon, is one of the world's finest examples of a caldera volcano: a stratovolcano that has partially collapsed (usually during the final phase of a catastrophic eruption), creating a gigantic, cauldron-like depression called a caldera. A caldera is sometimes described as a large crater, but it is actually a type of sinkhole, since it forms as a result of collapse rather than explosion. Over time, a caldera often fills with water, resulting in a deep lake like Crater Lake.

## Collapse of a stratovolcano

The ancient stratovolcano from which Crater Lake caldera formed is called Mount Mazama. Before its top part collapsed, Mazama stood at around 3,700 m (12,000 ft), making it one of the highest peaks in the Cascade Range. During its eruption (see panel, right), a column of pumice and ash reached some 50 km (30 miles) high. In the final eruption phase, a circular system of cracks, called a ring fracture, opened around the volcano's lower slopes, causing its top to disintegrate.

Following this event, volcanic activity continued for some time, creating relatively small volcanic features on the caldera floor. Sediments and material from landslides covered the floor, and the caldera gradually filled with water. Although no significant volcanic activity has been seen at Crater Lake for thousands of years, it is possible that in the future it may erupt again.

**◁ PINNACLES**
These needle-like formations, a little distance from the lake, were once gas vents under a mass of hot volcanic pumice. The vents hardened into tubes and now stand alone, because surrounding softer materials have been eroded away.

### HOW CRATER LAKE FORMED

Around 7,700 years ago, a large stratovolcano called Mount Mazama underwent a cataclysmic eruption that left its magma chamber partly empty. Much of the volcano's floor and its top part collapsed downwards. This left a huge, bowl-like depression – a caldera – with a raised rim. The caldera later filled with water, forming Crater Lake.

conduits to surface

vent

filled magma chamber

**MOUNT MAZAMA 7,700 YEARS AGO**

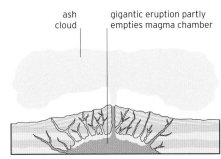

ash cloud

gigantic eruption partly empties magma chamber

**MAGMA CHAMBER PARTLY EMPTIES**

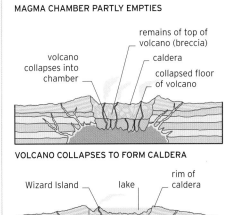

remains of top of volcano (breccia)

volcano collapses into chamber

caldera

collapsed floor of volcano

**VOLCANO COLLAPSES TO FORM CALDERA**

Wizard Island

lake

rim of caldera

**A FEW HUNDRED YEARS LATER**

At **594 m (1,949 ft)** deep, **Crater Lake** is the **deepest lake** in the **United States**

W. North
America

# The San Andreas Fault

*Part of the boundary along which two tectonic plates grind past each other, producing frequent earthquakes*

The San Andreas Fault is one of few places on Earth where the boundary between two tectonic plates can be seen running across land. Extending for approximately 1,300 km (810 miles) through California, the San Andreas is known as a continental transform fault. Along the fault, the Pacific Plate is slowly slipping past the North American Plate (see panel, below). Each slippage – which typically occurs only along a section of the fault – is accompanied by an earthquake. A large tremor that devastated San Francisco in 1906 is the most famous quake associated with movements on the fault.

## MOVEMENT ALONG THE FAULT

The Pacific and North American plates are moving about 4.6 cm (1³/₄ in) per year in relation to each other. This motion results in jerky movement on the fault itself, with periods of little or no displacement interspersed with sudden big slippages during major earthquakes.

relative motion of Pacific Plate

San Andreas Fault

relative motion of North American Plate

△ **LINEAR SCAR**
In parts of California, the San Andreas Fault visibly bisects the landscape, as seen here in the Carrizo Plains, around 120 km (70 miles) northwest of Los Angeles.

△ **COLOURPLAY**
Grand Prismatic Spring is the largest hot spring both in Yellowstone and in the USA. Its colours come from pigments produced by heat-loving microbes in the water.

▷ **SCALDING HOT**
The temperatures of these hot springs can be close to boiling point, which means they continually give off water vapour. This turns into dramatic plumes of steam when the air temperature above the spring's surface cools.

NW. North
America

# Yellowstone

*A national park in Wyoming, famous
for its spectacular geysers, hot springs,
and other geothermal features*

America's oldest national park, Yellowstone covers nearly
9,000 square km (3,500 square miles) of forest, grassland,
mountains, lakes, and canyons. It contains the greatest
concentration of geothermal features in the world,
including geysers, colourful hot springs, and mud pools.

## Heat supply

Yellowstone's geothermal activity results from the fact
that part of the area rests above a huge underground
chamber of magma, which serves as a constant heat
source (see pp.32–33). The features are arranged in
distinct areas called geyser basins. Upper Geyser Basin
contains around 150 geysers, including the famous Old
Faithful, and several colourful springs. Many of the
geysers erupt more than 30 m (100 ft) into the air.
Plentiful rainfall seeps through Earth's crust and is
heated by the magma chamber, providing a continuous
hot water supply for the springs and geysers.

## Geological diversity

Rugged mountain scenery, deep glacier-carved valleys,
waterfalls, canyons, and rich wildlife are among other
interesting features of Yellowstone. The Grand Canyon
of Yellowstone, as well as two waterfalls on the
Yellowstone river, constitute some of its most dramatic
attractions. Considered the world's largest intact
ecosystem in the northern temperate region, Yellowstone
is home to many large mammals such as American
bison, wolves, black and grizzly bears, and elk.

## Yellowstone has more than
## 300 geysers, almost one-third
## of all the geysers in the world

◁ **BLOWHOLE**
Every 10-12 hours, Castle
Geyser throws up a
fountain of hot water
nearly 27 m (90 ft) high
for roughly 20 minutes.
It is thought to have
been doing this for about
1,000 years.

Underneath a

# SUPERVOLCANO

There are a handful of volcanic sites in the world known as supervolcanoes: sites that have seen at least one eruption rating an 8 on a scale called the Volcanic Explosivity Index. These are the largest eruptions that geologists know about; intervals between them tend to be measured in tens or hundreds of thousands of years. These sites are also capable of future eruptions that could radically alter landscapes and

severely impact the world's climate. All supervolcanoes are calderas that sit above large, active magma chambers. Yellowstone National Park lies over a supervolcano that last erupted about 640,000 years ago. There are two large, interconnected magma chambers under Yellowstone caldera. The heat from the upper chamber drives geothermal features: hot springs, geysers, fumaroles, and boiling mud pools.

△ COLOURFUL POOLS

The colours in hot spring pools and around their edges come from pigments produced by heat-loving micro-organisms or from minerals.

Grand Prismatic spring

▷ HOT SPRING TERRACES

Hot springs can create terraces of travertine (a form of limestone) as seen at Mammoth Hot Springs, just outside the caldera boundary.

Old Faithful geyser

centre of caldera is about 10km (6 miles) east of Old Faithful geyser

Teton Range

rim of Yellowstone caldera

◁ MUDPOT

Acid hot springs, such as Sulphur Cauldron, dissolve nearby ground into a pool of boiling, sulphurous-smelling mud, called a mudpot.

Yellowstone Lake

the Absaroka Range is one of several mountain ranges of the Rocky Mountains that surround the caldera

Upper crustal magma reservoir

This contains hot, mostly spongelike solid rock and about 9 per cent magma at a temperature of around 760°C (1,400°F).

**► YELLOWSTONE CALDERA**
The heat source for the Yellowstone Caldera is a mantle plume – a mass of hot, semisolid rock moving slowly upwards in the mantle. This supplies heat and material to the magma reservoirs in the crust under Yellowstone.

**Lower crustal magma reservoir**
This contains hot rock and about 2 per cent magma at a temperature of about 1,000°C (1,830°F). The amount of material here would fill the Grand Canyon over 11 times.

**Dikes or vertical cracks**
The upward movement of hot rock or magma occurs through dikes.

**△ FUMAROLES**
Around 4,000 of these holes in the ground occur within the boundaries of the Yellowstone Caldera. They release steam and other gases – some of them foul-smelling, such as hydrogen sulphide.

**Mantle plume**
This plume rises at least 1,000 km (620 miles) and may even originate from Earth's core–mantle boundary, which is at a depth of 2,900 km (1,800 miles).

# The Yellowstone supervolcano erupts about once every 600,000–800,000 years

**EVIDENCE OF PAST VOLCANISM**

The North American Plate is moving to the southwest at the rate of about 22 km (14 miles) per million years, whereas the mantle plume currently under Yellowstone stays in approximately the same place. As a result, traces of past eruptions and calderas stretch for hundreds of kilometres to the southwest of Yellowstone.

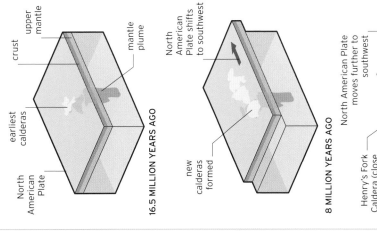

crust

upper mantle

earliest calderas

mantle plume

North American Plate

**16.5 MILLION YEARS AGO**

North American Plate shifts to southwest

new calderas formed

**8 MILLION YEARS AGO**

North American Plate moves further to southwest

Henry's Fork Caldera (close to Yellowstone Caldera)

**1.3 MILLION YEARS AGO**

# Ship Rock

*A twin-peaked volcanic plug with spectacular radiating dikes, which stands high above the deserts of New Mexico*

SW. North
America

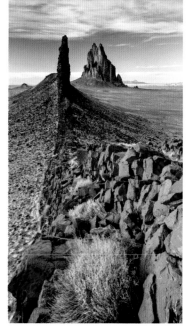

The most famous volcanic plug in the American Southwest, Ship Rock towers 483 m (1,585 ft) above a high-desert plain in the Navajo Nation – a Native American territory in Arizona and New Mexico. Also called a volcanic neck, this type of feature originates when a mass of erosion-resistant rock forms inside a volcano after it has stopped erupting (see panel, right).

## Rock ribs

Regarded as a sacred monument by the Navajo people, Ship Rock's Navajo name is *Tsé Bit'a'í*, which means "winged rock". This probably alludes to the presence of prominent walls of volcanic rock, called dikes, that radiate from the monolith. The three most prominent of these rock ribs run at roughly 120 degrees to each other. Seen from certain perspectives, the whole monument can have a bird-like appearance.

### HOW SHIP ROCK FORMED

Ship Rock is composed of a hard igneous rock called lamprophyre that solidified in the throat of an ancient volcano about 30 million years ago. The accompanying radiating dikes formed around the same time. The rest of the volcano, which consisted of softer rock, has since eroded away.

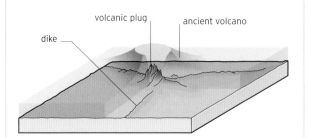

volcanic plug          ancient volcano

dike

△ **KNIFE EDGE**
Parts of Ship Rock's dikes have a knife-like form and tower at a height of 20 m (65 ft) above the high-desert plain.

▽ **NAVAJO MONUMENT**
Ship Rock and three of its dikes can be seen here. According to a Navajo legend, bird-like monsters once nested on the rock's high peaks, feasting on human flesh.

The **longest** of **Ship Rock's dikes** stretches for nearly **3 km (2 miles)**

W. North
America

# Devil's Tower

*A 40-million-year-old massive domed monolith in Wyoming with columnar fluting on its sides*

Devil's Tower is one of the most iconic US landmarks. Often referred to as a butte (an isolated hill with vertical sides and a small, flat top), its exact origins are debated – in particular, whether or not it formed inside a volcano. However, experts agree that it started off as a mass of magma that cooled and solidified underground. In the process, remarkably uniform polygonal columns formed in the cooling body, and this fluting is still visible today.

Regarded as a sacred spot by Native Americans, in 1906 it became the first site in the US to be declared a National Monument. It is also famous as the location for a climactic meeting between humans and visitors from space in the 1977 film *Close Encounters of the Third Kind*.

△ **STANDING TALL**
Devil's Tower rises 264 m (866 ft) from its rocky base to its relatively flat top, which is about the size of a football field.

## HOW DEVIL'S TOWER FORMED

One theory about the formation of Devil's Tower suggests that a volcano formed, and later, some magma within the volcano solidified into a hard mass. This was exposed as the tower when surrounding rock layers were eroded away.

volcano

magma

sedimentary
rock surface

mass of solidified
magma

Devil's Tower

---

# The Appalachians

*A complex and highly scenic series of mountain ranges, stretching from Alabama to southeastern Canada*

E. North
America

The Appalachians are the eroded remnants of an ancient mountain belt that formed about 500 million years ago through collisions between ancient continents. Over the last 65 million years, the whole belt has been uplifted and eroded to what is seen today – a vast area of gently rounded, mostly forested hills and mountains. It consists of many parallel provinces (see panel, below), including the Blue Ridge Mountains, noted for a bluish haze when seen from a distance. The colour is in part due to a chemical released into the air by trees in the region.

▽ **AUTUMN PALETTE**
In autumn, foliage creates an explosion of colour across the Appalachians, as seen here in the White Mountains range in the northern Appalachians.

## INSIDE THE APPALACHIANS

This west-to-east cross-section through the central part of the Appalachians shows folding, faults, and ridges. It also shows the provinces into which the whole belt is split.

Appalachian plateaus | Valley and Ridge province | Blue Ridge Mountains | Piedmont | Atlantic Ocean

sedimentary layers | fault | folding

# The Cave of Crystals

*A horseshoe-shaped underground cavern in Mexico containing some of the most enormous natural crystals ever discovered*

S. North America

The Cave of Crystals lies 300 m (980 ft) below a mountain at Naica in northern Mexico. In April 2000, workers at the Naica Mine were excavating a new tunnel when they discovered the Cave of Crystals. The chamber is filled with gigantic, beam-like crystals of the mineral selenite.

## Birth of giants

The cave, which has been eroded out of the limestone that forms much of Naica Mountain, has existed for several thousand years. At some point after its formation, when it was flooded with water, mineral-rich groundwater rising from below came into contact with cooler, oxygen-rich surface water in the vicinity of the cave. The oxygen diffused into the groundwater, causing

## The cave's **largest crystal** is **12 m (39 ft) long** and weighs **50 tonnes (55 tons)**

the selenite to be deposited in the form of crystals. This continued at an extremely slow rate for at least 500,000 years, forming the gigantic crystals that exist today.

## Terra incognita

The cave is relatively unexplored because of its depth and the continued presence of heated water below the surface. Air temperatures can reach 58°C (136°F), and at present, the cave lacks public access.

△ **GYPSUM TROVE**
The crystal reserves found within the Cave of Crystals are made of selenite. A variety of gypsum, selenite is a soft, light-coloured mineral containing calcium sulphate and water.

◁ **CRYSTAL FOREST**
The Cave of Crystals in Mexico houses some of the largest natural crystals ever found. Here, massive beams of selenite dwarf explorers in the cavern.

## LOCATION OF THE CAVE OF CRYSTALS

Naica Mountain harbours several caverns, including the Cave of Crystals. Some of these can be explored only because the Naica Mine Company pumps water out of the ground to keep the water level down.

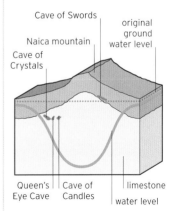

Cave of Swords

original ground water level

Naica mountain

Cave of Crystals

Queen's Eye Cave

Cave of Candles

limestone

water level

## HIGHEST PEAKS IN THE SIERRA MADRE

**HIGHEST PEAKS IN THE SIERRA MADRE**

1 Mt San Rafael 3,730 m (12,238 ft)
2 Mt El Potosí 3,720 m (12,205 ft)
3 Mt El Nacimiento 3,710 m (12,172 ft)
4 Sierra de La Marta 3,705 m (12,156 ft)
5 Mt Teotepec 3,550 m (11,647 ft)

# The Sierra Madre

*A majestic system of mountain ranges extending diagonally across Mexico and forming a part of the North American Cordillera*

S. North America

Mexico's Sierra Madre has three main parts – the Sierra Madre Occidental (in northwest Mexico), the Sierra Madre Oriental (in the northeast), and the Sierra Madre del Sur (in the south). Between the Occidental and Oriental ranges lies the Mexican Plateau. This consists of layers of sedimentary rock laid down in shallow seas 250–65 million years ago, which have subsequently been eroded into an angular, gouged landscape in the Sierra Madre Occidental and a more rounded terrain in the Oriental. Parts of the range also show signs of past volcanism in the form of igneous rock intrusions and layers of hardened lava poured out by long-disappeared volcanoes.

## Labyrinthine landscape

The western side of the Sierra Madre Occidental is a series of high escarpments (steep slopes that form due to faulting or erosion) whose sides drop into deep canyons called *barrancas*. The dimensions of some of these, notably the Copper Canyon, or Barrancas del Cobre, are comparable to those of the Grand Canyon (see pp.54–57) in the US.

The Sierra Madre Oriental is cut through by many steep-walled narrow valleys. Many of these are aligned in a north–south direction, but there are also several passages through these mountains that run from the lowlands of the Gulf of Mexico to the east. The Sierra Madre del Sur consists of a labyrinth of narrow ridges and steep-flanked valleys.

◁ **GREAT DESCENT**
A part of the Copper Canyon of the Sierra Madre Occidental, Urique Canyon is the deepest canyon in North America, descending 1,870 m (6,135 ft) from its highest point.

# Popocatépetl

*North America's second highest volcano and the second highest peak in Mexico, with a history of violent eruptions*

S. North America

Towering to a height of 5,636 m (18,491 ft), Popocatépetl, which means "smoking mountain" in the ancient Aztec language, is famous because of the ferocity of its eruptions, its tongue-twisting name, and its proximity to Greater Mexico City, the largest metropolitan area in the Americas.

## Violent nature

A stratovolcano, Popocatépetl, or "El Popo", can be seen erupting from most parts of Mexico City. Over the past few thousand years, it has had three Plinian eruptions – extremely violent blasts that produce huge plumes of gas and ash. Such explosions can cause large loss of life in populated areas. After a massive eruption in 1947, the volcano lay dormant until it came back to life in 1994. Ever since, smoke has been seen emanating almost daily from its huge summit crater. The year 2000 saw the most dramatic eruption in more than a thousand years. In April 2016, it erupted again, spewing out lava, ash, and rock, and it has continued to produce powerful explosions at irregular intervals.

▽ **HIGH CONE**
Popocatépetl's height, symmetrical shape, and snow-covered summit give it an imposing presence.

△ **MELTWATER STREAMS**
In summer, meltwater streams appear all over
the ice-sheet. An estimated 250 cubic km
(60 cubic miles) of ice is currently being
lost from the ice-sheet each year.

N. North
America

# The Greenland Ice-sheet

*The largest glacier in Earth's northern hemisphere, covering 80 per cent of the island of Greenland and holding a significant fraction of the world's fresh water*

The Greenland Ice-sheet is the world's second largest ice-sheet (and second largest glacier) after the Antarctic Ice-sheet (see pp. 306–307), covering 1.71 million square km (660,000 square miles) of Greenland. It has an estimated volume of 2.85 million cubic km (680,000 cubic miles), with an average thickness of about 1,670 m (5,480 ft) and a maximum thickness of 3,205 m (10,515 ft). Some of its ice is thought to be up to 110,000 years old.

## Shape and movement

The ice-sheet's surface is slightly domed, reaching a height of about 3,290 m (10,800 ft) above sea level at the top of the dome. From here, ice moves slowly down towards the ice-sheet's edges, where it is mostly constrained by coastal mountains; it only reaches the sea along a broad front in a few places. As a result, Greenland (unlike Antarctica) has no ice-shelves. In many places, the ice flows through gaps in the mountains in the form of outlet glaciers. When these reach the coast, they discharge enormous numbers of icebergs into the sea. One outlet glacier, the Jakobshavn Glacier in western Greenland, is the world's fastest-flowing glacier. At its sea end, the ice flows at about 1 m (3 ft) per hour.

## Threats to sea level and fresh water

The Greenland Ice-sheet is melting due to global warming. If it melted completely, the world's oceans would rise by about 7.2 m (24 ft), drowning many of the world's major cities and leading to loss of about 6.5 per cent of the world's freshwater reserves.

◁ **ARCTIC PREDATOR**
Although rarely venturing onto the ice-sheet itself, polar bears are active around its edges on the Greenland coast, where they hunt and feed on seals.

△ **ICEBERG MAKER**
A helicopter flies past the end of Graah Glacier, an outlet glacier that flows out of the Greenland Ice-sheet. From this wall of ice, huge icebergs break off, or calve, from time to time.

◁ **WASHED OUT**
From the snout of Russell Glacier in central western Greenland emerges a river of meltwater, skirting rock debris that has been carried there by the ice.

## ICE MOVEMENT AND ICE DIVIDES

Although ice covers most of Greenland, there is an ice-free zone of variable width around much of the island's coast. Snow falling on the ice-sheet compacts to ice, which then slowly flows down in the direction of the coast along the flow lines shown here. Imaginary lines called ice divides separate regions of ice moving towards different coasts.

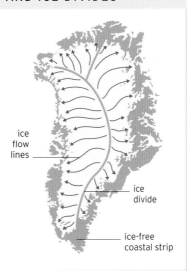

ice flow lines

ice divide

ice-free coastal strip

The **Jakobshavn Glacier** produces about **90 million tonnes (99 million tons)** of **icebergs** every day

# The Columbia Glacier

*An enormous Alaskan glacier that moves quickly but is also rapidly shortening in length*

NW. North America

The Columbia Glacier is one of Alaska's largest glaciers, with an area of about 920 square km (360 square miles) and a length of 40 km (25 miles). It descends from the Chugach Mountains in southern Alaska to Prince William Sound (an inlet of the Gulf of Alaska), where it calves large numbers of icebergs into the sea.

## Shrinking record

The Columbia Glacier is one of the fastest moving in North America, but its fast forward movement is offset by extremely rapid calving of icebergs at its terminus, at the rate of some 13 million tonnes (14 million tons) every day. As a result, since the 1960s, the glacier's terminus has moved backwards, and since 2011 it has become detached from some of its largest tributary glaciers. The retreat cannot be attributed solely to global warming, as other nearby glaciers are not shrinking at the same pace. Instead, experts believe it is connected to the shape of the bedrock channel under the glacier.

### THE RETREAT OF THE COLUMBIA GLACIER

Since the 1960s, the iceberg-calving front of the Columbia Glacier has retreated about 20 km (12 miles). Marked here are its positions at roughly 15-year intervals since 1969.

Chugach Mountains    Columbia Glacier

2016
2001
1985
1969

◁ **RETREATING TERMINUS**
About two-thirds of the area of the glacier visible in this photograph, which was taken in the mid-2000s, is now open water.

▽ **MILITARY FORMATION**
The higher part of the glacier is criss-crossed by crevasses, producing an army of ice pinnacles on its surface.

**SPRAWLING GIANT**
This satellite view shows the Bagley Ice-field with its linked arms at top right, the glacier's main descending trunk at centre right, and the terminal lobe at the bottom.

△ **ICE SHARDS**
Massive icebergs regularly break off the end of the glacier and may then float around for months in Vitus Lake.

# The Bering Glacier

*The longest glacier in North America – and the largest by surface area – which terminates
in a large lake and has thick vegetation growing on part of its surface*

NW. North
America

The Bering Glacier is around 178 km (111 miles) long
and flows down from Alaska's Chugach Mountains
towards the Gulf of Alaska. In parts, the ice is
about 800 m (2,600 ft) thick. The glacier has two
very different main parts. The upper part, known
as the Bagley Ice-field, lies at an altitude of 1.1–2 km
(¾–1¼ miles) and is a vast basin of clean ice formed
from compaction of fallen snow. This ice-field is about
90 km (56 miles) long and completely fills several
interconnecting valleys. The lower part is a broad trunk
of ice ending in a 42-km- (26-mile-) wide, roughly oval
terminal lobe, which is partly covered in rock debris
and vegetation. A large lake of meltwater, Vitus Lake,
fringes the glacier's terminus. From this lake and
the edge of the terminal lobe, streams flow a short
distance to the coast, which is on the Gulf of Alaska.

△ **GLACIER HERB**
Some areas on the
lower part of the glacier,
consisting of stagnating
ice, have a variety of
plants growing on them,
including this herb
called mistmaiden.

## ICE GAINS AND LOSSES

Ice from the glacier
descends to the
ablation area where
it is lost by melting
and evaporation. The
equilibrium line marks
the ice gain and ice loss
balance. The Bering
Glacier's accumulation
area is a vast ice-field
at an altitude of 1.1 km
(¾ mile), while the
ablation area comprises
much of the rest of
the glacier.

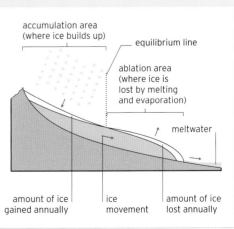

# The Kaskawulsh Glacier

*A vast glacier in Canada, which is notable for its sheer size and impressive medial moraines*

NW. North America

Around 1,000 square km (400 square miles) in area, the Kaskawulsh Glacier snakes for some 70 km (45 miles) through the St Elias Mountains of Canada's Yukon Territory. At its terminus, or snout, it produces meltwater that maintains the level of one of the Yukon's largest lakes, Kluane Lake.

### Bold stripes

All glaciers constrained by valleys have dark bands of material running along their sides, called lateral moraines (see panel, p.150). These contain rock fragments that have been plucked off the walls of the valley and are being carried along by the glacier. When two glaciers merge to form a larger glacier, the lateral moraines of the smaller glaciers typically combine to form a bold dark stripe running down the merged glacier. This stripe is called a medial moraine. The lower half of the Kaskawulsh Glacier has several bold medial moraines, formed through a series of mini-mergers of its different arms, or tributaries, upstream. The glacier has three main arms, each with its own tributaries.

The Kaskawulsh's **broadest medial moraine** is over **400 m (1,300 ft) wide**

▷ **MERGER OF EQUALS**
In this view, different arms of the glacier can be seen to merge in the distance. The merger produces the widest medial moraine (dark band) on the glacier's downstream surface.

# The Malaspina Glacier

*The world's largest and finest example of a piedmont glacier – a huge lobe of ice spread out over a broad, flat plain*

NW. North America

The enormous Malaspina Glacier is up to 65 km (40 miles) wide and extends up to 45 km (28 miles) forwards from where its ice "root" emerges from between a gap in some mountains near the coast of Alaska. Studies of earthquake waves passing through it have shown that its ice is up to 600 m (2,000 ft) thick.

### Patterned lobe

The glacier has complex patterns of light and dark bands on its surface – the dark areas are moraines, containing rock debris. The bands are arranged in zigzags and swirls, and are thought to have been caused by alternating, perhaps seasonal, patterns of fast and slow ice flow. Also on the surface are colonies of tiny creatures called ice worms.

△ **ICE SWIRL**
The glacier, with its vast swirling ice patterns, is named after an Italian explorer, Alessandro Malaspina, who visited it in 1791.

### PIEDMONT GLACIERS

A piedmont glacier occurs where the flow of ice within a valley spills out through a narrow opening onto a broad plain, often near a coast. The ice spreads out to form a wide lobe.

valley glacier | narrow outlet | broad, flat plain

wide lobe

# The Black Rapids Glacier

*An unusual glacier, the largest in the Alaska Range, with a history of phases of exceptionally rapid advance*

NW. North America

The Black Rapids Glacier flows down a valley in south central Alaska and is one of the world's more accessible glaciers, since its snout lies only a few kilometres from a road. It is about 40 km (25 miles) long and has several tributaries. In 2002, an earthquake caused a huge part of it to be buried in rock debris.

The Black Rapids is a surge type glacier, in which periodic rapid transfers of ice occur from the upper to the lower parts, which increase in thickness and then surge forwards. The Black Rapids Glacier made national news in the US for three months in 1937, when its ice suddenly began moving at 30 m (100 ft) a day – more than 20 times its normal rate – and its snout began to rapidly advance down its valley. During this period, it became known as the Galloping Glacier. Since this surge, however, the Black Rapids has steadily retreated to its current length, which is at a historic minimum.

◁ **SURFACE MELT**
Surface pools, and even streams, can develop on the higher parts of the Black Rapids when meltwater forms faster than it can be absorbed by the glacier.

▽ **RARE FIN**
An ice fin - caused by pressure squeezing ice upwards as a glacier grinds against hard bedrock - is a rare glacial feature. This fin was sighted on the Black Rapids in 2013.

NW. North
America

# The Kennicott Glacier

*A spectacular and accessible glacier in the Wrangell–
St. Elias National Park of south central Alaska*

The Kennicott Glacier flows down from the southern
Wrangell Mountains, its principal source of ice being
the flanks of Mount Blackburn – a snow-clad, eroded,
extinct volcano and the fifth highest peak in the US.
The glacier is about 42 km (26 miles) long.

## Arms and tributaries

The upper part of the Kennicott Glacier has two arms,
west and east, that flow around a ridge of rock and a
nunatak (a small rocky peak poking through the ice)
known as Packsaddle Island. After these two arms
have joined, the impressively striped main trunk of
the glacier flows south, where it merges with two
large tributaries – the Gates and Root glaciers – and
later passes an abandoned mining community called
Kennecott. It ends in a broad, rock-covered snout,
surrounded by shallow meltwater pools and lakes.
Several streams emerge from underneath the glacier
and merge to form the Kennicott River.

The Kennicott Glacier and Root Glacier are
accessible from a road built to supply the old mining
camp. Guided hikes onto the glaciers provide a chance
to view the ice formations, and vivid blue pools and
streams on the glaciers' surfaces, or to explore the ice
caves underneath them. Crossing the Root Glacier
gives access to Donoho Peak. A strenuous scramble
up this can give brilliant views of the whole area.

## The glacier has **lost** about **75 m (245 ft)** of its **thickness** since the 1960s

▷ **SUBGLACIAL CAVE**
Under the glacier's edge is
a world of glowing blue
ice, dripping water, and
light seeping through
cracks. These caves
can be safely visited
only at certain times
of year due to seasonal
flood risks.

▶ **KENNICOTT GLACIER**
Kennicott is a typical valley glacier - one
hemmed in by surrounding mountains. It
covers an area of around 250 square km
(100 square miles). The full length of the
glacier is shown here, flowing down from
Mount Blackburn to its terminus at the
head of the Kennicott River.

surface stream
of flowing
meltwater

**Englacial debris**
This consists of rocks
and dust carried inside
the glacier.

surface debris
made up of rocks
that have fallen on
to the glacier

**Medial moraine**
A line of debris is
formed by the union
of two lateral moraines
when glaciers merge.

**Lateral moraine**
Rock debris is
dragged along
and deposited at
a glacier's sides.

**Water channel**
Channels carry
meltwater along the
glacier's base.

moulin (vertical
water channel)

**Crevasses**
These are deep,
wedge-shaped clefts
in the ice surface.

△ **ANATOMY OF A VALLEY GLACIER**
A typical valley glacier like the Kennicott is hundreds of metres
deep at its centre and contains rock debris on its surface, inside
it, and dragged along at its sides and base. The ice is not solid
but fractured by crevasses and carved by meltwater channels.

**Mount Blackburn**
The highest peak in the Wrangell Mountains, Mount Blackburn reaches 4,996 m (16,390 ft).

Kennicott Glacier West Upper Arm

Kennicott Glacier East Upper Arm

Packsaddle Island (nunatak)

**The Gates Glacier**
A tributary of the Kennicott Glacier, the Gates Glacier is 13 km (8 miles) long.

△ **WINDING TRAILS**
This view of a middle section of the Kennicott Glacier shows its well-developed medial moraines (dark stripes containing rock debris). Some of these are more than 350 m (1,150 ft) wide.

Donoho Peak provides excellent views from its summit at 2,041 m (6,696 ft)

**The Root Glacier**
This 22-km- (14-mile-) long tributary consists of relatively clean, rock-free ice.

◁ **MELTWATER CHANNEL**
The surface of both the Kennicott Glacier and its tributary, the Root Glacier, feature numerous turquoise meltwater channels that snake across the surface.

Kennecott (old mining camp)

**Meltwater pools**
The glacier's terminus, or snout, which consists of a grey-coloured slush, is surrounded by meltwater pools.

**Snout edge**
Over the past several decades, the snout edge has retreated markedly, leaving mounds of moraine behind.

▷ **BLUEST BLUE**
Some meltwater pools on the glacier's surface are more than 6 m (20 ft) deep and an exceptionally vivid blue. The blueness is due to the depth and purity of the water, which naturally absorbs all other light wavelengths.

Kennicott River

NW. North
America

# The Mendenhall Glacier

*A partially hollow glacier close to the Alaskan city of Juneau, best known for its extraordinary blue ice caves*

The Mendenhall is one of 38 large glaciers that flow down off the Juneau Ice-field. The glacier is about 21 km (13 miles) long and ends in an iceberg-filled lake. It has been retreating and shrinking since at least the 1700s, with some phases of rapid retreat in recent years, and it is expected that it will recede out of the lake by about 2020.

## Exploring the glacier

By paddling a canoe across the meltwater lake, it is possible to reach the side of the glacier's terminus, clamber across its surface, and then explore its ice caves, with their deep blue ceilings and streams of water running over the rocks inside. Wildlife in the area includes beavers, bears, spawning salmon in season, and the occasional wolverine (a carnivore resembling a small bear).

▷ **BLUE DESCENT**
An ice climber descends through a moulin into one of the large ice caves in the glacier.

◁ **ICE PROWLER**
Occasionally, a wolverine is spotted in the area. It is thought that wolverines use glaciers for temporarily storing killed prey.

## MELTWATER DRAINAGE

The lower parts of many glaciers have extensive systems for draining meltwater. These include vertical channels, called moulins, and streams at the base. Changes in the drainage pattern can cause outlet channels to dry up, leaving them as internal ice caves, as in the case of Mendenhall.

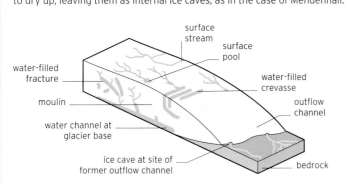

surface stream
surface pool
water-filled fracture
water-filled crevasse
moulin
outflow channel
water channel at glacier base
ice cave at site of former outflow channel
bedrock

NW. North
America

# The Margerie Glacier

*A glacier with a spectacular terminal wall that drops blue icebergs into Glacier Bay*

The Margerie Glacier originates from the eastern end of the St Elias Mountains, in southeastern Alaska, and discharges icebergs into Glacier Bay (an inlet of the Gulf of Alaska). Glaciers that flow down to the sea are called tidewater glaciers. The Margerie Glacier's terminal wall is in Tarr Inlet, at the northern end of the Bay, and is 80 m (260 ft) high. Unlike most other Alaskan glaciers, it has not retreated in recent years.

## History of the bay

Glacier Bay was entirely filled with a single massive glacier when the English naval officer Captain George Vancouver discovered it in 1794. But due to higher temperatures and lower average snowfall over the last few centuries, this has been transformed into a 100-km- (60-mile-) long fiord with many smaller glaciers feeding ice and meltwater into it, including the Margerie.

◁ **FALLING APART**
A huge quantity of ice drops from the 1.5-km- (1-mile-) wide terminus of Margerie Glacier into Glacier Bay. This occurs several times each day, accompanied by thunderous cracking noises.

## ICEBERG CALVING

Glaciers that flow to the coast extend only a short distance over the sea. The stress created by the glacier's weight causes chunks to break off and float away as icebergs. This is called iceberg calving.

glacier

direction of glacier flow

calving iceberg

iceberg floats out to sea

terminal moraine (pile of rock debris)

sea inlet

W. North
America

# The Yukon

*One of North America's longest watercourses, meandering through some of the continent's last wild frontiers*

The Yukon River is 3,185 km (1,980 miles) long, making it the third longest river in North America after the Missouri and Mississippi. Eight major rivers flow into the Yukon, and its tributaries drain an area of about 850,000 square km (328,000 square miles) – the fourth largest drainage basin in North America.

## Epic journey

From its mountain source in British Columbia, the Yukon flows slowly northwest through the lowland forests of the Yukon Territory before entering Alaska, where it widens into a vast wetland area called the Yukon Flats. From here, its lower course turns southwest to wind across the entire breadth of the relatively flat central Alaska landscape to its mouth in the Bering Sea.

The vast scale of the Yukon and the remoteness of the landscapes that it crosses have led to the romantic mythology of frontier life associated with its name. Once the principal means of transportation for the pioneers of the Klondike Gold Rush at the turn of the 19th century, the river remains an epic watercourse for those with an adventurous spirit.

▷ **MIGRATION ROUTE**
In early May, large groups of sandhill cranes can be seen at various points along the Yukon, migrating to nesting grounds in western Alaska and northwestern Siberia.

# Despite its great length, **only four bridges** cross the Yukon River

△ **HAZARDOUS PASSAGE**
One of the best-known landmarks on the Yukon River, the Five Finger Rapids is a point at which the river splits into channels around four large rock towers. It presented a hazardous obstacle to early gold prospectors.

◁ **WINDING COURSE**
For much of its lower course, the Yukon is a slow, winding river with many large meanders and oxbow lakes scattered throughout its floodplain.

## HOW AN OXBOW LAKE FORMS

Erosion and deposition make the shape of a river meander more and more accentuated over time. Eventually, the river cuts its course across the neck of the loop and follows a more direct route. The abandoned section forms a distinctive U-shaped oxbow lake.

erosion on outside of bend

steady bend (meander) in river

flow direction

deposition on inside of bend

**RIVER WITH MEANDER**

river straightens out

old loop cut off, forming oxbow lake

abandoned channel

**BEND IS CUT OFF**

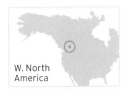

W. North America

# Lake Abraham

*A glacial lake with unusual natural sculptures captured within its frozen surface during winter*

Lake Abraham is an artificial lake created by the damming of the North Saskatchewan River in northern Alberta, Canada, in 1972. It stretches 32 km (20 miles), oriented north to south along the river valley, and is 3.3 km (2.0 miles) across at its widest.

## What lies beneath

For much of the year, Lake Abraham displays the turquoise waters seen in most glacial lakes of the Canadian Rockies. However, during winter, beneath its clear, icy surface, bizarre frozen stacks of cotton-wool-like bubbles appear. These amazing formations actually contain highly flammable methane gas produced as bacteria at the bottom of the lake decompose dead organic material. In summer, the methane simply rises to the surface and escapes, but in winter the gas is trapped – and held on dazzling display – until the spring thaw.

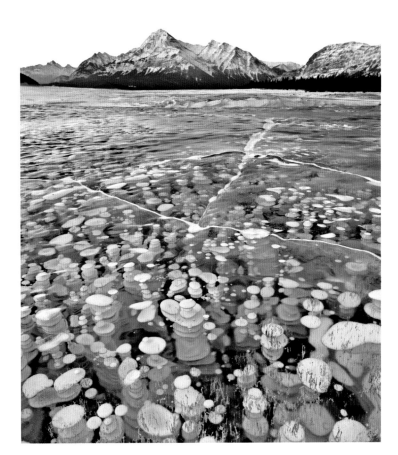

△ **BUBBLE STACKS**
Bubbles of methane gas rising from the bottom of the lake are trapped in ice as they reach the colder surface, with successive bubbles freezing in stacks.

## MAKING WAVES
Although unaffected by tides, the Great Lakes can take on sea-like features. Sustained winds may produce strong currents and huge waves, like these in Lake Erie.

## ELEVATIONS AND DEPTHS

This profile shows the water surface elevations and depths of the Great Lakes. Four of the lakes are at varied elevations, producing the effect of a stepped waterway flowing from west to east. Lake Superior, the deepest lake, holds more water than the other four lakes combined.

Lake Erie elevation: 173 m (569 ft)

Lakes Michigan and Huron elevation: 176 m (577 ft)

Lake Erie depth: 64 m (210 ft)

Niagara Falls

lock

Lake Ontario elevation: 74 m (243 ft)

Lake Superior elevation: 183 m (601 ft)

dams

Gulf of St Lawrence

Lake Superior depth: 406 m (1,333 ft)

Lake Huron depth: 229 m (750 ft)

Lake Michigan depth: 281 m (923 ft)

Lake Ontario depth: 244 m (802 ft)

△ WINTER FREEZE
During winter, large areas of the Great Lakes freeze over, causing much disruption to shipping. Here, on Lake Huron, distinctively shaped pancake ice has formed.

E. North America

# The Great Lakes

*Earth's largest system of freshwater lakes, often referred to as inland seas*

North America's Great Lakes have a combined surface area of 244,000 square km (94,000 square miles) – exceeding that of the UK – and contain one-fifth of the world's fresh surface water. The system comprises five interconnected lakes – Superior, Michigan, Huron, Erie, and Ontario – draining from west to east and emptying into the Atlantic Ocean via the St Lawrence River. Although the gradient between the lakes is generally shallow, the descent between lakes Erie and Ontario features the dramatic Niagara Falls (see p.52).

## Diverse shorelines

The Great Lakes began to form at the end of the last ice age, when moving ice sheets gouged huge basins that subsequently filled with water as the ice melted. The current shapes of the lakes started to become established about 10,000 years ago. Their shorelines vary considerably, from sandy beaches to rocky cliffs, and even wetlands. Also located on their shores are several major cities including Toronto, Chicago, Cleveland, Buffalo, and Detroit.

rows of protective bony plates (scutes)

barbels used to sense bottom-dwelling prey

shovel-like nose

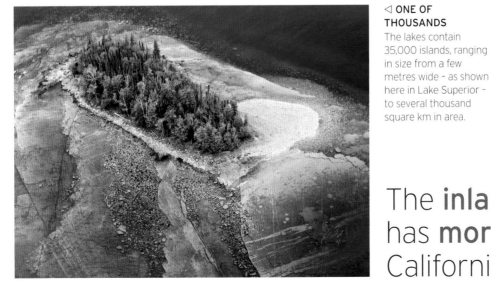

◁ **ONE OF THOUSANDS**
The lakes contain 35,000 islands, ranging in size from a few metres wide - as shown here in Lake Superior - to several thousand square km in area.

▷ **LARGEST NATIVE**
The lake sturgeon is the largest of the more than 150 native fish species in the Great Lakes, growing to about 2 m (6 ft) in length and weighing 90 kg (200 lb) or more.

The **inland** US state of **Michigan** has **more coastline** than California or **Florida**

# Niagara Falls

*A powerful and world-famous waterfall system with one of the highest flow rates of any waterfall on Earth*

E. North America

△ HORSESHOE FALLS
Visitors can experience the incredible power of Niagara Falls by taking a boat trip to Horseshoe Falls - one of the world's most accessible natural wonders.

Niagara Falls is a system of three waterfalls located on the Niagara River, through which water flows from Lake Erie into Lake Ontario. The falls lie on the Canada–US border and form the southern end of the Niagara Gorge.

The Horseshoe Falls, the largest of the three, has a dramatic crest line that is 670 m (2,200 ft) wide and a vertical drop of more than 57 m (188 ft) into a plunge pool 35 m (115 ft) deep. Separated by Goat Island, the American Falls and the diminutive Bridal Veil Falls crash down 20–35 m (66–115 ft) onto a talus – a slope of rocks – formed by a huge rock slide in 1954. At peak times, about 170 million litres (45 million gallons) of water pass over the three falls each minute.

## Cutting back

Niagara Falls was formed as the Niagara River plunged over a huge escarpment on its way to the Atlantic Ocean. In a process that continues today, the river eroded layers of soft rock sitting below a harder dolomite caprock. As the soft layers eroded, undercutting the dolomite above, great chunks of the caprock gave way, leaving a vertical cliff face.

▽ DISTINCTIVE GOOSE
The Canada goose is one of the most conspicuous birds in the Niagara Falls area. Some are resident, but large flocks of these geese also travel through during their spring and autumn migrations.

### THE RETREATING FALLS

The great erosive power of Niagara Falls is demonstrated by its fast rate of movement, tracked since the end of the 17th century. It is estimated that, at the present rate of erosion, the falls will reach Lake Erie in about 50,000 years from now.

direction of flow of river

2005
1886
1842
1819
1764
1678

edge of waterfall

## Niagara Falls has **retreated 11 km (7 miles)** over the last **12,500 years**

**△ MEANDERING WATERWAY**
The lower Mississippi River and its tributaries are slow-moving waterways. They wind through a gently contoured landscape that includes lowland forest habitats, as shown here, and open floodplains.

**◁ BIRD'S FOOT DELTA**
Also known as a bird's foot delta, due to its shape, the Mississippi River Delta deposits more than 550 million tonnes (606 million tons) of nutrient-rich sediment into the Gulf of Mexico each year, leading to the formation of algal blooms.

# The Mississippi-Missouri

*A vast river system that extends over much of North America, from the Rocky Mountains in the west to the Appalachians in the east*

C. North America

The Mississippi–Missouri river system is the largest watershed in North America, draining more than 40 per cent of the contiguous US (all of the country except for Alaska and Hawaii). The Mississippi River's source is Lake Itasca in northern Minnesota, which it leaves as a stream just 3 m (10 ft) wide. The Mississippi's western tributaries, including those of the Missouri, drain the Great Plains, while those to the east drain the Appalachian Plateau.

## Great River

In combination, the Mississippi–Missouri stretches 5,970 km (3,710 miles), making it the fourth longest river in the world after the Nile, Amazon, and Yangtze. Also known as the Big Muddy, the silt-laden Missouri discharges into the clearer Mississippi just north of St Louis in Missouri. However, it is only after its confluence with the Ohio River – the Mississippi's largest tributary by volume – at Cairo, Illinois, that the river takes on the full grandeur that led Algonquian-speaking inhabitants to name it *Misi-ziibi* – "Great River".

**▷ FALSE MAP TURTLE**
The endemic false map turtle is an inhabitant of large streams and rivers of the Mississippi-Missouri. It is often found basking on logs and rocks.

## SHIFTING DELTAS

The modern Mississippi Delta evolved after the last ice age, when water levels began to stabilize. Over the last 9,000 years, it has shifted across a distance of about 320 km (200 miles), in a sequence of at least seven episodes. The seven major delta lobes and changing river courses are shown here.

present course of Atchafalaya River

present course of Mississippi

**KEY**

| | |
|---|---|
| ■ | 4,600 years before present (BP) |
| ■ | 4,600-3,500 years BP |
| ■ | 3,500-2,800 years BP |
| ■ | 2,800-1,000 years BP |
| ■ | 1,000-300 years BP |
| ■ | 750-500 years BP |
| ■ | 550 years BP |

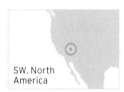

SW. North
America

# The Grand Canyon

*A vast gorge cut by the Colorado River, the Grand Canyon holds a record of 1.7 billion years of Earth history in its walls*

One of North America's most iconic landscapes, the Grand Canyon is a spectacular valley carved into the Colorado Plateau by the Colorado River over millions of years. Renowned for its dramatic rock formations, precipitous cliffs, and striking coloration, the canyon is named for its impressive scale. Averaging 1,220 m (4,000 ft) deep for its entire length of 446 km (277 miles), it drops 1,830 m (6,000 ft) below the rim at its deepest and reaches 29 km (18 miles) across at its widest.

### A record in the walls

An extraordinary demonstration of the powerful forces of erosion, the Grand Canyon is one of Earth's best-studied geological features. The layers of rock exposed on its walls present a remarkable record of geological history reaching back almost 2 billion years, although the canyon itself is much younger.

▷ **CLIFFS AT TUWEEP**
Sunlight pierces the early morning fog, illuminating the sheer face of the canyon wall and revealing the Colorado River far below.

◁ **GAINING TRACTION**
Forked hooves that spread apart to improve grip enable the desert bighorn sheep to move freely on steep slopes. It can survive for long periods without drinking water.

## In some places, **the canyon** is more than **1.8 km (1 mile) deep**

### THE SHAPE OF THE CANYON

The Grand Canyon extends from Lees Ferry in the east to Grand Wash Cliffs in the west. As well as the main river channel, there are hundreds of miles of tributaries contained within side canyons. The canyon's North Rim is about 300 m (980 ft) higher than the South Rim as the plateau slopes downwards from north to south.

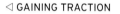

edge of
North Rim

upstream end
of canyon

Marble
Canyon

Great Thumb
Point

Colorado
River

Tuweep

Supai

Phantom
Ranch

North Rim

Angel's
Window

downstream end
of canyon

edge of
South Rim

Powell Plateau
(see pp.56-57)

# The formation of a
# RIVER CANYON

The Grand Canyon is a spectacular example of a gorge – a steep-sided mountain valley eroded by a river. Weathering processes have both shaped the inside of the canyon – producing features such as mesas, buttes, and spires that are often seen in arid and semi-arid places – and helped to make it wider (see panel, below).

A river requires a lot of erosive power to carve such a deep gorge. The Colorado owes its power to its steep gradient, which increases the speed at which the water moves along its channel, and to a relatively high sediment load, which enables the water to scour and deepen the channel.

The Grand Canyon is a relatively young canyon carved into old rocks. It seems likely that the Colorado gained the power to carve the canyon when the Colorado Plateau was pushed upwards by tectonic forces, steepening the slope of its bed. This sudden increase in a river's power is called rejuvenation.

## HOW THE GRAND CANYON WAS FORMED

The Colorado Plateau is made up of layers of sedimentary rock (or strata), including limestone, sandstone, and shale. The shape of the canyon is partly determined by how resistant these rocks are to processes of weathering and erosion.

### DOWNCUTTING
The Colorado River has cut down progressively into the plateau. The canyon widens significantly whenever the river encounters a layer of relatively soft rock.

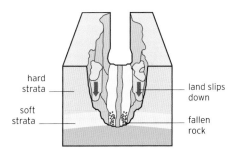

### CLIFF COLLAPSE
The river is able to cut relatively easily through soft strata, undercutting cliffs above made of harder rock and causing them to collapse.

▶ **THE GRAND CANYON**
This cross-section of the Grand Canyon, at a point near the Powell Plateau, displays an array of geological formations created by the cutting power of the Colorado River. Each layer of rock has put up a different level of resistance to the river.

**The South Rim**
The South Rim tilts away from the canyon, so water falling on its surface does not flow over the edge into the Colorado.

**Shallow slopes**
Shallow slopes are formed when a river encounters a layer of soft rock, such as shale.

◁ **COLORADO CONFLUENCE**
The Little Colorado River is one of the Colorado River's largest tributaries. Tributaries add both water and sediment to the main channel, often altering the colour of the water. The point where two rivers meet is called the confluence.

The **oldest rocks** exposed in the walls at the **bottom of the canyon** are almost **2 billion years old**

**Side canyon**
Side canyons are carved by tributaries of the main river channel. Those of the Grand Canyon have contributed to its width.

◁ **INCISED MEANDER**
Meanders are winding bends usually found in the flat lower courses of rivers. Those occurring on an entrenched river – one that is confined to a canyon – are known as incised meanders. These cut deep gorges, like those seen in the Grand Canyon, along pre-existing routes.

△ **KAIBAB PLATEAU**
The Grand Canyon cuts through the southern end of the Kaibab Plateau, which is capped with limestone.

**The North Rim**
More elevated than the South Rim, the North Rim experiences much colder temperatures and snowfall during winter.

**Steep cliffs**
Cliffs are often made of more durable rocks, usually strongly jointed sandstone and limestone.

**Rock layers**
Sedimentary rock strata are usually laid down in horizontal beds, with younger layers above older layers. This creates a geological record that is revealed through erosion.

**Colorado River**
The Colorado flows along the floor of the canyon and continues to cut through layers of rock.

**Oldest rock layer**
Unlike the sedimentary rocks above, the oldest rock layers are made up of igneous and metamorphic rocks. These hard rocks resist weathering and erosion, so the bottom of the canyon is narrow with steep sides.

▷ **ViSHNU BASEMENT ROCKS**
These very hard, crystalline rocks – a combination of schist, granite, and gneiss – exposed near the bottom of the Grand Canyon are among its oldest rocks, having formed more than 1.7 billion years ago.

▷ **NIGHT HUNTER**
Living in the aptly named Bat Cave at the start of the cavern complex are thousands of Mexican free-tailed bats, which roost there by day and leave each evening to feed.

tail extends beyond tail membranes

large external ears

S. North America

# Carlsbad Caverns

*A vast underground cave system containing a stunning array of calcite formations*

Beneath the Chihuahuan Desert in the Guadalupe Mountains, New Mexico, lies a series of massive underground chambers and passages decorated with stalactites and stalagmites. The caverns formed through an unusual process in which groundwater, rich in sulphuric acid, invaded and dissolved the limestone rock on a huge scale. Only later did percolating rainwater enter the resulting caves and furnish their interiors with the elaborate calcite formations seen today. The famous Big Room, a rambling chamber up to 191 m (625 ft) wide in places, contains many remarkable features including the Hall of Giants, a series of gigantic stalagmites rising almost 20 m (65 ft) from the cave floor, and the Painted Grotto, coloured by iron oxide and hydroxide pigments.

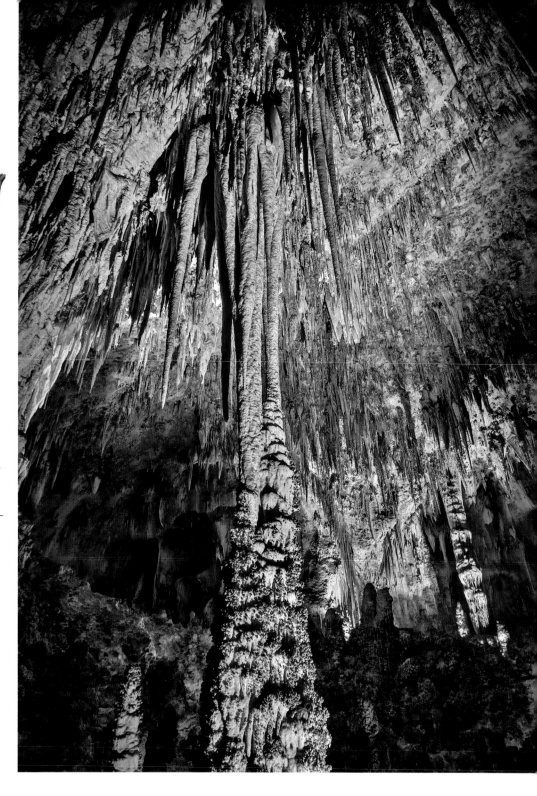

## STALACTITES, STALAGMITES, AND PILLARS

As mineral-saturated water seeps through the roof of a cave and drips down, a residue starts to build up and a stalactite forms. As drops of water hit the ground, they gradually create a mound of residue that grows into a stalagmite. Over time, the two may join to create a pillar.

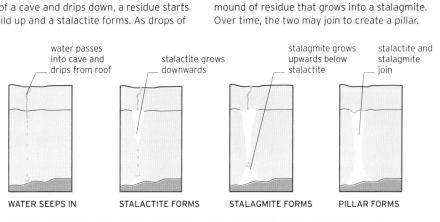

water passes into cave and drips from roof

stalactite grows downwards

stalagmite grows upwards below stalactite

stalactite and stalagmite join

**WATER SEEPS IN**    **STALACTITE FORMS**    **STALAGMITE FORMS**    **PILLAR FORMS**

◁△ **ELABORATE DECORATIONS**
Many of the caverns' roofs are covered with delicate tubular stalactites called straws, formed by water moving slowly through cracks in the rock. Curtain-like deposits of calcite, called flowstones (left), form when water runs down the cavern walls.

# The Okefenokee Swamp

*One of the largest swamps in North America, a diverse wetland
ecosystem displaying remarkable biodiversity*

SE. North
America

Straddling the border between Georgia and Florida,
the Okefenokee Swamp is a large, shallow wetland
area lying in a basin that was once beneath the Atlantic
Ocean. The swamp is dominated by peat-filled mires
but also contains lakes, flooded prairies, and mixed
woodlands. With an area of 1,770 square
km (685 square miles), it is considered
the largest "blackwater" swamp in
North America. Although its waters
are free of sediment, they are stained
a dark colour by tannins released by
peat and decomposing vegetation.

Over the 7,000 years that the swamp
has been forming, peat has collected to
depths of more than 4.5 m (15 ft). In some
areas, floating sections shudder when
trodden on; translated from the Hitchiti
language, Okefenokee means "water-
shaking". The swamp is drained by two
rivers: the Suwannee flows into the Gulf
of Mexico, while the St Marys empties
into the Atlantic.

## Flourishing wildlife

A great variety of animals and plants thrive in the
Okefenokee Swamp's mosaic of habitats. A haven for
more than 230 bird species, including many wading
birds and waterfowl, the swamp is well known for
its abundance of amphibians and reptiles including
alligators, snakes, and the alligator snapping
turtle, the world's largest freshwater turtle.
It is also a key habitat for a population
of Florida black bears and home to other
mammals, including bobcats and otters.
More than 600 plant species include the
carnivorous sundews, bladderworts, and
pitcher plants, which have developed
their unusual diets to compensate for
the nutrient-poor soils.

◁△ **ACIDIC WATERS**
The acidic waters and soils of the Okefenokee
Swamp have led to unusual plant adaptations. The
hooded pitcher plant is one of several varieties that
capture and digest insects in their modified leaves.

## HOW A MIRE FORMS

Mires (or bogs) form as flooded
areas are infilled by sediment.
Fen peat forms as mineral-rich
groundwater prevents dead
plant matter from decomposing.
As the surface rises above the
groundwater, mosses grow and
their remains become bog peat.

accumulating mud

aquatic plants
take root in shallows

lake clay  impermeable bedrock
**WETLAND SUCCESSION**

trees grow on new peat

fen peat collects
and fills up lake

**INTERMEDIATE STAGE**

bog peat, covered
with bog moss

**PEAT MIRE**

# The Everglades

*North America's only subtropical wetland system, dominated by iconic sawgrass prairies known locally as the river of grass*

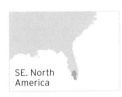

SE. North America

Located at the southern tip of the Florida peninsula, the Everglades is a vast wetland wilderness containing a variety of lowland habitats. Much of the inland area is covered by the famous sawgrass prairies that are permanently inundated with water. This so-called river of grass, which also contains lakes, waterways (sloughs), and cypress marshes at its edges, is dependent on fires – usually started by lightning strikes – for maintenance and regeneration. Islands of hardwood forests, known as hammocks, contain a mixture of trees including tropical mahogany and temperate oaks and maples, while some of the highest areas, including dry ridges, are covered in South Florida slash pine. In coastal areas, where fresh and salt water merge, salt marshes and tidal mangroves dominate. The combined mangrove system of the Everglades lining the coast, creeks, and estuaries is the largest in the western hemisphere.

## Critical habitats

A major migration stopover, the Everglades is home to more than 400 species of birds. Some 50 species of reptile live in the area, and it is the only place in the world where alligators and crocodiles coexist. The wetland provides critical habitats for many threatened and endangered species.

▷ **LOCAL WADER**
Roseate spoonbills roost in trees, then fly to their feeding grounds at sunrise, where they search for fish and other small prey.

# One in three people living in Florida **rely** on the Everglades for their water supply

△ **UNDERWATER GRAZERS**
A West Indian manatee mother and nursing calf graze in shallow waters on seagrasses and other aquatic plants.

◁ **WINDING SLOUGHS**
A series of slow-moving sloughs snake a route through sawgrass before emptying their fresh water into the brackish waters of an Everglades estuary.

▷ **CYPRESS SWAMP**
Found throughout the Everglades, cypress swamps are populated by trees that have adapted to flooded conditions. They are home to many species, including crocodilians.

## ECOLOGICAL SUCCESSION

This cross-section of the Everglades shows the relationship between different vegetation types and water inundation levels. The forested areas develop slowly as layers of organic material build up and elevate the land. The change from one type to the next, known as succession, may be affected by fire or environmental factors and may even be reversed.

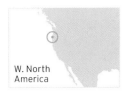

W. North
America

# Thor's Well

*A gaping hole in a rocky stretch of Oregon's coastline that acts like a giant drainpipe*

Situated on an area of coast called Cape Perpetua, Thor's Well is circular in shape, about 6 m (20 ft) deep, and has a hole at the bottom that connects with the open sea. As the waves crash in at high tide, the well fills up, overflows, and spouts seawater into the air. The water is then sucked back for several seconds, making it appear as if the sea is being drained.

## Churning chasm

Thor's Well churns most spectacularly around the time of high tide. The stirring action depends on the height of the high tide and the direction and size of the waves. The wind can also be a factor. During storms, water can burst out of the well extremely violently, making it too dangerous to approach closely.

△ **NATURAL DRAIN**
Every 10 seconds or so, during high tide, it seems as if the ocean is being drained away down a bottomless hole. This can be a dramatic sight.

A large number of **mussels** gamely **cling on** to the well's **internal walls** as the water churns

W. North
America

# Monterey Bay

*A diverse ecosystem, with an underwater canyon and kelp forests on the sea bottom*

Monterey Bay occupies a large stretch of the coast of central California. Underwater is a thick forest of kelp and the shore end of one of the world's longest undersea canyons, which runs for 100 km (60 miles) to the southwest. Around the bay are some pristine beaches and jewel-like tide pools.

## Marine sanctuary

Offshore, the Monterey Bay National Marine Sanctuary covers a much larger area of around 15,700 square km (6,100 square miles). It is home to a fantastic diversity of sea life including more than 30 species of mammals, over 300 species of fish, and nearly 100 bird species.

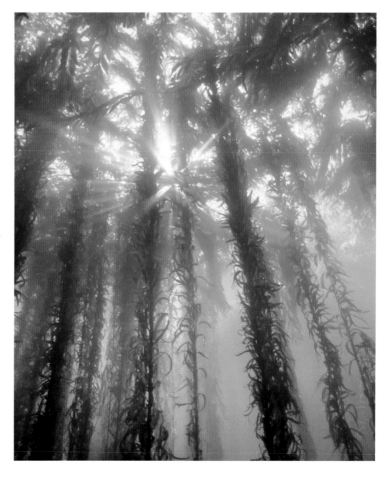

◁ **ALGAL FOREST**
Several species of kelp, a form of seaweed, grow in the bay's nutrient-rich waters. Kelp can grow up to 50 cm (20 in) a day, with some specimens becoming as tall as trees.

# Big Sur

*A long, rugged stretch of the central California coastline, with steep mountains and stunning views*

W. North America

Widely regarded as one of the most spectacular meetings of sea and land in the world, Big Sur is a roughly 160-km- (100-mile-) long section of undeveloped, wild coastline. Some sections of the hillsides just above the shoreline have been shaped into flat areas called marine terraces. These are former beaches lifted up by tectonic activity. A winding, narrow road extends down its length, affording spectacular views of the Pacific. The road and Big Sur in general, however, are susceptible to landslides as a result of wave action, weakening of the cliffs by faulting and fracturing, the destruction of vegetation by summer fires, and heavy winter rainfall.

## BIG SUR'S COASTAL SLOPE

One area of the Big Sur coastline, to the west of Cone Peak (in the Santa Lucia Range), has the steepest coastal slope in the USA outside of Alaska and Hawaii. Here, the land rises sharply in a series of steps to Cone Peak, which is just 5 km (3 miles) from the Pacific Ocean.

Villa Creek Canyon • Cone Peak 1,572 m (5,158 ft) • coastal road • marine terrace • Pacific Ocean

quartz crystal — diopside crystal

### ▷ DEEP-GREEN JEWEL
The mineral diopside is common in the Big Sur region. Here, typical dull green diopside crystals are embedded with quartz crystals in a rock groundmass.

### ◁ MISTY HEADLANDS
Big Sur, seen here at sunset, winds in and out for much of its length, forming a near-endless array of small bays and headlands, often shrouded in mist.

### ▽ WAVE-BATTERED ROCKS
Around Big Sur's headlands, small rocky islets typically extend up to 100 m (330 ft) offshore. These are continuously pounded by waves surging in from the Pacific.

# The Bay of Fundy

*A deep inlet into North America's eastern seaboard, best known for its huge tides but also for its dinosaur fossils and semi-precious minerals*

E. North America

Located on Canada's east coast, the Bay of Fundy is a unique coastal environment with a landscape of breathtaking cliffs, sea caves, and unusual rock formations. The rock layers and outcrops around the bay display an enormous geological diversity and range of ages. They tell a story of hundreds of millions of years of natural history with, for example, dinosaur discoveries from the Triassic Period and fossils of marine invertebrates from the Carboniferous Period.

## Massive tides

High cliffs channel the tidal waters flowing into the bay until they separate into two narrow arms (called Chignecto Bay and the Minas Basin) at its northeastern end. This channelling effect and the bay's overall shape give rise to some of the world's fastest-running tides and highest tidal ranges (the difference between high and low tides) – the highest ever being a range of 21.6 m (71 ft) recorded one night in October 1869 at Burntcoat Head in the Minas Basin. The tidal surge in the northern arm, Chignecto Bay, produces a bore up to 1.8 m (6 ft). The twice daily tides transform the shorelines, tidal flats, and exposed sea bottom as they flood into the Bay and its harbours. They also create unique conditions for whales and seabirds.

analcime crystals

◁ **MINERAL CLUSTER**
Cape Blomidon, which borders the Minas Basin, and other sites in the bay are important source areas for the white mineral analcime.

## 160 billion tonnes (176 billion tons) of seawater flow in and out of the bay twice a day

### TIDAL RANGES

At a typical Fundy location like Cape Blomidon (see graph below), the average tidal range is about 12 m (39 ft). But at some locations, it can be as high as 20 m (65 ft). As well as depending on location, tidal ranges are affected by the monthly lunar cycle.

Tidal range

| M | FT |
|---|---|
| 8 | 25 |
| 6 | 20 |
| 4 | 12 |
| 2 | 6 |
| 0 | 0 |
| -2 | -6 |
| -4 | -12 |
| -6 | -20 |
| -8 | -25 |

tidal period: 12 hours 25 mins

Time (hours): 0 6 12 18 24 30 36 42 48

E. North America

# Acadia Coastline

*An area of rugged coastal scenery, interspersed with tranquil inlets and harbours*

Forming part of the coast of Maine, Acadia National Park consists principally of one large island, Mount Desert Island, and several smaller islands nearby. It is best known for its distinctive rocky coastline – the result of a complex series of geological processes over hundreds of millions of years. These include the intrusion of magma into existing sedimentary rock layers about 360 million years ago to form granite masses. From about 2 million years ago, a huge ice-sheet blanketed the area and sculpted out a series of mountains separated by U-shaped valleys. Today, waves and tides are major agents of change at Acadia, gradually eroding cliffs and depositing the debris as sediments around the coastline.

## The **base rocks** of the Acadia coastline are up to **500 million years old**

△ **MEANDERING CHANNELS**
The bay's vast intertidal zone (the area covered at high tide and exposed at low tide) has a web of deep channels scoured out by tidal currents.

▽ **LOW TIDE**
At St Martins, on Fundy's northern shore, it is possible at low tide to "walk the ocean floor" out to sea caves. At high tide, the sea fills the caves.

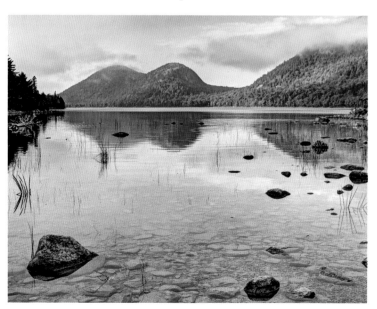

△ **SHELTERED INLET**
In some areas, wave-pounded cliffs give way to tranquil inlets. Sea-level rise since the last ice age has resulted in the flooding of these glacier-sculpted valleys.

# North American Boreal Forest

*The largest intact forest on Earth, one of only five that contain expanses of untouched wilderness, and a breeding ground for billions of birds*

N. North America

Boreal forest, or taiga, is the world's largest terrestrial biome, consisting of vast expanses of mostly conifer species. In North America alone, 6 million square km (2 million square miles) of boreal forest stretches from Alaska to Newfoundland, and from the Great Lakes as far as the Arctic tundra.

## Cold-climate forest

The boreal forest is also known as a snow forest due to its climate. Short, humid summers of 50–100 frost-free days are followed by long, cold winters with up to six months of temperatures below freezing. Precipitation falls mainly as snow. Conifers, such as spruce, are well adapted to survive such rigorous conditions (see panel, below), and they dominate the landscape. Lakes, rivers, and wetlands are numerous, making the North American boreal forest Earth's largest source of unfrozen fresh water. Because it is still relatively intact, this is a vital habitat for wildlife such as moose and brown bears, and a breeding ground for up to 3 billion North American birds.

▷ **BOREAL SPECIALIST**
The mainly solitary Canadian lynx roams the boreal forest in all seasons, following its primary prey species, the snowshoe hare.

◁ **DENSE COVER**
Conifers such as spruce, pine, and fir grow close together, forming an almost unbroken canopy that allows little light to reach the forest floor.

### SHEDDING SNOW

A conifer's conical shape and flexible, downward-pointing branches prevent heavy snow accumulation. Narrow, needle-like leaves also prevent snow build-up and have a waxy outer coat that prevents water loss by evaporation in freezing conditions.

snow slides off branches when too much builds up ____

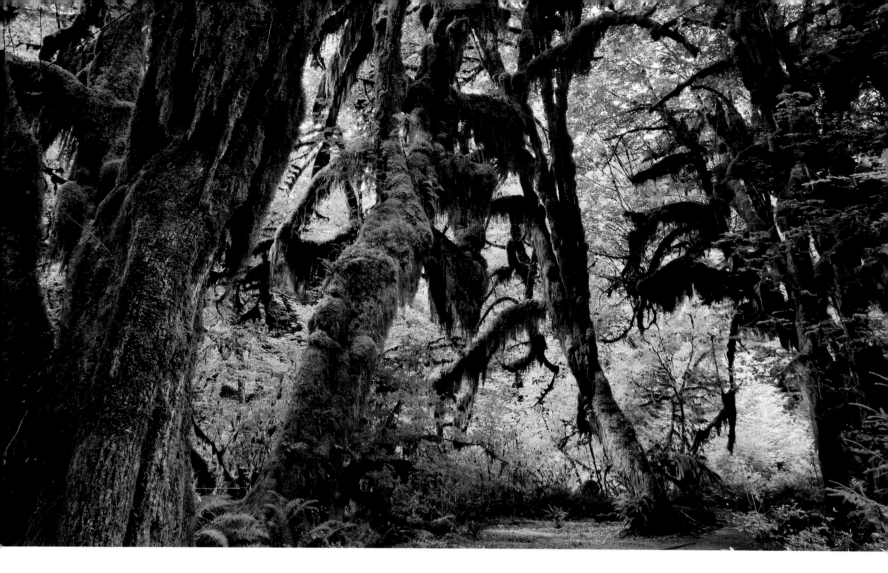

△ CANOPY OF MOSSES
The continually wet conditions of the coastal temperate rainforest provide an ideal habitat for moss. Over 100 varieties can be found draped over the branches of the conifer trees that dominate this environment.

# Pacific Northwest Rainforest

*The world's largest undeveloped coastal temperate rainforest, with the greatest amount of wood – living and dead – of any forest on Earth*

W. North America

△ FOREST FERNS
The rich forest floor provides nutrients for many fern species, including lady, deer, and sword ferns.

Coastal temperate rainforests are a very rare forest type. Today, they are believed to cover just 302,200 square km (116,700 square miles), which is less than 0.2 per cent of Earth's land area. Oceans, mountains, and high rainfall are features of these forests, and the largest intact areas stretch from northern California to Canada and the Gulf of Alaska. The Pacific Northwest measures annual rainfall in metres; 2.5–4.2 m (8–14 ft) is not unusual. Constantly decaying cedar, Sitka spruce, Douglas fir, and coast redwood needles and deciduous leaves create rich soil that supports hundreds of moss, fern, and invertebrate species, while rivers, lakes, and streams are home to salmon and trout, which in turn attract bears and bald eagles.

▷ WASTE MANAGEMENT
A vital part of the ecosystem, the Pacific banana slug breaks down rotting plants, spreading seeds and spores in the process.

## THE WORLD'S COASTAL TEMPERATE RAINFORESTS

Apart from North America's Pacific Northwest, coastal temperate rainforests occur in southwest South America; the northeastern Atlantic region (including Iceland, Ireland, Scotland, Norway, and northern Spain); the eastern Black Sea; southwest Japan; New Zealand, Tasmania, and Australia's New South Wales; and the tip of South Africa.

W. North
America

# The Giant Forest

*An ancient, sky-scraping forest of more than 8,000 giant sequoias, home to half of the largest and longest-lived trees on Earth*

Growing at the heart of central California's Sequoia National Park, in the southern Sierra Nevada mountains, the Giant Forest is home to more than 8,000 giant sequoias (*Sequoiadendron giganteum*). While they may not be the world's tallest trees, they are easily the largest by volume.

## A tale of two redwoods

Giant sequoias are often called redwoods and vice versa, but while both species are native to California and share similar cinnamon-coloured bark and heady heights, they are decidedly different. The trees of the Giant Forest generally grow at elevations of 1,500–2,100 m (4,900–6,900 ft) on western Sierra Nevada slopes, making them much more cold-hardy than coast redwoods (*Sequoia sempervirens*), which

grow at lower coastal elevations. Giant sequoias live for around 3,000 years (as opposed to the redwoods' 2,000), and their bark is up to 60 cm (2 ft) thick. Sequoias do not grow as tall as redwoods – the maximum is around 95 m (310 ft) compared to the redwoods' 115 m (380 ft) – but what they lack in height is made up for in girth. The largest tree in the world, the Giant Forest's General Sherman, stands just 84 m (275 ft) high but is more than 30 m (100 ft) wide and weighs an estimated 1,225 tonnes (1,350 tons). At 1,487 cubic meters (52,513 cubic feet), it is the single largest living organism by volume in the world.

◁ **OPPORTUNISTIC SCAVENGER**
The Steller's jay's raucous call often breaks the silence of the Giant Forest. As omnivores, jays eat anything from seeds and nuts to hikers' leftovers.

## The **high tannin** content in giant sequoia **bark prevents rotting**

▷ **MAMMOTH PROPORTIONS**
The President is 75 m (245 ft) tall, making the scientists shown with it ant-like by comparison.

### CONIFER HEIGHTS

Because they lack the spread of deciduous trees, conifers grow much taller than the latter, but their heights vary greatly according to species. Coast redwoods are the tallest trees on Earth.

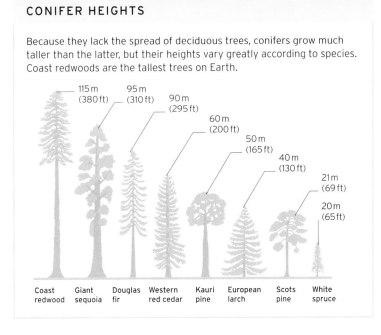

115 m (380 ft) — Coast redwood
95 m (310 ft) — Giant sequoia
90 m (295 ft) — Douglas fir
60 m (200 ft) — Western red cedar
50 m (165 ft) — Kauri pine
40 m (130 ft) — European larch
21 m (69 ft) — Scots pine
20 m (65 ft) — White spruce

△ **SHADE TOLERANT**
Although the trees of the Giant Forest cast giant shadows, small trees such as the Pacific dogwood survive well in the shade.

# Pando Aspen Grove

*One of the oldest organisms in the world, a quivering grove with a surprising genetic heritage*

W. North America

△ **SHIVERING TREES**
Quaking or trembling aspens are so named because their leaves, which are attached via long, flattened stalks, flutter in even a slight breeze, making the trees appear to shiver.

It contains what seem to be around 47,000 individual trees, but the Pando grove of central Utah is in fact one giant living organism – a clonal colony of quaking aspen. All of its trees are genetically identical, and all began as ramets, or suckers, growing from and sharing a single root system.

## Ancient roots

Aspen clonal colonies are common in North America. What makes Pando – Latin for "I spread" – special is its size: it sprawls over 44 hectares (108 acres), with an estimated weight of 6,000 tonnes (6,615 tons). Pando's accepted age is also unusual. Although individual trunks live around 100–150 years, the clonal root system is thought to be at least 80,000 years old. Some scientists, however, believe that the figure could be closer to 1 million. Recently, Pando's survival has been threatened because deer and elk eat its ramets. Parts of the forest are now fenced off so that new suckers grow undisturbed.

**SHARED ROOT SYSTEM**

Aspen clonal groves begin with a single seed. As a "parent" tree grows, its lateral root system spreads. Suckers sprout from these roots, forming genetically identical trees that send out their own lateral roots, and the process continues.

parent tree

ramet

root system shared by trees

# The Cherokee National Forest

*At the heart of an ancient mountain range, a mist-shrouded forest that is one of the most biologically diverse in North America*

E. North America

The Cherokee National Forest follows Tennessee's state border with North Carolina, covering around 2,630 square km (1,015 square miles) of the Appalachian Mountains' eastern slopes. Because the southern Appalachians were not covered by glaciers during the last ice age, diverse plant and animal species flourished, and the forests that covered the mountains provided a relatively isolated habitat until the mid-19th century. The Cherokee forest is a natural sanctuary, with more than 20,000 plant and animal species. Beech, hickory, sassafras, birch, maple, and numerous types of oak are just some of the forest's common trees, attracting more than 120 bird species including the threatened red-cockaded woodpecker.

▷ **MORNING FOG**
The Cherokee forest is bisected by the Great Smoky Mountains National Park, and the thick, blue-tinged fogs are common to both areas.

▽ **ON THE PROWL**
Around 1,500 black bears roam the forest, relying on the dense cover its trees provide.

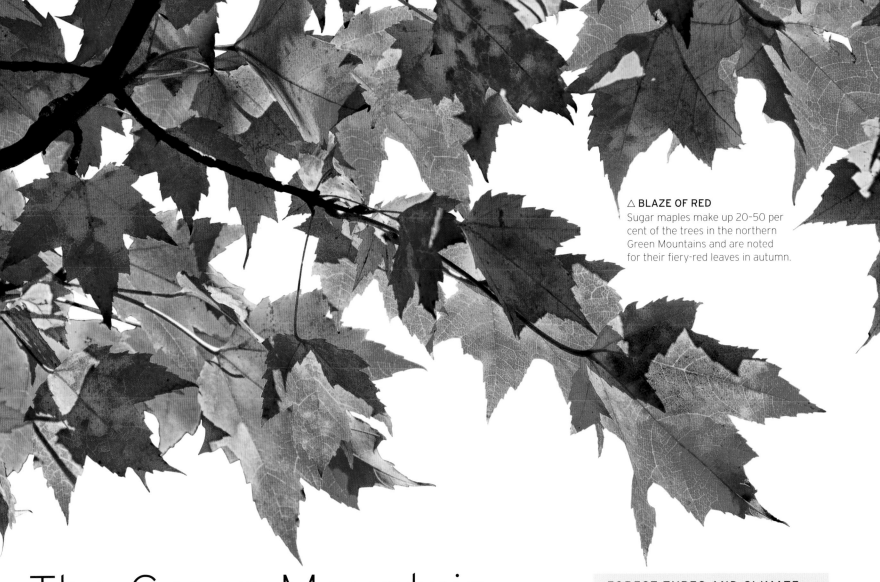

△ BLAZE OF RED
Sugar maples make up 20-50 per cent of the trees in the northern Green Mountains and are noted for their fiery-red leaves in autumn.

# The Green Mountain National Forest

*One of two national forests in northeastern North America, covering the Green Mountains with many colours in autumn*

E. North America

Vermont's Green Mountain National Forest consists of more than 1,620 square km (625 square miles), spread over and around the mountains that share its name. Many of the peaks in the forest are over 900 m (2,950 ft), and the spruce, maple, beech, and paper birch that cover them make the region a popular tourist destination, particularly in autumn, when the forest becomes a blaze of colour. Softwood as well as hardwood tree species flourish in the region, and these, coupled with a dense understorey fed by mountain streams and winter snowfall, provide food and shelter for a wide variety of bird and other animal species. Beavers, elk, deer, and wild turkeys are common, as well as grey wolves and black bears.

△ AUTUMN COLOURS
The wide variety of Green Mountain trees produces a multihued landscape in the autumn, ranging from light gold to deep red.

## The **Green Mountain Bear Corridor** serves as a vital **north-south passage** for bears

## FOREST TYPES AND CLIMATE

Forest types vary with latitude. Closest to the equator are tropical forests, which include evergreen rainforests or those with two seasons - dry and wet. At higher latitudes lie temperate forests, which consist of broadleaf rainforests as well as deciduous forests with defined seasons. At latitudes of 50° to 60° north, where summers are warm and short and winters cold and long, coniferous boreal forests flourish.

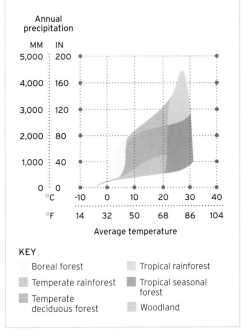

Annual precipitation

| MM | IN |
|---|---|
| 5,000 | 200 |
| 4,000 | 160 |
| 3,000 | 120 |
| 2,000 | 80 |
| 1,000 | 40 |
| 0 | 0 |

°C  -10  0  10  20  30  40
°F  14  32  50  68  86  104

Average temperature

KEY

Boreal forest

Temperate rainforest

Temperate deciduous forest

Tropical rainforest

Tropical seasonal forest

Woodland

# North American Tundra

*Stretching across the continent's northern reaches, a cold, treeless landscape where plants huddle to the Earth to survive*

N. North America

△ POLYGONAL PATTERN
The continual Arctic freeze-thaw cycle creates ice wedges that eventually shape tundra soil into polygons. These become freshwater ponds during the summer thaw.

In North America, tundra covers northern Alaska in the US, extends eastwards through northern Canada, and forms a narrow band around the coast of Greenland. North of the Arctic Circle, this ecoregion is known as Arctic tundra, while in the mountains above the timberline, it is classed as alpine tundra.

## Challenging conditions

Arctic tundra conditions are the coldest: summer high temperatures are just 4°C (40°F), plunging to -32°C (-25°F) or below in winter. The alpine tundra climate is less extreme: summers may reach 12°C (54°F) while winters rarely fall below -18°C (0°F). However, in both cases, this is mainly a treeless landscape, where vegetation hugs the ground to survive freezing winds that contribute to blizzard conditions in winter, especially on alpine tundra. Elsewhere, the constantly frozen underlying ground

layer called permafrost (see p.175) blocks the movement of water through the soil during the summer thaw, turning some low-lying Arctic tundra areas into bogs. Despite these challenges, almost 2,000 plant species grow in the tundra, primarily grasses, sedges, mosses, and shrubs. All are shallow-rooted to cope with the tundra's thin active soil layer and to survive the region's freeze-thaw cycle.

Flowering plants such as saxifrage bloom vigorously during the brief summer, while mosses and lichens provide vital nourishment in lean months for mammals such as reindeer and Arctic hares. In areas where the soil is slightly deeper and richer, mat-forming shrubs such as bilberry and crowberry produce fruit in summer that may last beneath winter snows until the following spring.

▽ SUMMER COAT
The Arctic fox is well-adapted to its environment. In summer, its white winter coat thins, changing to brownish-grey to match the tundra's rocks and low-growing vegetation.

# The Great Plains

*The largest expanse of temperate grassland in North America, a fragile landscape shared between two nations and increasingly under threat*

C. North America

At around 4,800 km (3,000 miles) long and 500–1,100 km (310–680 miles) wide, the vast grassland known as the Great Plains covers the heart of the North American continent. It slopes gently downwards from the Rocky Mountains east to the Mississippi River, and stretches from southern Canada as far south as Texas.

## Types of prairies

Once immense seas of grass known as prairies, much of the Great Plains today consists of seemingly endless fields of grain – primarily corn (maize) and wheat – planted on the most fertile soil. Other sections have been converted to cattle ranching. The natural grassland that remains consists of shortgrass and bunchgrass in the drier west, giving way to medium-grass and tallgrass prairie in central and eastern regions that receive more rainfall. Tallgrass prairie is among the world's rarest ecosystems. It is dominated by up to 60 different grass species, some of which reach heights of up to 2.4 m (8 ft), towering over an understorey of flowers, lichens, and liverworts. These support a wide range of wildlife, from tiny western harvest mice and shrews to coyotes. All prairies are fragile habitats, and various programmes are underway to preserve them.

## Less than **1 per cent** of the **original tallgrass prairies** that existed prior to European settlement **remains**

### NORTH AMERICA'S PRAIRIE TODAY

Land surveys estimate that up to 70 per cent of the original grasslands that once constituted the Great Plains have disappeared, due to conversion of land for agriculture or for oil and gas exploration. Losses have been greatest in tallgrass prairies, while around 52 per cent of shortgrass prairie remains intact.

greatest extent of prairie

present-day extent of open prairie

△ **SUPER STORM**
Huge, violent supercell thunderstorms are common summer occurrences in the Great Plains. Their rotating updrafts often spin off tornadoes.

▷ **RETURN OF THE HERDS**
American bison were hunted almost to extinction by the late 19th century. Today, around 500,000 roam the Great Plains.

# The Great Basin Desert

*North America's largest desert, where moisture arrives as snow – if it arrives at all*

W. North America

With an area of about 518,000 square km (200,000 square miles), the Great Basin is the largest desert in North America. It lies mainly in Nevada and western Utah, but stretches into California and Idaho.

Classed as a cold desert, the Great Basin's elevations range from 1,200 to 3,000 m (3,900 to 9,800 ft). The Sierra Nevada and the Rocky Mountain ranges block most moisture from the Pacific Ocean and Gulf of Mexico, forming a rain-shadow area. So the average annual precipitation is only 30 cm (12 in), and it usually arrives as snow. A single "wet year" can often be followed by many years of drought.

Habitats here vary greatly, with mountaintop "islands" of bristlecone or pinyon and juniper forests separated by sagebrush-filled valleys or arid salt lakes – in some places, the soil is so salty that no plants can survive.

## Bristlecones can live for more than 5,000 years

### ▷ BONNEVILLE SALT FLATS
The flats are covered with water in winter, which evaporates in summer, leaving behind crystallized salt.

### ▽ BRISTLECONE LANDSCAPE
Bristlecones are found at high altitudes and in harsh conditions. They are more disease-resistant and at lesser risk of wildfires than many other desert plants.

### ▷ MOJAVE ICON
The Joshua is actually a tree-sized succulent that grows up to 12 m (40 ft) in height. It takes about 50 years to mature and can live for 150 years or more.

### HOW STONES MOVE

Shallow water covers the playa (dry lakebed) surface. It freezes overnight but begins to melt in sunlight, separating the ice into panels. Light winds move these panels over the water beneath, pushing the stones along. Once the water evaporates, the "push trails" of the stones are left behind in the mud.

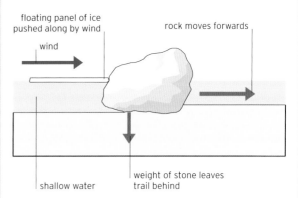

floating panel of ice pushed along by wind

rock moves forwards

wind

shallow water

weight of stone leaves trail behind

sharp, thin, spiky leaves
minimize water loss

branching stem
covered in fire-resistant
fibrous bark

SW. North
America

# The Mojave Desert

*Home of the lowest, hottest point in North America, giant yuccas, and moving stones*

North America's smallest, driest desert is sandwiched between the cold Great Basin Desert to the north and the hot Sonoran Desert to the south. Due to its proximity to two contrasting desert areas, experts rarely agree on the Mojave's boundaries, size, or type.

## Levels of extremes

Badwater Basin in the Mojave's Death Valley lies 86 m (282 ft) below sea level – the continent's lowest point – yet Telescope Peak in the Panamint Range soars 3,366 m (11,043 ft) above it. North America's highest air temperature of 57°C (134°F) was recorded here, at Furnace Creek in 1913, yet winter temperatures regularly drop below freezing. Winds above 80 kph (50 mph) often buffet the Mojave's western edge but are rare in the east. In the north, low-growing scrubland resembles the Great Basin Desert, but to the south, creosote bush takes hold.

The Joshua tree, the largest North American yucca, is a symbol of the Mojave and is often used by ecologists as an indicator of the desert's southern boundary. Death Valley is also home to "sailing stones" – boulders that appear to move unaided (see panel, left).

◁ **DEVIL'S GOLF COURSE**
In Death Valley, growing crystals push against one another, forming an inhospitable, jagged landscape called the Devil's Golf Course.

△ **MYSTERIOUS MOVERS**
Stones, some as large as 320 kg (700 lb), move across a dry lakebed in Death Valley. They leave behind them trails up to 460 m (1,500 ft) long.

SW. North America

# The Sonoran Desert

*One of the wettest deserts on Earth, with an abundance of plant species*

A place of transition between temperate and tropical climates, the Sonoran Desert merges into the Chihuahuan Desert to the southeast and the Mojave to the northwest. Calderas bear witness to the prehistoric volcanic eruptions that created the desert's mountain ranges, valleys, and alluvial fans known as *bajadas*. The Sonoran's annual rainfall of 7–50 cm (3–20 in) – the most in any North American desert – supports over 2,000 plant species. These must survive average summer temperatures of 40°C (104°F), but winters are relatively mild at 10°C (50°F), with few frosts.

## ▽ DESERT IN BLOOM

Winter rains transform the Sonoran with patches of colour, as plants such as dune evening primrose and pink sand verbena bloom in profusion.

## ROOT NETWORKS

Sonoran plants use varied strategies to survive drought conditions. Within hours of rainfall, saguaro cacti send out new roots close to the surface to store water in their expanding stems. Velvet mesquite roots burrow 50 m (165 ft) downwards to tap into the water table.

saguaro cactus

velvet mesquite

shallow, spreading roots in subsoil

deep-growing roots in water table

## ▽ HANDS UP

The valley's distinctive West and East Mitten Buttes have intrigued visitors for decades. Viewed from the south, they resemble two giant mittens, held thumbs-upwards.

# Monument Valley

*A remote valley, where thousands of acres of wide, red plain are broken only by isolated stone monoliths, formed naturally over millennia*

W. North
America

Part of the Colorado Plateau, Monument Valley is a high-elevation desert, averaging about 1,700 m (5,600 ft) above sea level. It is located in one of the Southwest's driest and least-populated areas, across Utah and Arizona.

During the last 50 million years, wind and water have eroded what was once an even higher plateau, peeling away layers of shale, sandstone, and conglomerate. What remains are incredible stone formations, 120–300 m

(400–980 ft) high. Settlers in the 19th century considered the desert terrain hostile, and this, along with its harsh climate and isolated location on Navajo Nation-owned land, saved it from exploitation. Today, all of Monument Valley is owned by the Navajo Nation, which restricts visitors to a 27-km (17-mile) road tour and one hiking trail.

## The valley's majestic **rock formations** began as **sand** on a seafloor **270 million years ago**

△ **RESIDENT RATTLER**
The prairie rattlesnake is one of several species that inhabit the desert. It often takes over burrows abandoned by prairie dogs or burrowing owls.

◁ **ICONIC LANDSCAPE**
Monument Valley's buttes, mesas, and spires are familiar to millions via Hollywood's classic westerns. Each formation has a descriptive name, such as (from left to right) Brigham's Tomb, Castle Rock, Bear and Rabbit, and Stagecoach.

## MESAS AND BUTTES

Once part of a plateau, both mesas and buttes have flat tops and sheer sides. Their top layers are hardened rock, more resistant to erosion, but the layers beneath are made of less-resistant sedimentary rocks. Over time, these sedimentary layers are eroded by the action of wind and water, separating mesas from the parent plateau and forming canyons between them. Eventually, the larger mesas are worn down into smaller buttes.

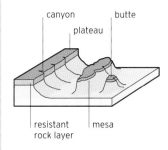

canyon
butte
plateau
resistant
rock layer
mesa

# Bryce Canyon

*A natural amphitheatre filled with thousands of strange stone spires, columns, and towers – more than anywhere else on Earth*

W. North America

△ BRYCE AMPHITHEATRE
The rim of Bryce Canyon is 2,400-2,700 m (8,000-9,000 ft) high – perfect for overlooking the national park's "amphitheatre" area.

Bryce is not a conventional canyon, because it was not created by flowing water cutting through rock. However, an ancient lake that once covered much of what is now southwestern Utah did play a role in its formation. The process began 55–35 million years ago.

## A sea of stone totems

Mineral-rich sediment from surrounding mountains was washed down by rainfall and deposited on the lake's base, eventually turning into the pink limestone that geologists today know as the Claron Formation. Geological upheaval forced the lakebed upwards by nearly 1.6 km (1 mile) to create a plateau. Millennia of weathering and erosion subsequently carved the Claron Formation first into cliffs, then into tall columns of rock, some a mere 1.5 m (5 ft) tall, others rising as high as 45 m (150 ft) (see panel, right). Known as hoodoos, these multicoloured towers contain sections of varying thicknesses and are composed mainly of limestone but with silt and mudstone as well. Minerals such as iron oxide and manganese oxide add pink, purple, and blue tints. The different hues and lumpy, unequal shapes of the stone spires remind some visitors of wooden totem poles.

## Freezing and thawing

While hoodoos occur on every continent, they are more common in the northern part of Bryce Canyon than anywhere else on Earth. This is partly due to the region's climate. On average, Bryce Canyon undergoes 200 freeze-thaw cycles a year; temperatures rise above freezing point during the day but drop below it at night, causing any meltwater that has seeped into the rock to freeze and expand. Known as frost-wedging, this process shatters stone, reshaping it into the hoodoos that the Paiute Indians, who settled in the area around 1200 CE, named the Legend People. As frost-wedging continues, many Legend People collapse each year.

◁ THOR'S HAMMER
Top-heavy hoodoos, such as Thor's Hammer, are fragile and at greatest risk of collapse, especially during spring and autumn, when frost-wedging is at its peak.

## HOW A HOODOO FORMS

Rainwater washes debris away from the plateau's edges, forming fins. Frost-wedging creates ice windows in the fins, separating them into hoodoos. Rain then rounds off the edges, lending a lumpy appearance.

rainwater erodes slopes, carrying away debris

FIN

frost-wedging creates ice window in fin

WINDOW

section separates from fin, forming hoodoo

HOODOOS

# Meteor Crater

*The best-preserved meteor impact site on Earth and evidence of an explosive
50,000-year-old collision whose effects can still be seen several miles away*

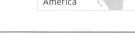
W. North
America

A massive, bowl-shaped cavity carved into the
Arizona Desert, Meteor Crater is one of the most
studied meteorite craters on Earth. Its shape has been
remarkably well preserved, and it is comparatively young
for an impact site. At nearly 1.6 km (1 mile) across, more
than 168 m (550 ft) deep, and with a circumference of
3.9 km (2½ miles), the depression known as Meteor
Crater was initially believed to be an extinct volcano. It
was not until the early 20th century that it was identified
as having been created by an extraterrestrial impact,
and details about its origins are still being discovered.

Around 50,000 years ago, a meteorite estimated
at 40 m (130 ft) across and weighing 272,000 tonnes
(300,000 tons) hurtled into Earth's atmosphere at
43,120 kph (26,800 mph) – 10 times faster than a rifle
shot. Half of it slammed intact into what is now the
northern Arizona Desert, forming the crater; the rest
fragmented before the collision, falling to the surface
as a cloud of debris. Iron-rich meteorites weighing
0.5–454 kg (1–1,000 lb) have been found in a 10-km
(6-mile) radius around the main impact site.

△ **TITANIC IMPACT**
Scientists estimate that the force released by the
meteorite impact was greater than that generated
by 18 million tonnes (20 million tons) of TNT.

## CRATER FORMATION

The impact created when a
single meteorite hits Earth
often destroys the meteorite
itself in a shock-wave-inducing
explosion that fractures bedrock
and melts surface rock. It leaves
a bowl-shaped depression with a
raised outer rim of ejected
debris and a base lower than
the surrounding landscape.

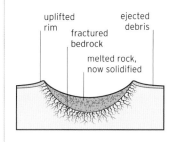

uplifted
rim     fractured     ejected
debris
bedrock

melted rock,
now solidified

△ **CLIFFTOP ARCH**
Mesa Arch stretches across the top of the Island in the Sky
mesa, a headland that rises 600 m (2,000 ft) above the base
of Canyonlands National Park. Below it, the mesa wall drops
vertically for more than 150 m (490 ft).

W. North
America

# Mesa Arch

*A dramatic natural archway formed on
the edge of a high mesa*

It looks like a bridge, but the 27-m (88-ft) long Mesa
Arch, in southeastern Utah's Canyonlands National
Park, is a pothole arch, created not by running water
but by weathering and erosion. Water accumulated
in a small depression in a sandstone mesa known
as the Island in the Sky. Over time, the depression
was enlarged by weathering and erosion into a
larger pothole, undercutting more resistant rock
at the cliff edge and eventually forming what is
now thought to be the most-photographed natural
arch in the world. Although pothole arches survive
for thousands of years, they do eventually collapse.
However, seismic research in 2015 indicated that
Mesa Arch is stable – for now.

SW. North America

# Antelope Canyon

*Beautiful chambers sculpted by flash floods, hidden beneath the Arizona desert*

In northern Arizona lies one of the American Southwest's most-visited canyons, but few passing drivers know that it exists. This is because, unlike most other canyons, whose broad gorges open to the sky, Antelope Canyon is a slot canyon. Also called narrows, slot canyons have nearly vertical, twisting walls divided by fissures that can be just a handspan's width at the surface yet plunge downwards for hundreds of metres.

### A tale of two canyons

Antelope Canyon was discovered in 1931 on land owned by the Navajo Nation. It twists and turns beneath the desert for 8km (5 miles); some sections are only a few feet wide, but can be 36m (120ft) or more deep. The canyon has two parts: Lower Antelope Canyon, known in Navajo as *Hazdistazí* ("spiral rock arches"), and Upper Antelope Canyon, or *Tse' bighanilini* ("the place where water runs through rocks").

The spiral-patterned chambers have been created by running water, but the process has been intermittent. Over millennia, the local red Navajo sandstone has been subjected not to constant scouring from a river but to rare yet violent flash floods. Whenever the nearby Antelope Creek overflows, water forces its way through the desert surface via tiny cracks; rocks and other debris carried by the floods help to carve out the rainbow-coloured corridors. Where the canyon widens enough to allow in sunlight, winds blow in sand, abrading the walls and changing the patterns and the floor levels.

Were it not for **floods** periodically **washing the sand away,** the canyon would have **disappeared long ago**

△ **SANDSTONE SPIRALS**
Shafts of light, created briefly when the Sun is directly overhead, illuminate the chambers of Upper Antelope Canyon. The lighting effect transforms the sandstone into a multicoloured spectacle.

◁ **RIBBON IN THE SAND**
Antelope Canyon's snake-like shape is best seen from above. Antelope Creek periodically forces its way through the canyon, its waters cleansing and altering the winding course before draining into nearby Lake Powell.

# The Chihuahuan Desert

*One of North America's largest deserts and the most species-rich in the western hemisphere*

SW. North America

Bisected by the Rio Grande, the Chihuahuan Desert is bordered to the west and east by the Sierra Madre Occidental and Oriental (see p.37). It is a rain-shadow desert, formed because the mountains block the flow of moisture-laden air from coastal waters.

## High life

Most of the desert lies at an altitude of about 1,370 m (4,500 ft), making the summers slightly milder than in other deserts, although daytime temperatures still exceed 38°C (100°F). The nights can be cold, with snow at higher elevations. Despite such extremes, more than 3,500 plant species flourish here, including more cacti species than in any other desert. While higher slopes are characterized by thick-leaved, frost-hardy yuccas, shrubs such as mesquite and creosote bush dominate the lower region. Such varied habitats support a wide range of wildlife, including bats, roadrunners, and tarantulas. A surprising number of fish are able to survive in pools and oases.

△ **WORLD'S LARGEST GYPSUM DUNEFIELD**
White gypsum deposits punctuate the Chihuahuan Desert, with the largest concentration in northern New Mexico's Tularosa Basin. Its 712 square km (275 square miles) include the White Sands National Monument.

## STORING WATER

Cacti such as the golden barrel are adept at surviving on a limited water supply. The plant's pleated "barrel" is actually a fleshy stem that stores water, expanding and contracting as the supply waxes and wanes. Instead of having leaves that lose moisture through transpiration, it is covered in thorns.

rounded stem swollen with water

spines protect plant from being eaten by thirsty animals

roots absorb moisture from soil

◁ **SPIKY MEAL**
The Chihuahuan Desert contains over one-fifth of the world's estimated 1,500 cacti species, including this prickly pear. Many animals, such as desert tortoises, eat its pads and fruit – barbs and all.

**Mist rising**
Water vapour collects above the canopy of the Amazon
Rainforest on the eastern slopes of the Peruvian Andes.
Hot, wet, and humid all year round, rainforests reach their
greatest extent in the tropical zones of South America.

# Central & South America

# ANDES TO AMAZON

## Central and South America

South America is roughly triangular in shape, with a curved hook of land at either end: the narrow neck known as the Isthmus of Panama in the north; and the aptly named Cape Horn and the islands of Tierra del Fuego in the south.

The soaring Andes mountains, South America's sturdy spine, extend 7,200 km (4,500 miles) down the continent's western edge, making them the world's longest mountain range by far. They are also the second highest, after the Himalayas, and are crowned by the 6,961 m (22,838 ft) peak of Aconcagua in Argentina. To the northeast, high plateaus rise above the vast, dense tropical rainforests, which are watered by the mighty Amazon River and its tributaries – together, they carry huge quantities of water east into the Atlantic. Southwards, the continent narrows to a point, and wedged between the Andes and the sea is flat pampas grassland.

### KEY DATA

▲ **Highest point** Aconcagua, Argentina: 6,961 m (22,838 ft)
▼ **Lowest point** Laguna del Carbón, Argentina: -105 m (-344 ft)
● **Hottest record** Rivadavia, Argentina: 49°C (120°F)
● **Coldest record** Sarmiento, Argentina: -33°C (-27°F)

### CLIMATE

Much of Central and South America lies within the tropics, but the south extends close to icy Antarctica. The Amazon forests are warm and moist, but plateaus in the mountains are dry.

AVERAGE TEMPERATURE

| °C | °F |
|---|---|
| -40 | -40 |
| -30 | -20 |
| -20 | 0 |
| -10 | 20 |
| 0 | 40 |
| 10 | 60 |
| 20 | 80 |
| 30 | |

AVERAGE RAINFALL

| MM | IN |
|---|---|
| 0 | 0 |
| 2,500 | 100 |
| 5,000 | 200 |
| 7,500 | 300 |
| 10,000 | 400 |

ATLANTIC OCEAN

Planalto da Borborema

Serra dos Carajás

Mouths of the Amazon

Ilha de Marajó

Amazon

Serra do Cachimbo

Xingu

Tocantins

Araguaia

SOUTH AMERICA

Guiana Highlands

Amazon

Madeira

Rio Negro

Amazon Basin

Orinoco

Apure

Pampas

Juruá

Amazon

Llanos

Lake Maracaibo

Cordillera Oriental

Ucayali

Nevado Huascarán 6,768 m

Cordillera Real

Cordillera Occidental

Isthmus of Panama

West Indies

Windward Islands

Greater Antilles

Lesser Antilles

Puerto Rico

Hispaniola

Caribbean Sea

Bahamas

Cuba

Jamaica

Gulf of Mexico

Lake Nicaragua

Galápagos Islands

## ECOSYSTEMS

South America's centre is dominated by the Amazon rainforest with its huge diversity of life. The Andes have unique mountain ecosystems, while the south of the continent is mostly dry grassland.

KEY

- Tropical broadleaf forest
- Tropical dry broadleaf forest
- Tropical coniferous forest
- Temperate broadleaf forest
- Mediterranean woodland and scrub
- Tropical and subtropical grassland
- Temperate grassland
- Wetland
- Montane grassland
- Desert and scrub

## GEOLOGY

South America's core, the high tablelands of the northeast, is made of ancient rocks. The Andes are comparatively young, forming 50 million years ago. Between the tablelands and the Andes is a series of young river basins.

KEY

- Precambrian (pre-541 million years ago)
- Paleozoic (541-252 million years ago)
- Mesozoic (252-66 million years ago)
- Cenozoic (66 million years ago to present day)

ATLANTIC OCEAN

PACIFIC OCEAN

Brazilian Hig

Serra do Espinhaço

Serra da Mantiqueira

São Franc

Serra

Serra do Mar

Serra Geral

Lagoa dos Patos

Mirim Lagoon

Uruguay

Paraná

Paraná

Mato Grosso

Serra do R

Serra de Maracaju

Serra do Caiapó

Paraná

Pantanal

Mesopotamia

Río de la Plata

Gran Chaco

Pampas

Rivadavia 49°C

Río Grande

Altiplano

Lago Poopó

Sierras de Córdoba

Cerro Ojos del Salado 6,893m

Aconcagua 6,961m

Sarmiento -33°C

Laguna del Carbón -105m

Patagonia

Andes

Tierra del Fuego

Cape Horn

Lake Titicaca

Salar de Uyuni

Atacama Desert

Falkland Islands

ELEVATION

M    FT

5,000 ● 16,000

2,000 ● 6,600

0 ● 0

A

B

Andes

Altiplano

Serra Monte Cristo

Planalto de Mato Grosso

Planalto da Borborema

CROSS-SECTION

△ ANCIENT PLATEAU
This plateau in Venezuela is one of hundreds of dramatic tabletop mountains, or tepuis, formed from the remains of a sandstone plateau that built up on top of the ancient Guiana Shield.

# THE SHAPING OF CENTRAL AND SOUTH AMERICA

Attached to Central America, South America is a huge continent – with high plateaus of old rocks in the east and the towering Andes to the west. It became an independent continent about 90 million years ago, when it split away from Africa.

## Shields and tablelands

To the east, a large portion of South America's land is dominated by three giant slabs of ancient rock, typically made of gneiss and granite, forming the Patagonian Shield, the Brazilian Shield, and the Guiana Shield. These slabs survived and moved together as a group for billions of years, even when South America was part of the southern supercontinent Gondwana. South America, along with Africa, split from Gondwana about 180–170 million years ago. Then, 140–90 million years ago, South America slowly split from Africa and the shields eventually became the bedrock around which the rest of South America has been gradually built.

Today, the shields form a base under some of the world's largest high plateaus. In places, these old plateaus, or tablelands, end so abruptly in high cliffs that people once believed they were lost worlds where ancient creatures might survive.

The world's highest waterfall, Angel Falls, plunges over the edge of the Guiana Shield, dropping 979 m (3,212 ft). The slabs are by no means intact. In places, they have cracked and slipped to open up huge valleys – for example, the Parana, Sergipe, and São Francisco rivers now flow through such valleys in the Brazilian Shield.

## The rise of the Andes

After it separated from Africa and became a separate continent, South America drifted slowly west, allowing the South Atlantic to expand. Its relentless movement brought it into direct collision with the Nazca Plate under the Pacific Ocean. The collision forced the eastward-moving, smaller but heavier Nazca Plate to subduct under the oncoming, larger but lighter South American Plate. As the

# Central and South America have only been linked by land for 3 million years

## KEY EVENTS

**80-70 million years ago** Complex tectonic activity in a shallow ocean north of South America causes uplifting magma to cool in an arc of new islands – forming the Caribbean.

**50 million years ago** The Nazca Plate under the Pacific Ocean subducts beneath the South America Plate to the west of the continent, and this causes the Andes to start rising. Most of the mountains have grown over the last 10 million years.

**15-10 million years ago** The Andes rise even higher and block the Amazon River from flowing into the Pacific. It flows eastward and pools into a lake. About 10 million years ago, the Amazon starts to flow into the Atlantic, lowering the lake's water level, and the lush Amazon Rainforest grows in its place (see opposite page).

**180 million years ago** The supercontinent Gondwana comprises South America, Africa, India, Australia, and Antarctica. South America and Africa separate from the rest about 180-170 million years ago, and South America starts its own journey when it splits from Africa 140-90 million years ago.

**16-13 million years ago** The area that will become the Andean Altiplano undergoes its first bout of elevation, with two more instances occuring 13-9 and 10-6 million years ago, which bring the area closer to its present-day height.

**14 million years ago** Desertification starts to occur in the area that will form the Atacama Desert.

**3 million years ago** A land bridge called the Isthmus of Panama forms, connecting South America with North America.

◁ **HIDDEN CAVE**
This underground lake in Chapada Diamantina National Park, Brazil, contains some of the continent's oldest rocks.

◁ **CONTINENTAL SPINE**
The Andes are the world's longest mountain range, stretching 7,200 km (4,500 miles) – almost along the entire western edge of South America.

◁ **RUPTURING VOLCANO**
The 5,023-m- (16,480-ft-) high Tungurahua volcano in Ecuador is one of many towering volcanoes that have burst through the Andes.

◁ **CHANGED COURSE**
The Amazon River is the world's largest river, carrying huge amounts of water through its enormous, lush lowland basin. Until 10 million years ago, it used to be a huge inland lake.

Nazca Plate was driven down into the mantle, a deep trench opened up in the ocean floor along the west coast of South America. The descending Nazca Plate dragged against the South American Plate, fracturing and deforming it.

The Andes started to form 50 million years ago as rocks were crumpled and piled on top of each other. Meanwhile, the Nazca Plate, melting as it was pushed into the mantle, caused magma to punch through the young mountain range as volcanoes, adding a series of towering volcanic peaks such as Cotopaxi in Venezuela, and Chimborazo and Tungurahua in Ecuador.

### River basins

Between the three ancient shields and the newer fold mountains of the Andes are lowland river basins, filled over time with layer upon layer of sediment. There are three main basins, forming a linked lowland belt running north to south down the middle of the continent. The Orinoco in the north flows from the Venezuelan Andes into the Atlantic south of Trinidad. The sprawling Amazon River in the centre flows right across South America from the Andes to emerge in the Atlantic at the equator. Until about 10 million years

ago, the Amazon pooled into a giant, inland lake and drained westwards into the Pacific Ocean. The Chaco-Pampa in the south is a large basin through which the Parana, Paraguay, and other rivers flow into the Rio de la Plata.

highlands block flow to Atlantic

giant lake forms

lake starts to drain to Atlantic

lake drains to Pacific

Amazon River forms

Andes block river's flow

**40–20 MYA**    **10–7 MYA**    **5,000–2,000 YA**

**HOW THE AMAZON BASIN FORMED**
The Amazon River flowed from the Guiana Highlands into a huge lake, which drained into the Gulf of Mexico and the Pacific Ocean. Eventually, the rising Andes forced the lake to drain into the Atlantic Ocean instead.

**BUILDING THE CENTRAL AND SOUTH AMERICAN CONTINENT**    KEY ▬ Convergent boundary ▬ Divergent boundary

Nazca Plate subducts under South American Plate

South America starts to split from Africa

Gondwana supercontinent

Central American Seaway

three shields form basis of continent

Mid-Atlantic Ridge spreads, South Atlantic opens

Africa drifts east

Central and South America linked by land

**94 MILLION YEARS AGO** South America begins to break away from Africa and move west. The South American Plate abuts the Nazca Plate, which subducts beneath it.

**50–40 MILLION YEARS AGO** The Mid-Atlantic Ridge spreads, forcing South America and Africa further apart and creating the Atlantic Ocean. The Central American Seaway separates the Americas.

**18,000 YEARS AGO** The Central American Seaway, which once separated the continent from Central America, has now closed and South and North America have become linked by land.

W. Central
America

# Santa María

*A dangerous Central American volcano that produced the third largest eruption of the 20th century*

Santa María is one of the most prominent of a chain of large volcanoes that rise high above the Pacific coastal plain of Guatemala. This chain is the result of the process of subduction as the Cocos oceanic plate slides beneath the Caribbean Plate. For much of recorded history, forest-covered Santa María has lain quiet, but in October 1902 it erupted, tearing a huge crater in its southwestern flank and killing more than 5,000 people.

## Activity today

In 1922, a lava dome – a huge, mound-shaped mass of grey lava – began growing in the massive flank crater formed in the 1902 eruption. This has since developed into a complex of four overlapping domes. Right up to the present day, frequent eruptions have occurred from these domes, sometimes generating dangerous pyroclastic flows (see panel, p.89) or blasting plumes of ash to heights of a few kilometres. Smaller explosions occur almost daily and can be viewed from the summit of Santa María's quieter main cone.

△ **BURSTING FORTH**
Most current activity at Santa María consists of regular, often spectacular eruptions from the lava domes on its southern flank.

◁ **DAWN VIEW**
In this view at sunrise, Santa María's main volcanic cone, standing high in the background to the left, overlooks its group of dangerous lava domes.

### SANTA MARÍA'S LAVA DOMES

Santa María's four lava domes, collectively known as the Santiaguito Complex, are called El Caliente ("the hot one"), La Mitad ("middle one"), El Monje ("the monk"), and El Brujo ("the sorcerer").

El Brujo

El Monje

La Mitad

El Caliente

Santa María main cone 3,772 m (12,375 ft)

Santa María's **1902 eruption** produced about **5.4 cubic km (1.3 cubic miles)** of **ash and pumice**

W. Central
America

# Masaya Volcano

*A huge volcano in Nicaragua that sometimes features a rare, spectacular lake of fiery lava*

Masaya is an example of a complex volcano: it has more than one volcanic cone as well as multiple vents and craters. Its two main cones lie within a larger, roughly oval, structure called the Masaya Caldera, which formed during a cataclysmic eruption some 2,500 years ago. The caldera is about 11 km (7 miles) long, 5 km (3 miles) wide, and has a rim that in places is 300 m (980 ft) high.

At the summit of Masaya's main cones are several deep holes called pit craters. One of these, the Santiago Crater, has been the main site of volcanic activity at Masaya for many years and sometimes contains a red-hot pool known as a lava lake.

### THE MASAYA CALDERA

Masaya Volcano lies within the much larger Masaya Caldera. The volcano itself consists of two main overlapping volcanic cones. One contains at its summit two pit craters - the Santiago and San Pedro craters - while the other is indented by the larger San Fernando Crater.

adjoining volcanic cones of Masaya Volcano

rim of Masaya Caldera

lake

San Pedro Crater | Santiago Crater | San Fernando Crater

△ HELLISH PIT
The Santiago Crater sometimes fills with a lava lake that produces spectacular bursting bubbles as it boils off noxious gases. It is one of just a handful of lava lakes in the world.

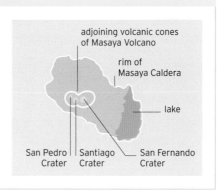

---

### PYROCLASTIC FLOWS

Also known as a nuée ardente (glowing cloud), a pyroclastic flow is an avalanche of volcanic gases, ash, and lava fragments. It often forms from the collapse of a volcanic eruption cloud or a lava dome. Most pyroclastic flows travel for about 5-10 km (3-6 miles), typically moving downhill at around 100 kph (60 mph) or faster, burning, flattening, and burying everything they encounter.

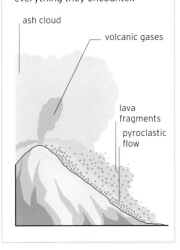

ash cloud

volcanic gases

lava fragments

pyroclastic flow

# Soufrière Hills

*A Caribbean volcano that has devastated over half the island of Montserrat since 1995*

Caribbean

There are several volcanoes named Soufrière in the Caribbean – soufrière is French for "sulphur outlet" – but the most famous is on the island of Montserrat, called Soufrière Hills. Since 1995, several destructive events called pyroclastic flows (see panel, left) have made it the most studied volcano in the world.

## Collapsing lava domes

Soufrière Hills produces a viscous lava that, instead of flowing away, piles up to form a lava dome. This emits a lot of steam and sometimes glows at night. The danger with lava domes, though, is that they can disintegrate to produce pyroclastic flows. At Soufrière Hills, a series of dome collapses and flows between 1995 and 2000 destroyed Montserrat's capital city, Plymouth, obliterated many settlements, and caused several deaths. Two-thirds of the population had to leave the island. However, since 2010, the volcano has quietened, and Montserrat remains a compelling place to visit because of its scenic beauty and volcanic history.

△ DANGEROUS DOME
The grey conical object in the background here is a lava dome. Part of its flank has begun to break up, triggering a pyroclastic flow, which resembles an avalanche.

N. South
America

# The Guiana Highlands

*A spectacular area of South America with astonishing flat-topped mountains and some of the world's most impressive waterfalls*

The Guiana Highlands extend across the northern part of South America, occupying much of southern Venezuela as well as parts of Guyana and northern Brazil. The Highlands are famous for their ancient and magnificent flat-topped mountains called tepuis. The word tepui means "house of the gods" in the language of the local Pemon people. Many of these landforms appear otherworldly, with their sheer cliffs and flat, vegetated, seemingly inaccessible tops. Some of the most famous include Auyan Tepui, Mount Roraima, and Kukenan Tepui. Mount Roraima is said to have inspired the Scottish author Arthur Conan Doyle to write his novel *The Lost World*, about the discovery of a remote plateau where prehistoric creatures still exist.

## Auyan Tepui and Angel Falls

Auyan Tepui is the largest of the tepuis, with a surface area of 700 square km (270 square miles), partly covered by cloud forest. The world's highest uninterrupted waterfall, Angel Falls, drops from a cleft in its summit. The River Churun, a tributary of the Orinoco, flows over the falls, descending a total of 979 m (3,212 ft) to the valley floor. The main plunge, of 807 m (2,648 ft), is followed by a series of sloping cascades and rapids.

◁ **ENDEMIC FLORA**
A long period of isolation means that many animal and plant species, such as the example from the genus *Orectanthe* shown here, have evolved on tepui summits and are found only on tepuis.

## The **rock** from which **Mount Roraima** was **carved out** was laid down about **1.8 billion years ago**

△ **LOST WORLD**
One of the highest tepuis, Mount Roraima is surrounded by stepped cliffs up to 400 m (1,300 ft) high and is the source of several rivers.

◁ **LONG DROP**
In Angel Falls' uppermost few hundred metres, the water barely makes contact with the sheer sandstone cliffs behind.

### HOW TEPUIS FORM

The Guiana Highlands were once an extensive sandstone plateau. Over a period of some 1.5 billion years, weak sections of the plateau were eroded away by rivers and streams, leaving behind the tepuis.

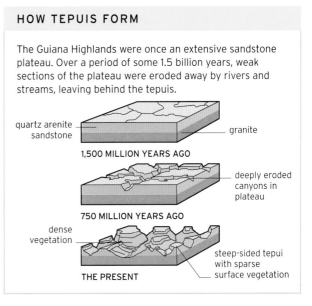

quartz arenite sandstone — granite
**1,500 MILLION YEARS AGO**

deeply eroded canyons in plateau
**750 MILLION YEARS AGO**

dense vegetation
steep-sided tepui with sparse surface vegetation
**THE PRESENT**

# The Andes

*The world's longest continental mountain range, comprising a series of impressive snow-capped peaks, volcanoes, and high plateaus*

W. South America

The Andes extend for 7,200 km (4,500 miles) north to south along the western edge of South America, passing through Venezuela, Colombia, Ecuador, Peru, Bolivia, Argentina, and Chile. The range comprises several distinct subranges that formed in the last 50 million years. Some of these sit on either side of high-plateau regions, one of which, the Altiplano (see pp.96–97), is the world's second highest plateau after the Tibetan Plateau.

## Deserts and glaciers

The Andes form a formidable physical and biological barrier between the Pacific coast and the rest of South America, ranging in width from 100 km (60 miles) wide at the southern tip of the continent to 700 km (440 miles) in the central region. A variety of physical features can be found across the range, including one of the most arid places on Earth – the Atacama Desert (see pp.124–25) –

and some of the world's deepest canyons. In the Altiplano, there are vast salt flats and lakes. The southern sector of the Andes has been heavily glaciated in the past, and glaciers still extend down to sea level in southern Chile, where there is an extensive network of fiords (see p.117).

## Volcanoes and earthquakes

Interspersed within the Andes are about 180 active and numerous extinct volcanoes. They are the result of tectonic activity (see pp.94–95) that results in the production of magma under the Andes, where it rises to create substantial volcanic activity. This tectonic activity also generates earthquakes, including the most powerful ever recorded, which had its epicentre off the coast of Chile in 1960.

▽ **MOUNTAIN PACK ANIMAL**
Along with alpacas, vicuñas, and guanacos, llamas are one of four relatives of camels found in the Andes. While llamas and alpacas are domesticated, the others are wild.

## THE SOUTHERN ANDES RAIN SHADOW

In the southern Andes, the prevailing winds blow from the west and push moisture-laden air from the Pacific Ocean towards the mountains. As this air meets the Andes, it rises, expands, and cools, dropping rain on the mountains' western flanks. However, it then continues over and down the eastern side as a dry wind, creating arid conditions (a rain shadow) in the southeastern part of South America.

rising air cools and condenses

rain falls on western slopes

rain shadow on eastern slope

prevailing wind

evaporation

ocean

△ ROOF OF THE AMERICAS
With its highest point at 6,961m (22,838 ft), Mount Aconcagua, seen here at the centre, is the highest peak not just in the Andes but in the entire world outside of Asia. It lies in Argentina, just east of the country's border with Chile.

△ DIRTY THUNDERSTORM
Lightning in and around a volcanic ash plume – as in this 2015 eruption of the Calbuco volcano in Chile – is referred to as a dirty thunderstorm. The flashes of lightning result from friction between ash particles and the atmosphere.

## Cotopaxi, in Ecuador, is one of the **highest active volcanoes** in the world. It has **erupted 50 times** since **1738**

# Building a
# MOUNTAIN CHAIN

Most mountain chains form where two tectonic plates – at least one of which carries continental crust – move towards each other, or converge, driven by convection in Earth's mantle. If both carry landmasses, these collide and mountains are pushed up; the Himalayas formed in this way. But a mountain chain can also form if there is a landmass on just one of the plates – the formation of the Andes provides a prime example.

## The creation of the Andes

The process that built the Andes unfolded over tens of millions of years and began with the development of a new convergent plate boundary to the west of South America. The ocean-dominated plates to the west of the boundary, which include the precursor of today's Nazca and Cocos plates, as well as the Antarctic Plate, began to subduct, or move under, the South American Plate to their east. This movement squeezed the lithosphere (the crust and top layer of mantle) of the western part of South America, causing major fracturing, called thrust faulting, and uplift of blocks of lithosphere. In parts of the region, a single mountain chain was created; in others, parallel subranges were pushed up in different phases. The subduction process led to the production of magma at depth, which rose up to create volcanism in many parts of the chain.

### ANDEAN VOLCANIC ZONES

The Andean volcanoes are clustered in four regions – the northern, central, southern, and austral volcanic zones. Between these belts lie large gaps containing no volcanoes. For example, there are around 30 volcanoes in southern Peru, but none in the rest of the Peruvian Andes. The reasons for these gaps are not fully known. One theory is that the tectonic plates that subduct under South America do so at a narrower angle of dip in these "gap" regions.

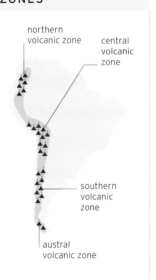

northern volcanic zone

central volcanic zone

southern volcanic zone

austral volcanic zone

**▶ THE CENTRAL ANDES**
This is a section through the central Andes of Bolivia and northern Chile, which also extends in the north into Peru. In this region, two roughly parallel subranges - the western and eastern cordilleras - are separated by a high, wide plateau called the Altiplano.

**▷ WESTERN SLOPES**
This region, the Atacama Desert, is extremely dry. Winds in these latitudes blow mainly from the east. As they push moist air up the eastern side of the Andes, rain falls on that side, leaving the western slopes and coastal settlements dry.

**Ocean trench**
The Peru-Chile Trench marks the place where one plate moves under the other. It is 7-8 km (4-5 miles) deep.

**Nazca plate**
This plate is about 50-60 km (30-40 miles) thick.

the Nazca plate is slowly moving towards and under the South American Plate at a rate of around 7.5 cm (3 in) a year

**Lithospheric mantle**
This is the uppermost layer of the mantle. With the overlying crust, it comprises a relatively strong layer called the lithosphere, which is divided into plates.

**Asthenosphere**
This relatively deformable layer of Earth's upper mantle is warmer than the overlying lithosphere.

**Oceanic crust**
The top part of the descending plate is oceanic crust. Made of rocks such as basalt and gabbro, it holds substantial amounts of water.

# As the Andes formed, South America shortened from west to east

◁ **SNOW-CAPPED VOLCANO**
Recent volcanic activity in the Western Cordillera is confined to an arc of over 40 active volcanoes with many adjacent extinct volcanoes, such as Nevado Sajama (pictured here).

the Western Cordillera is a subrange of the central Andes

Lake Titicaca

◁ **SALINE LAKES**
There are numerous salty lakes on the Altiplano (such as Lake Poopo, seen here). Their salinity is due to the fact that all rivers in the region drain inland, which means that salts produced by weathering of the nearby mountains are trapped here.

the Eastern Cordillera is broad in parts, with topography resulting from multiple thrust faults and the deformation of crustal layers

**Thrust fault**
Faults or fractures along which movement and uplift have occurred are called thrust faults.

South American Plate moving west relative to subducting Nazca Plate

**South American Plate**
In the Andean region, the South American Plate is around 150 km (90 miles) thick. The upper 40 km (25 miles) are continental crust; the rest is lithospheric mantle.

coastal block of continental crust being pushed east relative to stable South American Plate

**Magma**
Pockets of molten magma are generated by lower melting temperature.

continental crust squeezed and pushed upwards

the magma flow often follows fractures and other lines of weakness in the crust on its way to the surface

water is released from descending oceanic crust and moves into the mantle, where it lowers the melting temperature

▷ **HIGH PLATEAU**
Overlying a crustal block that has risen about 3 km (2 miles) in the past 25 million years, the Altiplano is a windswept region around 150–200 km (90–120 miles) wide of plains, lakes, and salt flats.

W. South
America

# The Altiplano

*A vast, high plateau of the central Andes with stunning, otherworldly scenery and harsh living conditions*

Set against a backdrop of volcanoes, the Altiplano is a windswept landscape of sparsely vegetated high plains, lakes, and salt flats, occupying about 200,000 square km (77,000 square miles) of the central Andes. Much of it lies in Bolivia, but parts of it extend into Peru, Chile, and Argentina in the south. It contains one large urban area – the adjacent cities of La Paz and El Alto in Bolivia.

## Land of fire, ice, salt, and wind

The Altiplano is bordered on the west by a subrange of the Andes, the Western Cordillera, which includes some massive volcanoes. The Altiplano is mostly arid, especially in the south, where the annual rainfall is only about 200 mm (8 in). It also has some large salt flats, such as the Salar de Uyuni (see p.126).

Because of the high altitude, the air is thin and the climate is quite harsh. Although temperatures during the day can reach 10–25°C (50–77°F), at night they plunge to as low as -20°C (-4°F) in the southern Altiplano. The local wildlife, which is adapted to these conditions, includes vicuñas (llama relatives), viscachas (rabbit-like rodents), and the Andean fox.

◁ **PROTECTIVE COAT**
The viscacha's thick, fine fur allows it to withstand year-round freezing nighttime temperatures.

# The **average height** of the Altiplano is **3,750 m** (12,300 ft)

## BENEATH THE ALTIPLANO

In the central Andes, two or more phases of mountain building pushed up the Eastern and Western Cordilleras, which are separated by the Altiplano. The details of how the Altiplano formed are fiercely debated, although recent research suggests that the crust under it was lifted about 3 km (2 miles) in a series of spurts in the last 25 million years.

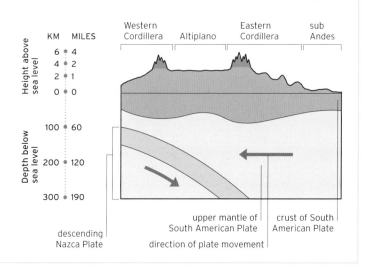

| | KM | MILES |
|---|---|---|
| Height above sea level | 6 | 4 |
| | 4 | 2 |
| | 2 | 1 |
| | 0 | 0 |
| Depth below sea level | 100 | 60 |
| | 200 | 120 |
| | 300 | 190 |

Western Cordillera | Altiplano | Eastern Cordillera | sub Andes

descending Nazca Plate

upper mantle of South American Plate

crust of South American Plate

direction of plate movement

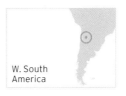

W. South
America

# El Tatio Geysers

*One of the highest concentrations of geysers in the world, high in the Chilean Andes*

El Tatio is the third largest geyser field in the world and by far the highest-altitude large field, at 4,320 m (14,170 ft) above sea level. It lies in a volcanic region on the edge of Chile's Atacama Desert (see pp.124–25) and can be reached from San Pedro de Atacama – one of the nearest towns – via a minor road.

The geysers here do not shoot up to enormous heights – the tallest eruptions observed are to around 6 m (20 ft). However, the sheer density of geysers makes up for this: more than 20 of them are within a few minutes' walk. Interspersed between the geysers are hot springs and fumaroles (gas-emitting volcanic vents). Visitors have to be warned not to approach the springs and geyser pools too closely as fatalities have occurred when people have fallen into the scalding water.

### LAYOUT OF THE GEYSER FIELD

El Tatio contains around 80 geysers spread over an area of about 5 square km (1.9 square miles). Most are concentrated within three main areas: the upper, middle, and lower geyser basins.

lower basin    stream    upper basin

middle basin

▽ **SCALDING WATER SPOUT**
The water spouted out by this and other geysers at El Tatio is extremely hot - up to 85°C (185°F), which is close to the boiling point of water at this altitude.

△ **CHILEAN ALTIPLANO**
This part of the Altiplano in Chile features a dark blue lake, Laguna Miscanti, with brackish water. Behind it towers an extinct volcano, Cerro Miscanti. It is surrounded by yellow bunch grass.

◁ **SCARLET LAKE**
Situated near the Chile–Bolivia border, and known locally as Laguna Roja, this is actually a hot spring with a temperature of 40-50°C (104-122°F). A colony of bright red, heat-loving algae thrives in the water.

# The Sugarloaf Mountain

*A monolith made from granite and quartz that rises vertically from the water's edge in Rio de Janeiro*

E. South America

Rising to a height of 396 m (1,299 ft) above Rio de Janeiro, the Sugar Loaf is situated on a peninsula that juts out into the mouth of Guanabara Bay – a large sea inlet that includes Rio's harbour.

The rock from which the Sugar Loaf and other nearby dome-shaped mountains are composed was formed millions of years ago from magma rising from deep underground and then solidifying beneath the surface. Starting from around 100 million years ago, South America separated from Africa to form the southern Atlantic Ocean, resulting in fractures in Earth's crust around the eastern coast of South America. The fractures created weaknesses in the rocks surrounding the granitic masses, allowing them to be broken down by agents such as rain (see panel, right).

▷ **RISING STRAIGHT UP**
From the top of the Sugarloaf (in the centre in this image), there is a bird's eye view of both Rio's harbour, seen in the foreground here, and Guanabara Bay, visible on the mountain's left.

## HOW THE SUGAR LOAF FORMED

The Sugar Loaf began as a mass of granite underground. It was exposed at the surface as the softer rock above and around it was weathered away. Further erosion created its shape.

area of fast weathering · granitic mass
**SOFT ROCK WEATHERS**

exfoliation sheet (thin surface layer) · debris from accelerated erosion
**GRANITIC MASS EXPOSED**

emerging Sugar Loaf · broken exfoliation sheets on sides
**STEEP DOME SHAPE FORMS**

# The Hornocal Mountains

*A small mountain range in northwest Argentina with a spectacular zigzag structure and incredible colours*

E. South America

▽ **DELUGE OF COLOUR**
The hues of the Hornocal Mountains' rock layers range from vivid reds to creams and grey-greens.

The Hornocal Mountains are part of a formation of sedimentary rocks that extends for some 500 km (310 miles) southwards from Peru. Known as the Yacoraite formation, it is composed mainly of limestones as well as some sandstones and siltstones, and was formed in a shallow sea between 72 and 61 million years ago (when this part of South America was underwater).

Variations in the mineral composition of the sediments deposited into the sea, the chemistry of the seawater, and the life-forms present at the time account for the striking variation in the rock colours. Over the years, these rock layers were lifted up and tilted as a result of tectonic plate movements, then weathered and eroded, resulting in the breathtaking sight that is visible today. The riot of colours lends a slightly surreal appearance to the location.

# The Torres del Paine

*An area of astonishing scenic beauty in Patagonia with snow-capped mountains, glaciers, waterfalls, lakes, and rivers*

S. South America

Some 1,960 km (1,220 miles) south of Santiago, the capital of Chile, lies one of South America's most spectacular national parks, the Torres del Paine (Towers of Paine). Paine, pronounced "pie-nay", means "blue" in the native Tehuelche language – a reference to the fact that the park's mountains appear slightly azure from a distance.

## Granite spires and sharp peaks

One of the most iconic features of the Torres del Paine is a group of three huge granite spires – the torres, or towers – rising 1,500 m (4,900 ft) above the Patagonian steppe. Close by, and part of the same massif, are the Cuernos del Paine (Horns of Paine) – three

enormous mountains surmounted at their tops by sharp black "tusks" of rock. For many visitors, these are the outstanding attraction because of their spectacular forms, sharp edges, and different rock colours. Other high peaks in the park include Cerro Paine Grande at 2,884 m (9,460 ft) and Mount Almirante Nieto at 2,640 m (8,660 ft).

One of the world's last remaining wildernesses, the Torres del Paine has a unique range of wildlife, including guanacos, foxes, rheas, and Andean condors. The park is adorned with beautiful vegetation, including evergreen shrubs, such as Chilean Fire Bush.

▽ **BLACK PEAKS**
The black-topped Cuernos del Paine, which soar to more than 2,500 m (8,200 ft) are visible on the left here, while Mount Almirante Nieto can be seen on the right.

◁ **THE THREE TOWERS**
Torre Sur on the left, Torre Centrale, rising to 2,800 m (9,200 ft), and Torre Norte on the right form the three-tower granite centrepiece of the park.

## HOW THE TOWERS FORMED

The Torres del Paine was formed partly from granite that originated from magma rising into the subsurface 13 million years ago. The magma cooled to form a granite block, which was later exposed when the area was eroded by glaciers.

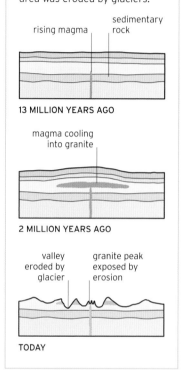

rising magma — sedimentary rock

**13 MILLION YEARS AGO**

magma cooling into granite

**2 MILLION YEARS AGO**

valley eroded by glacier — granite peak exposed by erosion

**TODAY**

# The Quelccaya Ice-cap

*The largest glaciated area in the world's tropical regions, located in the Cordillera Oriental of the Peruvian Andes*

W. South America

The Quelccaya Ice-cap covers an area of 44 square km (17 square miles) of the High Andes and – reflecting how high the freezing zone is in the tropics – all of it lies above an altitude of 5,000 m (16,000 ft). The ice-cap is of great importance to the surrounding region since its meltwater irrigates fields in the valleys below and provides drinking water for some large Peruvian cities, like Cusco. However, due to global warming, the ice-cap has lost about one-fifth of its area over the past 30 years, and the rate at which it is shrinking is accelerating. The recent appearance of 5,700-year-old plant remains from under the ice-cap indicate that more than 50 centuries have passed since it was smaller than it is today.

▷ **KEYHOLE LAGOON**
This unusual-looking keyhole-shaped feature is a meltwater lake – temporarily iced-over - that developed on one edge of the ice-cap in the 2000s.

The **ice-cap** lies close to the **Amazon rainforest**, where **temperatures** are about **27°C (80°F) higher**

# The Pastoruri Glacier

*A small Peruvian glacier that is shrinking but still a remarkable sight and a magnet for enthusiasts of various snow- and ice-related sports*

W. South America

The Pastoruri Glacier lies in the southern part of the Cordillera Blanca in the Andes of north central Peru, occupying one flank of a 5,240-m- (17,200-ft-) high Andean peak called Nevado Pastoruri. It is a cirque glacier – a type of glacier that occupies a bowl-shaped depression high up on a mountain.

## Steep sides

The Pastoruri Glacier has steep, cliff-like edges, making it a popular destination for ice climbers. Although parts of its surface are heavily crevassed, others are often covered in deep, soft snow, which attracts skiers and snowboarders. The glacier has lost over 15 per cent of its ice over the past 35 years, and it is expected to continue to shrink.

▷ **BLUE ICE CAVE**
Inside the Pastoruri glacier, at a height of around 5,200 m (17,000 ft), there is an ice cave. Visitors can hike around the glacier and visit the glowing blue interior.

◁ **HIGH-ALTITUDE CAP**
In this false-colour satellite image, part of the ice-cap and some small lakes surrounding it are visible. Vegetation in nearby valleys appears red.

▽ **MOUNTAIN CACTI**
Various species of cacti grow in the high Peruvian Andes near the ice-cap, including this variety in the *Tephrocactus* genus, often described as a mounding cactus.

# The Southern Patagonian Ice-field

*The world's second largest continuous ice-field, stretching almost 370 km (230 miles) down the southern Andes*

SW. South America

An ice-field is a single, extensive mass of ice at high altitude that is partly hemmed in by mountains (in contrast, ice-caps and ice-sheets almost completely submerge a region). The Southern Patagonian Ice-field is a remnant of the last glacial period, around 18,000 years ago, when all of southern Chile and Argentina was covered in a thick sheet of ice, over an area estimated to be about 480,000 square km (185,000 square miles). Today, it occupies an area less than 3 per cent of that size. Mount Fitz Roy at 3,359 m (11,020 ft) and the volcano Lautaro at 3,623 m (11,886 ft) are two of the main peaks surrounded by the ice-field.

▷ **OUTLET GLACIER**
The Grey Glacier is one of dozens of outlet glaciers that carry ice down from the ice-field. Some end in glacial lakes, while others discharge icebergs into Chilean fiords (see p.116).

### THE ICE-FIELD AND ITS OUTLET GLACIERS

The Southern Patagonian Ice-field is long and thin, with an average width of only about 35 km (22 miles). Some of the largest of its numerous outlet glaciers are marked here.

Occidental Glacier

Bruggen Glacier

Jorge Montt Glacier

O'Higgins Glacier

Viedma Glacier

Lago Argentino

Perito Moreno Glacier

Grey Glacier

# The Perito Moreno Glacier

*An impressive Patagonian glacier that can be seen discharging huge quantities of ice into an Argentinian lake*

S. South America

The Perito Moreno is an outlet glacier from the Southern Patagonian Ice-field (see p.101), descending to the east of the ice-field, where it discharges icebergs into Argentina's largest lake, Lago Argentino. The glacier is about 24 km (15 miles) long, has an area of about 200 square km (80 square miles), and terminates in a vast wall of ice that is some 5 km (3 miles) wide. Unlike most of the world's glaciers, the Perito Moreno is not retreating, although glaciologists are uncertain why that is.

## Calving face

Situated in Los Glaciares National Park, the glacier is one of Patagonia's most visited sites. It is possible to go on a guided trek over the glacier's surface. Many sightseers enjoy witnessing huge shards of ice falling off the glacier's calving face. The best time to view this is in the afternoon when, due to the Sun's warmth, disintegration events occur about every half an hour.

### THE BRAZO RICO ICE DAM

The glacier periodically advances and blocks off Brazo Rico, an arm of Lago Argentino. An ice dam forms, and Brazo Rico may then rise as much as 60 m (200 ft) before its waters break through the dam, discharging about a billion tonnes of water in 24 hours.

Lago Argentino

glacier advances to block off Brazo Rico

ice dam

Perito Moreno Glacier

Brazo Rico

# The Perito Moreno's **iceberg-calving** wall of ice towers **74 m (240 ft)** above Lago Argentino

▷ **CLOSING IN**
Here, the glacier is forming a dam between Brazo Rico (at the bottom) and Lago Argentino (at the top). Icebergs can be seen calving into Brazo Rico.

◁ **SAILING BY**
Massive blocks and pinnacles of ice slowly move downhill. In the foreground are boulders and other rock debris deposited at the glacier's side.

△ **CRASHING DOWN**
The ice dam eventually breaks down as water pressure forces a tunnel through it. Soon afterwards, the ice bridge over the tunnel dramatically collapses.

# The Llanos

*A huge expanse of savanna plains, transformed by seasonal flooding and supporting a unique collection of animals*

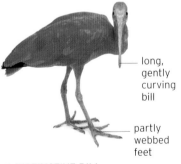

N. South America

△ **WATCHFUL EYE**
There are more than 100 species of reptile living in the Llanos, including many different turtles, snakes, and crocodilians. One of the most abundant is the spectacled caiman.

The Llanos, or "Plains", is a sprawling tropical grassland region, stretching across Colombia and Venezuela from the foothills of the Andes to the Orinoco Delta, where it meets the Caribbean. Although the Llanos contains a mosaic of different habitats, including woodland and swamps, most is treeless savanna, which is subject to seasonal flooding by the Orinoco River's tributaries.

### Unique wildlife
The Llanos is home to many species not typically found in savanna ecosystems. These include the critically endangered Orinoco crocodile, which is only found in the Llanos, and the largest rodent in the world, the semiaquatic capybara. The region supports several hundred species of birds – both residents and migrants – with a large proportion of waterfowl and waders.

long, gently curving bill

partly webbed feet

△ **DISTINCTIVE BILL**
The Llanos contains the world's largest population of scarlet ibis. The bird uses its long, thin bill to probe for food such as insects in soft mud and under plants.

### SEASONAL FLOODING

The Llanos is a region of climatic extremes with seasonal flooding and droughts. More than 90 per cent of its annual rainfall occurs between April and October. By May, the tributaries of the Orinoco have started to overflow, and at the height of the season much of the Llanos is wetland.

MAY        JUNE        OCTOBER

## The Llanos is home to the **world's most powerful snake**, the green anaconda

N. South America

# Caño Cristales

*Colourful aquatic plants and clear water combine to create a liquid rainbow*

On the Serrania de la Macarena tableland in central Colombia runs a 100-km- (60-mile-) long river that for five months of the year turns a combination of red, green, yellow, and blue. The Caño Cristales' colourful display is due to the blooming of an endemic plant called *Macarenia clavigera*, which requires a precise combination of water and sunlight levels to take on its bright red hues. Other green plants, the sandy riverbed, and the clear water reflecting the sky provide the other colours. Despite being rich in aquatic plants, the river is thought to contain no fish.

△ **CRYSTAL CLEAR WATERS**
The Caño Cristales is a relatively fast-flowing river with waterfalls, rapids, and deep, swirling pools. The water lacks sediment and nutrients, and as a result is extremely clear.

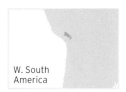

W. South America

# Lake Titicaca

*The world's highest navigable lake, and South America's largest*

The largest freshwater lake in South America, Lake Titicaca covers 8,370 square km (3,230 square miles) straddling the border between Peru and Bolivia. It is situated in the Altiplano basin (see p.96), high in the Andes, and at 3,812 m (12,507 ft) above sea level is the world's highest commercially navigable lake.

The lake's two sub-basins are connected by the Strait of Tiquina. The larger section to the north, Lago Grande, is much deeper than the smaller Wiñaymarka. More than 25 rivers from five major systems drain into the lake, but it has just one outflow, the Desaguadero River, at its southern end. The river accounts for only about five per cent of Lake Titicaca's water loss; the rest is lost by evaporation in its arid, windy environment.

◁ **FALLING LEVELS**
In the last few years, the water level of Lake Titicaca has been slowly falling. Lack of rainfall has been compounded by reduced levels of summer meltwater as glaciers shrink due to climate change.

**ALTIPLANO LAKES**

This cross-section of the northern and central drainage basins of the Altiplano shows the current levels of Lake Titicaca and Lake Poopó alongside levels of ancient lakes that have dried up, leaving salt flats behind. By the end of 2015, Lake Poopó had also completely dried up, partly due to low levels of water in Lake Titicaca, which ultimately feeds it.

Lake Titicaca
3,812 m
(12,507 ft)

Lake Poopó
3,685 m (12,090 ft)

Desaguadero River

Salar de Coipasa
3,656 m
(11,995 ft)

Salar de Uyuni
3,653 m
(11,985 ft)

# The Amazon

*The world's largest river, snaking through the most extensive rainforest on Earth and supporting a unique ecosystem*

N. South America

The greatest river on Earth, measured by both the size of its catchment area and the volume of water flowing through its channels, the Amazon carries one-fifth of all the fresh water entering Earth's oceans. Its vast basin drains more than 40 per cent of South America. Around two-thirds of its main stream and by far the largest portion of its catchment is in Brazil. The Amazon and its more than 1,000 tributaries support the great biodiversity of the world's largest rainforest, with which it shares its name (see pp.120–21). Although the exact source of the Amazon (see p.108) and its total length have been the subject of much debate, it is widely regarded as the world's second longest river, after the Nile (see pp.190–91).

The Amazon rises high in the Andes of southern Peru, less than 200km (125 miles) from the Pacific Coast, and flows more than 6,400km (4,000 miles) to the northeast coast of Brazil, where it discharges an average of 200 billion litres (53 billion gallons) every second into the Atlantic Ocean. During flooding, some parts of the Amazon can reach a width of more than 50km (30 miles). The river's estuary spans an area more than 300km (185 miles) wide.

◁ **FIERCE REPUTATION**
Despite its reputation as a ferocious predator, the red-bellied piranha mainly forages and scavenges for food and travels in shoals for protection against larger predators.

## The Amazon carries **more water** than the **next seven largest rivers** combined

### EARTH'S LARGEST RIVER BASINS

A river's drainage basin, also known as its catchment area, is the total area of land that is drained of water by the river and all its tributaries. The Amazon has by far the largest drainage basin of any river on Earth. Its area is about twice the size of the next largest, that of the Congo in Central Africa (see p.192).

Amazon:
7 million square km
(2.7 million square miles)

Congo:
3.5 million square km
(1.4 million square miles)

Nile:
3.4 million square km
(1.3 million square miles)

Mississippi-Missouri:
3.2 million square km
(1.2 million square miles)

Ganges: 1 million square km
(0.4 million square miles)

△ **VIGOROUS GROWER**
The water hyacinth is fast-growing, and outside its native range of the Amazon Basin it can be highly invasive, blocking up waterways and inhibiting other wildlife.

▷ **WIDE LOOPS**
A large part of the Amazon Basin is flat, and for much of its course the river and many of its tributaries meander slowly through vast areas of pristine rainforest.

# The features of a
# RIVER BASIN

A river basin is a geographic area in which all of the water that enters, usually by precipitation, drains to a common outlet through a river and its network of tributaries. River basins vary greatly in shape, complexity, and size – with the Amazon Basin being by far the largest on Earth. However, all river basins share a set of common features.

## Source to mouth

A basin's watershed marks its perimeter, separating it from the basins of adjacent river systems. Naturally, watersheds run through areas of high ground, such as mountain ranges and plateaus. The source of a river system – which may be a spring, glacial meltwater, or a marsh – is the farthest point of a river's course from its eventual outlet, measured along the river's course itself. The elevated sections of a river basin feature streams that are usually fast-flowing, with rapids and waterfalls, that descend onto lower, flatter terrain and become slower and wider, and often meander.

Most channels in a basin are tributaries that empty into the main river. Tributaries, such as the Amazon's Rio Negro, may themselves be substantial rivers, and waters from disparate sources, and of different content and colour, may come together at their meeting points, or confluences. A mature river in its lower course usually creates a floodplain by dropping sediment on its valley floor during floods. It meanders across this plain, its course shifting, either gradually as it erodes its banks or abruptly after severe floods. Most river basins, but not all (see p.197), empty into an ocean or a lake. A river's mouth usually features an estuary where fresh and salt water mix, and often a delta, formed by the accumulation of sediments carried by the river.

◁ **SOURCE**
Locating the source of a great river is often contentious. Since 2014, scientists have believed the Amazon's source to be in the Mantaro river basin, high in the Peruvian Andes.

**Watershed**
The boundary of a river basin, called the watershed, commonly runs along a mountain ridge. The Andes form most of the western watershed of the Amazon.

Pacific Ocean

**Upper course**
In the upper course of a river, it tends to descend steeply and is joined by many small tributaries. In the Amazon's upper course, rivers and streams cascade down steep Andean mountain slopes.

region of seasonal flooding

▶ **CATCHMENT AREA**
A river basin, also called its catchment area, encompasses the steep upper courses of a river system as well as the gentle gradients of the river's lower courses. In the Amazon's lower course, the river flows over such flat terrain that it falls only 70 m (230 ft) from the city of Manaus to the sea – a distance of some 1,400 km (820 miles).

## REGIONAL FLOODING

Every year, vast areas of the Amazon's floodplains become inundated with water, creating the most extensive system of river-flooded forests in the world. The weight of water is so great that GPS sensors show it depressing the land by 7.5 cm (3 in) during peak flood. Due to the Amazon Basin's vast size, and the uneven distribution of its seasonal rainfall, different regions in the basin are affected at different times.

water confined to river channels

**LOW-WATER PERIOD**

flooded forest

higher ground never flooded

**PEAK FLOODING**

△ **AQUATIC ACROBAT**
Unlike oceanic dolphins, the Amazon river dolphin has a flexible neck, which helps it negotiate vegetation in the flooded forest. The dolphins are seen lifting their heads clear of the water to check the surroundings, and occasionally they jump up to 1 m (3 ft) above the water's surface.

# It takes Amazon river water at least one month to flow from the Andes to the Atlantic

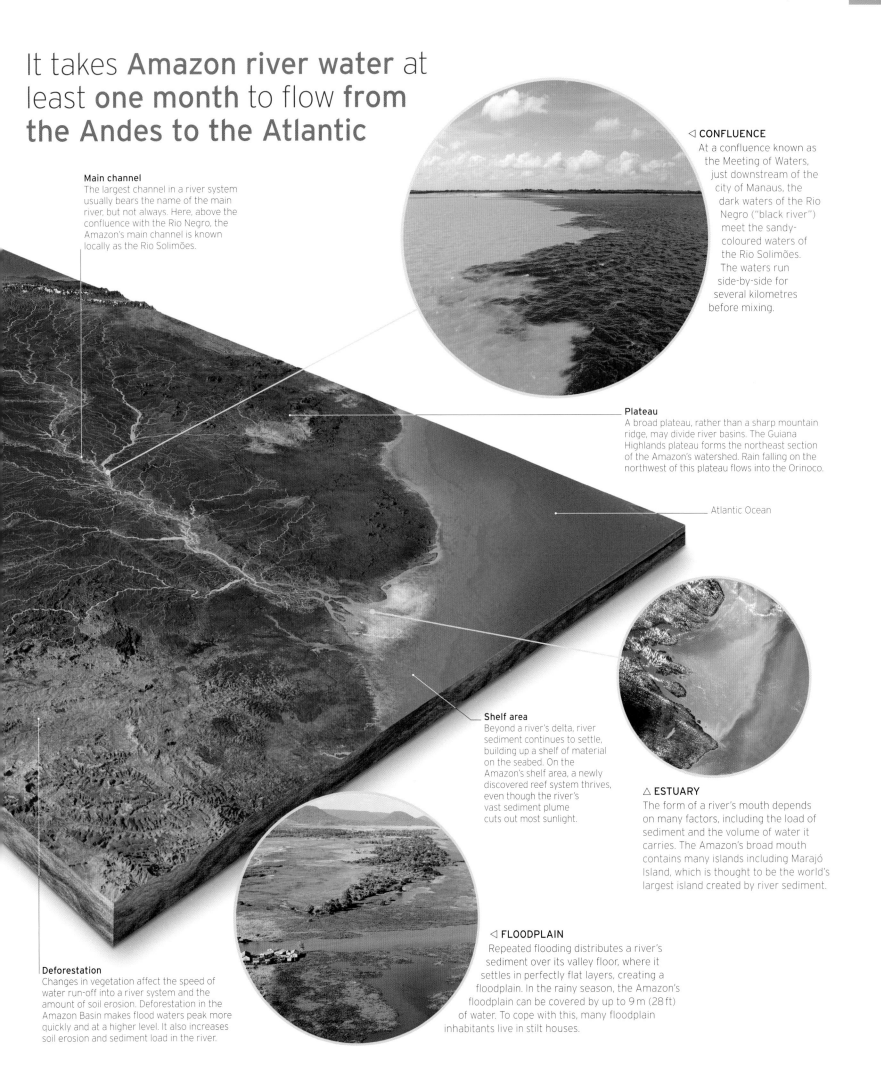

**Main channel**
The largest channel in a river system usually bears the name of the main river, but not always. Here, above the confluence with the Rio Negro, the Amazon's main channel is known locally as the Rio Solimões.

◁ **CONFLUENCE**
At a confluence known as the Meeting of Waters, just downstream of the city of Manaus, the dark waters of the Rio Negro ("black river") meet the sandy-coloured waters of the Rio Solimões. The waters run side-by-side for several kilometres before mixing.

**Plateau**
A broad plateau, rather than a sharp mountain ridge, may divide river basins. The Guiana Highlands plateau forms the northeast section of the Amazon's watershed. Rain falling on the northwest of this plateau flows into the Orinoco.

Atlantic Ocean

**Shelf area**
Beyond a river's delta, river sediment continues to settle, building up a shelf of material on the seabed. On the Amazon's shelf area, a newly discovered reef system thrives, even though the river's vast sediment plume cuts out most sunlight.

△ **ESTUARY**
The form of a river's mouth depends on many factors, including the load of sediment and the volume of water it carries. The Amazon's broad mouth contains many islands including Marajó Island, which is thought to be the world's largest island created by river sediment.

**Deforestation**
Changes in vegetation affect the speed of water run-off into a river system and the amount of soil erosion. Deforestation in the Amazon Basin makes flood waters peak more quickly and at a higher level. It also increases soil erosion and sediment load in the river.

◁ **FLOODPLAIN**
Repeated flooding distributes a river's sediment over its valley floor, where it settles in perfectly flat layers, creating a floodplain. In the rainy season, the Amazon's floodplain can be covered by up to 9 m (28 ft) of water. To cope with this, many floodplain inhabitants live in stilt houses.

Upon seeing Iguaçu Falls for the first time, **US First Lady Eleanor Roosevelt** is reputed to have exclaimed **"Poor Niagara!"**

◁ **WATER VOLUME**
The amount of water in a waterfall determines its erosive power. On average, more than 100 million litres (26.4 million gallons) of water flows over Iguaçu Falls every minute – one of the largest volumes of any waterfall on Earth.

Devil's Throat

▷ **FEEDING OPPORTUNITY**
Waterfalls may be home to a variety of plants and animals. Black-fronted piping guans live in the Atlantic Forest and sometimes venture out onto Iguaçu Falls to search for molluscs and insects.

viewing gallery on Brazilian side of falls

△ **BASALT OUTCROPS**
The hard caprock forming the crest of the falls was created during flood basalt eruptions that covered the region in layers of lava more than 120 million years ago.

underlying layer of soft rock

plunge pool

**Main flow of falls**
This cross-section downstream of the Devil's Throat shows the channel through which half of the Iguaçu's waters flow. A further series of falls lie below and to the right, on the Argentine side.

E. South
America

# Iguaçu
Falls

*One of the world's greatest waterfall systems, situated within a lush South American tropical forest*

---

Spanning part of the border between northeastern Argentina and southern Brazil, Iguaçu Falls is one of the largest waterfall systems on Earth, rivalled in size only by Africa's Victoria Falls (see pp.194–95). It is located on a tight horseshoe bend of the Iguaçu River and drops a total of 82m (269 ft).

## Complex system

Iguaçu Falls is a vast and complex series of up to 300 separate waterfalls – depending on the water level of the river – stretching 2.7km (1¾ miles) along a fragmented basalt escarpment. Above the falls, many islands split the river into the flows that feed the waterfalls, which vary in size. At the heart of the falls is the Devil's Throat, a narrow, semicircular chasm down which half of the river's waters plunge. Iguaçu Falls produces huge clouds of spray that constantly soak the surrounding area, creating a humid microclimate in which a diverse group of plants and animals flourish.

**Retreating falls**
Waterfalls retreat upstream as the caprock is eroded. The edge of the basalt cap of Iguaçu Falls is receding by 3mm (⅛in) a year.

**◁ IGUAÇU RIVER**
The Iguaçu River is one of the major tributaries of the Paraná River, South America's second longest river after the Amazon, which it joins 23km (14 miles) downstream of the falls.

viewing gallery on
Argentine side of falls

**◀ GREAT WATER**
Iguaçu Falls is about three times the width of North America's Niagara Falls (see p.52) and around 50 per cent taller. The name of the falls and river is from a word in the indigenous Guaraní language that translates as "great water".

**Atlantic Forest**
The tropical forest surrounding the falls is a remnant of the once vast Atlantic Forest. The Iguaçu National Park protects the thousands of species of plants and animals living within its borders.

top layers of hard rock

**△ DROUGHT-AFFECTED FALLS**
Waterfalls can experience great fluctuations in water flow. During the dry season, when long-periods of droughts may occur, Iguaçu Falls can shrink to a fraction of its size, completely transforming its appearance.

## HOW A WATERFALL FORMS

Waterfalls form where a river flows from hard to softer rock. The water erodes the soft rock faster than the hard rock, creating a step then a plunge pool. Undercut sections of the hard rock eventually collapse, and as this process continues, the waterfall retreats upstream.

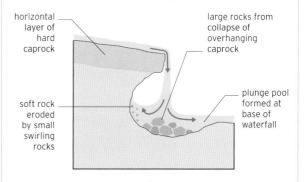

horizontal layer of hard caprock

large rocks from collapse of overhanging caprock

soft rock eroded by small swirling rocks

plunge pool formed at base of waterfall

C. South
America

# The
# Pantanal

*An immense freshwater wetland
system with one of the greatest
concentrations of plant and animal
species on the planet*

The world's largest freshwater wetland, the Pantanal covers more than 180,000 square km (70,000 square miles) of western Brazil and parts of Bolivia and Paraguay. Sitting in the upper Paraguay River basin, it is a gently sloping floodplain comprising a mixture of habitats including grasslands, forests, marshes, and rivers that are subject to seasonal flooding.

### Thriving in floods

A wide variation in water levels across the Pantanal has led to a great diversity in plant life. In higher areas, drought-resistant woodland is common, while in lower areas plants adapted to seasonal flooding flourish. Areas that are permanently under water support a diverse community of aquatic plants, including lilies and hyacinths. Many thousands of waterbirds feed on the great numbers of fish that migrate and spawn each year when the floods arrive, as do the estimated 10 million caimans that make up the largest population of crocodilians in the world.

rosettes on
coat assist
camouflage

short,
powerful
legs aid
climbing
and
swimming

△ **BIG CAT**
The Pantanal is a vital sanctuary for the jaguar, the world's third largest feline after the tiger and the lion. It is estimated that only 15,000 jaguars remain in the wild.

S. South America

# General Carrera Lake

*A large glacial lake in the shadow of the Andes with a remarkable complex of carved marble caves on its shores*

Also known as Lake Buenos Aires, General Carrera Lake spans the border between Chile and Argentina in Patagonia. With a surface area of 1,850 square km (715 square miles), it is the second largest freshwater lake in South America, after Lake Titicaca (see p.105). General Carrera Lake is surrounded by the Andes, and its glacier-fed water drains into the Pacific Ocean through the Baker River, the largest river in Chile. Its most famous feature, known as the Marble Cathedral, or Marble Caves, is a series of pure marble cliffs that have been carved into elaborate caves, tunnels, and columns by 10,000 years of wave action. The smoothly formed light-grey rock takes on a blue tint from the surrounding water, which is coloured turquoise by glacial silt.

▷▽ **SCULPTURED MARBLE**
The Marble Cathedral system includes an island monolith known as the Marble Chapel. The clarity of the water of General Carrera Lake reveals how the rock formations continue below the surface.

△ **FLOODED PLAINS**
Around 80 per cent of the Pantanal's vast plains are inundated during the wet season, as water levels rise by up to 5 m (16 ft).

## CLIMATE

The Pantanal has a tropical climate with a hot rainy season and a drier season during the cooler months. The extent and depth of floodwaters vary greatly throughout the year.

| Temperature | | | Rainfall | |
|---|---|---|---|---|
| °F | °C | Corumba (Brazil) | MM | IN |
| 104 | 40 | | 180 | 7 |
| 86 | 30 | | 135 | 5¼ |
| 68 | 20 | | 90 | 3½ |
| 50 | 10 | | 45 | 1¾ |
| 32 | 0 | | 0 | 0 |

Jan Feb Mar Apr May Jun Jul Aug Sep Oct Nov Dec

KEY ▮ Temperature ▮ Rainfall

C. America

# The Great Blue Hole

*A stunningly beautiful submerged sinkhole in a reef off the coast of Belize*

The Great Blue Hole is situated on Lighthouse Reef, an atoll lying 55 km (35 miles) east of the huge Belize Barrier Reef. It is a large, almost perfectly circular sinkhole within the limestone mass of the reef.

## Formation and exploration

Analysis of stalactites found in the Great Blue Hole shows that it formed in stages going back at least 153,000 years, during periods when sea level was lower than it is today. Erosion by streams during these times produced a complex of air-filled caves and tunnels in a block of limestone on dry land. Eventually, the ceiling of one cave collapsed, creating what is now the entrance to the Great Blue Hole. Since its formation, sea level has gone up and down, most recently rising to flood the blue hole and a cave system linked to it. Today, both the sinkhole and caves are accessible only by adventurous scuba divers but are regarded as one of the world's most exciting dive sites. Diving into the hole is not recommended for novice divers since perfect buoyancy control is needed. Caribbean reef sharks are often seen swimming around the entrance.

### PREVIOUS SEA LEVELS

The Great Blue Hole has not always been underwater. It formed over several tens of thousands of years through erosion of a mass of limestone on dry land. The sea level has risen over the past 18,000 years, since the last ice age, causing the Great Blue Hole to become submerged.

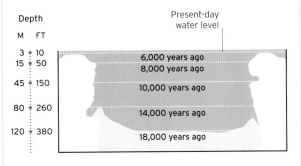

| Depth | | Present-day water level |
|---|---|---|
| M | FT | |
| 3 | 10 | 6,000 years ago |
| 15 | 50 | 8,000 years ago |
| 45 | 150 | 10,000 years ago |
| 80 | 260 | 14,000 years ago |
| 120 | 380 | 18,000 years ago |

◁ **STALACTITE EVIDENCE**
The presence of stalactites on several locations around the walls of the hole are proof that it formed above sea level, since stalactites and stalagmites cannot form underwater.

**South grotto cave**
Large stalactites hang from the cave roof and stalagmites project from its floor, giving the entrance the appearance of an open shark's mouth.

**Sediment dune**
This is created by coral fragments raining down from the lip of the shelf some 110 m (360 ft) higher up.

△ **WALL CAVITY**
The sinkhole's walls are pockmarked with small cavities. These provide ideal hiding places for some marine animals, except at great depth, where there is a lack of oxygen in the water.

▲ **THE GREAT BLUE HOLE**
With a diameter of 318 m (1,043 ft), the Great Blue Hole is 124 m (407 ft) deep and has rough-textured limestone walls indented by ledges, grottoes, and caves. Its entrance is almost completely surrounded by a ring of reef projecting slightly above the sea surface.

▷ **AT THE BOTTOM**
The floor of the blue hole is covered in chunks of limestone and fallen, broken stalactites, as first reported by the underwater adventurer Jacques Cousteau, who used a small submersible to explore the sinkhole in 1971.

**Surrounding reefs**
Smaller coral reefs surround the Great Blue Hole and are home to a great variety of marine life.

**West wall cave**
At a depth of 50 m (165 ft), a cave system extends horizontally for more than 46 m (150 ft) and contains sea turtle skeletons.

**Steep shelf**
This slopes down to a depth of 15–18 m (50–60 ft), to the rim of the sinkhole where there is a vertical drop-off.

△ **CORAL RESERVE**
The rim of the hole, close to the surface, is covered in a profuse array of both stony, reef-building corals (hexacorals), such as brain coral, and many colourful soft corals (octocorals).

**Notch**
This indentation is the result of dissolved hydrogen sulphide in the water at the bottom of the hole dissolving the limestone walls.

**Thermocline level**
At a depth of around 27 m (90 ft), warm surface water abruptly gives way to colder, clearer water lower down.

**Sediment layers**
At the bottom of the hole are layers of carbonate deposits – the remains of plankton and corals – and sediment washed from outside the reef and into the sinkhole.

**Hydrogen sulphide layer**
A turbid, brownish layer of water at about 95 m (310 ft) depth contains hydrogen sulphide gas from decomposing organic matter.

△ **GIANT SINKHOLE**
Like the pupil of an eye, the blue hole appears from above as a dark inky-blue circle. This contrasts with the mottled turquoise shades of the surrounding shallow water covering Lighthouse Reef.

# Stalactites hanging over the entrance to a cave on the sinkhole's wall are up to **8 m (25 ft) long**

# Lençóis Maranhenses

*A region of sweeping white dunes in northeastern Brazil, seasonally adorned with thousands of crystal-clear turquoise surface pools*

NE. South America

Covering some 1,500 square km (600 square miles), Lençóis Maranhenses is one of the most unusual areas of coastal sand dunes in the world. For part of the year, it looks like a desert, but from March to June it receives large amounts of rain, which collects in depressions between the dunes and creates many aquamarine pools. From the air, the vast expanse of dunes and blue pools looks like white bedsheets hung out to dry on a windy day. In fact, the word *lençóis* is Portuguese for "linen".

▷ **SLEEPING FISH**
In the dry season, tiger fish, or trahira, burrow down into the sand and lie dormant, re-emerging in the pools during the next rainy season.

## Blue lagoons

The pools, or lagoons, reach their maximum size around the end of the rainy season, in July, and remain so until September. Some are over 90 m (295 ft) long, with temperatures of around 30°C (86°F). Although they exist for only a few months each year, the pools are not devoid of life. Fish can usually access some of the pools via temporary channels that connect them to nearby rivers, such as the Rio Negro. The watery world of Lençóis Maranhenses is a brief one, however. Once the dry season returns, the hot equatorial sun heats the region, causing the pools to quickly dry up.

In places, the **dunes** stretch **50 km (30 miles)** back **from the coast**

# Chilean Fiordland

*A maze-like network of fiords, islands, and magnificent, glacier-sculpted rocky landscapes in southern Chile*

S. South America

▽ **GLACIAL SKYLINE**
In the Beagle Channel, the southern Andes soar behind a lonely lighthouse. Clear skies are rare here as moisture-laden Pacific air cools and forms clouds as it rises.

The Chilean Fiordland is a large region of southern Chile, including part of Tierra del Fuego, which was covered by a massive ice-sheet 10,000 years ago. Today, following the retreat of that ice-sheet, it is an extensive network of fiords – long, narrow sea inlets formed by the flooding of formerly glaciated valleys – and intervening islands. In total, it covers an area of about 55,000 square km (21,500 square miles) to the west of the southern Andes. At the terrestrial ends of some of the fiords, descending from ice-fields that still remain in the southern Andes, are the tongues of huge glaciers, which continuously calve enormous icebergs into the water. On the edges of the fiords, waterfalls cascade down granite walls, while hundreds of bird species feed around the often mist-shrouded coast and islands. Mammals that live along this coast include marine otters, sea lions, and elephant seals.

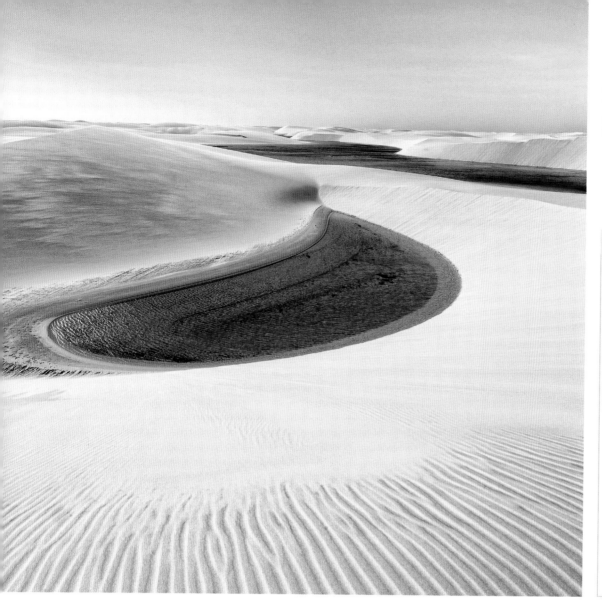

◁ **CLEAR WATERS**
The pools are typically crescent-shaped with crystal-clear water. They can be up to 3 m (10 ft) deep.

## SEASONAL CHANGES

In the dry season (September to November), strong winds blowing over Lençóis Maranhenses dry its surface and form sand dunes, with small valleys in between. In the rainy season, the valleys fill with water, which is prevented from percolating downwards by a layer of impermeable rock lying under the sand.

valleys   wind   rain
small dunes
Dry season
Rainy season
surface pool
impermeable rock
dried-out sand
water-saturated sand

# Cape Horn

*A headland at the southern tip of South America, notorious for the atrocious weather conditions that often develop nearby*

S. South America

◁ **BREEDING AT THE CAPE**
Some breeding colonies of Magellanic Penguin are found on coasts around the cape.

◁ **HIGH POINT**
The rocky, roughly pyramid-shaped landmass known as Cape Horn stands 425 m (1,390 ft) tall at its most elevated point.

Cape Horn is a rocky headland at the southern end of a small island called Hornos, in the Chilean part of Tierra del Fuego. It was named after the city of Hoorn in the Netherlands, the birthplace of the Dutch navigator Willem Corneliszoon Schouten, who rounded it in 1616.

Its importance and mystique derives from the fact that it is a navigational landmark, marking the northern edge of the Drake Passage, the strait between the southern tip of South America and the northern tip of the Antarctic Peninsula. "Rounding the Horn" is one of the main ways of moving between the Atlantic and Pacific oceans, and because the sea around the headland is noted for its bad weather – including powerful winds, large waves, and a risk of icebergs – sailors rightly consider it a major challenge. The number of ships rounding Cape Horn from the Atlantic Ocean to the Pacific Ocean dropped sharply when the Panama Canal opened in 1914.

# The Monteverde Cloud Forest

*One of the world's most biodiverse and protected habitats, where sea, moisture, and mountains create a rainforest constantly shrouded in clouds*

C. Central America

Characterized by fog-like clouds and humidity levels of close to 100 per cent throughout the year, cloud forests make up just 1 per cent of the world's woodlands. Costa Rica's Monteverde Cloud Forest, encompassing 105 square km (40 square miles), is nonetheless one of the most biodiverse habitats on Earth.

### A safe haven

The highly protected Monteverde Cloud Forest Reserve is located around 1,440 m (4,720 ft) above sea level in the Cordillera de Tilarán, part of a larger mountain chain running along the centre of Costa Rica.

Rainfall is extensive, with an average of 2,590 mm (102 in), while temperatures range between 14°C and 23°C (57°F and 73°F). Almost constant cloud covers the eastern forest, formed by warm Atlantic air ascending the mountains. The result is an astonishingly rich habitat containing more than 3,000 plant species, thousands of insect, 400 bird, and around 100 mammal species.

◁ **SURVIVAL TACTIC**
Algae-covered fur helps the brown-throated three-toed sloth to blend in with its lush forest surroundings – vital for such a slow-moving mammal.

△ **ONE FOREST, MANY TREES**
At least 755 tree species grow in the Monteverde reserve. They are taller on the cloud-covered eastern slope and shorter on the drier western side.

## Costa Rica is home to **4 per cent** of Earth's **plant** and **animal species**

# The Andean Yungas

*A unique band of dense forest, home to many species-rich ecosystems on the side of a mountain range*

W. South America

The Yungas covers the eastern slopes of the Andes from northern Argentina through Bolivia and into Peru. At an altitude of 1,000–3,500 m (3,300–11,500 ft), it forms a transitional area between the Andean highlands and the eastern lowlands. In this area are many temperature-sensitive climatic zones, ranging from high, subtropical cloud forests to moist temperate woodlands. Because many animal species are confined to "shoestring distributions" – areas that extend for several kilometres horizontally but only a few hundred metres vertically – the biodiversity of the Yungas is very high.

▷ **LIFE ON THE EDGE**
Steep-sided ravines like those near the ruins of Peru's Machu Picchu create many complex microclimates. The resulting ecosystems are vital for native species.

## TRADE WINDS AND CLOUD FORESTS

Warm, moisture-laden trade winds from the Atlantic form Costa Rica's cloud forests. As warm air rises, its moisture condenses into fog clouds, creating a rainy tropical climate. Mountains prevent the clouds from entering the country's Pacific side, so the climate is drier and less humid there.

northeasterly trade winds bring warm air from the Atlantic Ocean

clouds form on eastern side of mountains

rain-shadow area to west receives little rainfall

# The Valdivian Temperate Forest

*Earth's second-largest temperate rainforest, where almost 90 per cent of plant species are found nowhere else*

SW. South America

South America's only temperate rainforest is also one of the largest in the world, second only to the coastal temperate rainforest of the Pacific Northwest. Bordered to the west by the Pacific Ocean and to the east by the Andes, the Valdivian runs mainly along the Chilean coast, with parts stretching into Argentina. Having evolved separately from plants in the northern hemisphere, many species are unique to the region, including the rare Patagonian cypress (a giant conifer) and the prehistoric monkey puzzle tree.

▷ **PREHISTORIC SURVIVOR**
The monkey puzzle is a primitive evergreen conifer species thought to have been in existence for 200 million years.

N. South
America

# The Amazon Rainforest

*Earth's largest tropical rainforest, home to
390 billion trees and 2.5 million insect species*

Covering an area of almost 6 million square km
(2 million square miles), the Amazon is the world's
largest tropical rainforest, situated within the planet's
biggest river drainage basin. The bulk of the rainforest
is located in northern Brazil, but parts extend into eight
other South American countries. Rainfall is high – as
much as 3,000 mm (120 in) per year – with humidity
averaging over 80 per cent, and temperatures rarely
fall below 26°C (79°F).

This constantly moist, warm environment hosts an
incredibly rich ecosystem, where an estimated 16,000
tree species coexist alongside 1,300 bird species and
a staggering 2.5 million insect species. Scientists
believe even more species have yet to be discovered.
However, deforestation, due mainly to cattle ranching
and agriculture, remains a huge threat to the Amazon
Rainforest. From 1970 to 2016, 768,935 square km
(296,887 square miles), or roughly 20 per cent, of the
Brazilian rainforest was destroyed.

◁ **CANOPY FEEDING**
Many rainforest plants
depend on animals to
disperse their seeds. The
scarlet macaw spreads
seeds of the açai palm by
eating its fruit.

**Kapok**
This giant of the rainforest
towers above the canopy,
reaching heights of up to
60 m (200 ft). Kapok flowers
are pollinated by bats.

▶ **RAINFOREST DYNAMICS**
A tropical rainforest is both
highly structured and densely
populated – 300 species of
woody plant species can be
found in just 1 hectare
(2.47 acres). The many
species of plants and
animals interact and
depend on each other
to survive.

**Açai palm**
A slender palm of the
canopy layer, the açai
grows up to 15-30 m
(50-100 ft) tall.

## More than **half** of the world's **species live** in **tropical rainforests**

**Ground level habitat**
Larger animals search the forest
floor for food. Vines growing
from the floor create "highways"
to upper layers.

**Fallen tree**
If an old tree falls over, it creates
a gap in the canopy, allowing in
sunlight. Nearby trees undergo a
growth surge.

### TROPICAL RAINFOREST LAYERS

Tropical rainforests grow in five
main layers. The forest floor is
the hottest, darkest, and most
humid. Small shrub-layer plants
come next, competing for light
and food. In the understorey,
ferns and vines grow over taller
trees, where most birds nest.
Eighty per cent of rainforest life
lives in the hot, dry canopy at
heights of around 30 m (100 ft).
Only the tops of the tallest trees,
the emergents, project above it.

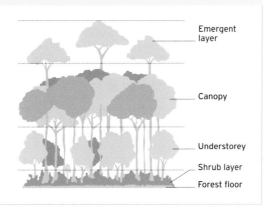

Emergent
layer

Canopy

Understorey

Shrub layer

Forest floor

## ▷ LIFE ON AIR

Epiphytes, or air plants, like this Cattleya orchid, require no soil. Instead, they anchor themselves to trees, drawing water and nutrients from moist air.

## ▷ CLOSED CANOPY

Most rainforest plants and animals live in the canopy, where trees compete for sunlight via maximum leaf growth and spread. The overlapping branches form an almost impenetrable "roof".

**Brazil nut tree**
This emergent tree reaches heights of 40–50 m (130–165 ft). Most brazil nuts come from wild rainforest trees.

**Rubber tree**
A rubber tree grows very fast so will soon establish itself if a gap appears in the canopy.

**Cacao tree**
The cacao tree grows in the understorey in the shade of other trees. Each of its pods contains up to 60 cocoa beans.

**Leaf litter**
Microorganisms feed on and recycle decaying plant material lying on the forest floor.

**Shallow roots**
An extensive shallow root system means a tree can absorb vital topsoil nutrients before they are washed away by rain.

## ▷ FOOD ON THE FLOOR

Some rainforest trees, like the Brazil nut, rely on ground-dwelling animals to spread their seeds. The agouti gnaws holes in tough, fallen seed pods to reach the nuts inside. It buries some but often forgets them, so new Brazil nut trees grow instead.

**Topsoil**
Rainforests have just a thin layer of topsoil. The soil contains all available organic matter and nutrients for plant growth.

**Buttress roots**
These large roots grow at angles from the trunk to stabilize a tall tree in poor-quality soil.

## CLIMATE

The Pampas has a temperate climate, with almost constant breezes that help to offset high humidity levels of 75 per cent or more. Although winter temperatures occasionally drop below freezing, the yearly average is 12–18°C (54–64°F). Precipitation is higher in coastal areas, decreasing inland, and annual rainfall ranges from 500 mm to 1,800 mm (20 in to 71 in), decreasing towards the southwest.

Temperature · Olavarria, Argentina · Rainfall

| °F | °C | | MM | IN |
|---|---|---|---|---|
| 104 | 40 | | 140 | 5½ |
| 86 | 30 | | 105 | 4¼ |
| 68 | 20 | | 70 | 2¾ |
| 50 | 10 | | 35 | 1¼ |
| 32 | 0 | | 0 | 0 |

Jan Feb Mar Apr May Jun Jul Aug Sep Oct Nov Dec

KEY ■ Temperature   ■ Rainfall

### △ UMBRELLA TREE

The evergreen ombu is one of the few tree species that grow in more arid Pampas regions. Its umbrella-like canopy spreads up to 15 m (50 ft) across.

SE. South America

# The Pampas

*Wide, flat grasslands stretching across the southeastern corner of South America, forming some of the richest grazing land in the world*

South America's huge expanse of grassland known as pampas extends from the foothills of the Andes east to the Atlantic coast. Because *pampa* means "flat surface", the term is used for many smaller plains all over the continent, but the largest and most well known is the Argentine Pampas.

Covering around 760,000 square km (295,000 square miles), the Pampas consists mainly of a flat, unbroken plain that slopes gently from an elevation of 500 m (1,640 ft) in the northwest to just 20 m (65 ft) in the southeast. This expanse of land can be divided climatically into two main sections. The larger western dry zone of central Argentina is dominated by coarse grass species such as stipa and is dotted with marshes. The smaller, more fertile, but heavily populated humid zone in the east features a combination of grassland and dry woodland. While much of the Pampas' original coarse grass has been replaced with grass species more suitable to domestic livestock that have grazed here for two centuries, the region remains a vital habitat for endangered species such as the Pampas deer and the Argentine tortoise.

▷ **STEALTHY MOVEMENT**
Geoffroy's cat is a small, efficient, and opportunistic predator, hunting throughout the Pampas for rodents, reptiles, birds, and fish.

△ **FEATHERY TUFTS**
Pampas grass reaches heights of up to 3.6 m (12 ft) or more. The feather-like flower heads hold up to 100,000 seeds each.

▷ **HIGH VIEWPOINT**
Standing 1.5 m (5 ft) tall, the greater rhea cannot fly, but it can run as fast as 60 kph (37 mph). It is often seen grazing with herds of Pampas deer.

# Few trees can **survive** the frequent **wildfires** of the Pampas, but **grasses** regenerate easily

### HIGH DUNES
Vast sand dunes collect at the base of the Andean foothills. One, called Cerro Medanoso, rises as high as 550 m (1,800 ft).

### △ PENITENTES
Snow rarely melts in the Atacama. Instead it forms clusters of jagged, ice-encrusted penitentes, named after the spiky hats of Spanish penitents.

### ▷ CAPTIVE CLOUDS
The Pacific's Humboldt Current pushes cold water towards the ocean's surface, creating fog banks. When trapped near the coast, these bring moisture ashore.

W. South
America

# The Atacama Desert

*The world's highest desert and one of the driest places on Earth, a sterile, stony plateau full of salt flats, sand, and active volcanoes*

Lying to the west of the Andes and parallel to the Pacific coastline, the Atacama forms a narrow strip of arid desert in northern Chile, averaging less than 160 km (100 miles) in width but extending about 1,000 km (620 miles) south from the border with Peru. With an average elevation of about 4,000 m (13,000 ft), this high plateau is, according to NASA studies, the driest place in the world. Some weather stations have never recorded a single drop of rain; even at lower levels, a dry spell of four years is common. The average rainfall in the desert's estimated 105,200 square km (40,600 square miles) range is just 1 mm (1/16 in) a year.

## In mountain shadows

The Atacama is a double rain-shadow desert, as two high mountain ranges, the Chilean Coast Range and the Andes, block incoming moisture from both Pacific and Atlantic oceans. In this hyper-arid climate, nothing decays; Earth's oldest mummies, dating from around 7020 BCE, were found here, still beautifully preserved. However, on the eastern side, where the foothills of the Andes rise up, there is a small amount of rainfall in winter.

Although parts of the Atacama are considered "absolute desert", other areas are more hospitable. Fog zones called *lomas*, created at lower levels when clouds are trapped by mountains or coastal slopes, bring in enough moisture to sustain lichens, algae, and hardy cacti and shrubs. Birds, which are able to fly in and out as conditions dictate, are the most abundant wildlife: three species of flamingo are found in the Atacama, feeding on algae growing in salt lakes, while hummingbirds and Peruvian song sparrows are seasonal visitors.

However, due to the high content of sulphur, copper, and mineral salts such as sodium nitrate in its soils, the desert is hostile to most animals apart from lizards, insects, and rodents – and the South American grey fox, which feeds on them. In fact, the Atacama's soil composition is so similar to samples taken from Mars that NASA uses it for testing space rovers.

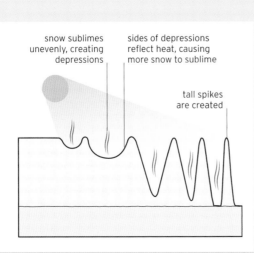

◁ **HARDY MAMMAL**
The highly adaptive South American grey fox, or *chilla*, copes with the Atacama's challenging conditions by hunting small animals in lower-lying fog zones.

## HOW PENITENTES FORM

Penitentes are ice structures that usually form at altitudes of 4,000 m (13,000 ft) or more. Their formation is believed to be due to a combination of factors. When the Sun shines on snow, freezing temperatures prevent it from melting. Instead, it sublimes (changes directly from solid to gas without melting). Troughs form and reflect sunlight, raising the temperature enough to melt the surrounding snow. Strong winds freeze the melting snow into spikes.

snow sublimes unevenly, creating depressions

sides of depressions reflect heat, causing more snow to sublime

tall spikes are created

Studies of sediments indicate that parts of the Atacama have had **no rain for** more than **20 million years**

# The Salar de Uyuni

*The largest salt flat on Earth, the legacy of a huge prehistoric South American lake*

W. South America

## HOW SALT CRUSTS FORM

Salt pans form when the rate of evaporation of mineral-rich water is higher than the rate of precipitation. As the water dries up, the previously dissolved salt ions solidify, forming a hard white crust, which thickens as the process continues.

evaporation

saline lake

**SALINE LAKE** — lake bed

dry saline pan

**SALINE PAN** — water drains through bed, forming brine layer

layered salt crust

**SALT CRUST**

Southwestern Bolivia's Salar de Uyuni is considered the planet's largest and least-polluted salt flat. Situated on a high plateau, around 10,600 square km (4,100 square miles) of salt, so fine it is almost flour-like, floats in a crust on top of a lake of brine thought to contain 50–70 per cent of the world's lithium reserves. While around 22,700 tonnes (25,000 tons) of salt per year are harvested, a staggering 9 billion tonnes (10 billion tons) remain.

### Islands in a white sea

The Salar de Uyuni's salt crust varies from a few centimetres to 10 m (33 ft) thick, reflecting the 2–20 m (6–65 ft) depth of the saltwater lake below it. Despite these variations, the crust's surface is remarkably flat, with less than a metre's difference in level from one side to the other. The vast expanse of white is not entirely unbroken, however. A few elevated areas, such as the cactus-covered Isla Incahuasi (Inca House Island), pierce the crust in places. For the most part, though, there is little vegetation. Some wildlife is present, mainly in the form of flamingos that migrate here to breed.

▽ **ROCKY OASIS**

One of the few oases in an almost pure-white landscape, the rocky Isla Incahuasi supports the growth of tree-sized pasacana cacti, which can reach 10 m (33 ft) in height.

▽ **NATURAL GEOMETRY**

When saline water evaporates undisturbed, hexagonal patterns are formed on the salt crusts. Widespread similar hexagons indicate calm conditions throughout the salar.

## Ponds formed by rain **dissolve bumps** on the **salt flat's** surface

W. South America

# The Valle de la Luna

*A maze of canyons and badlands sculpted into bizarre and otherworldly rock formations by millions of years of wind and water erosion*

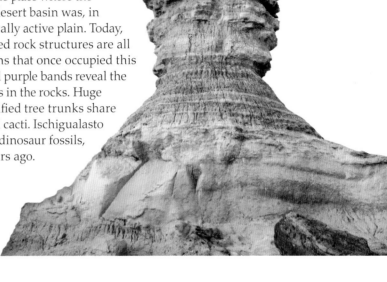

The region in northwest Argentina popularly known as the Valley of the Moon is actually a series of sandstone and mudstone rock formations. Otherwise known as Ischigualasto – "the place where the Moon rests" – this rugged desert basin was, in prehistoric times, a volcanically active plain. Today, stone monoliths and contorted rock structures are all that remain of the mountains that once occupied this landscape. Pink, yellow, and purple bands reveal the mineral content of sediments in the rocks. Huge mudstone spheres and petrified tree trunks share space with sparse scrub and cacti. Ischigualasto also has some of the oldest dinosaur fossils, dating from 230 million years ago.

## HOW ROCK PEDESTALS FORM

A pedestal-like rock formation is created by abrasion, when wind blows sand particles against a rocky outcrop. The force of the particles erodes softer sedimentary layers swiftly. The bottom section wears away faster than the top layers.

mushroom-shaped rock

prevailing wind

rock base eroded by wind-blown sand

neck

▷ **THE MUSHROOM**
One of Ischigualasto's most well-known rock formations is *El Hongo*, Spanish for "the Mushroom".

---

# The Patagonian Desert

*Argentina's largest desert, a cold, windswept region that undergoes seven months of winter and five of summer*

coarse hairs extend beyond bands of skin-covered bony armour

SE. South America

△ **HAIRY ARMOUR**
The centre of the desert is too harsh for most animals, but some, such as the big hairy armadillo, survive in vegetated border areas.

Bordered by the Atlantic to the east and the Andes to the west, the Patagonian Desert covers 673,000 square km (260,000 square miles) of southern Argentina and just crosses the border into Chile. The region has two distinct climatic zones: in the semiarid north, top temperatures average 12–20°C (54–68°F), while the south is cooler, with highs of 4–13°C (39–55°F). Average precipitation is 200mm (8in). The centre is mainly barren rocky plateau, but near the desert borders, mixed northern shrubland changes to sparser grassy vegetation in the south.

▷ **PETRIFIED FOREST**
Once a vast prehistoric forest, the Patagonian Desert today contains only the fossilized remains of huge trees at its centre - some up to 27m (89ft) long.

**Transport network**
Rivers deliver huge volumes of summer meltwater from Iceland's glaciers to the Atlantic Ocean. On reaching flat ground close to the coast, the rivers branch out to form complex networks of channels called distributaries.

# Europe

# LAND OF RIVERS AND PLAINS
## Europe

Europe is the world's second smallest continent; only Australia is smaller. Yet it has an astonishing variety of landscapes, sliced through by the continent's plentiful rivers, from the east-flowing Danube to the north-flowing Rhine.

Mainland Europe is attached to Asia all the way down its eastern boundary along the line of the ancient Ural mountains in Russia. To the north, it is fringed by the chilly Arctic Ocean, to the west by the Atlantic Ocean, and to the south by the warm, almost landlocked Mediterranean Sea. The Mediterranean south of the continent is warm and the landscape dry, but it is separated from the damper, cooler north by chain after chain of jagged, high mountains, including the Alps and the Caucasus, which have been thrust up by the northward pressure of the subducting African Plate. To the north of the mountains are rolling, sediment-covered plains, and further north still are ancient mountains worn smooth and sculpted with deep valleys by glaciers during the last ice ages.

## KEY DATA

▲ **Highest point** Mount El'brus, Russia: 5,642 m (18,510 ft)

▼ **Lowest point** Caspian Depression, Russia: -28 m (-92 ft)

● **Hottest record** Murcia, Spain: 47°C (117°F)

● **Coldest record** Ust'Shchugor, Russia: -58°C (-73°F)

## CLIMATE

Europe lies mostly within the mild temperate zone, with plentiful rains brought by winds off the Atlantic. Temperatures range from the warmth of the Mediterranean to the chill of the Arctic north.

AVERAGE TEMPERATURE

AVERAGE RAINFALL

°C 30 20 10 0 -10 -20 -30 -40

°F 80 60 40 20 0 -20 -40

MM 10,000 7,500 5,000 2,500 0

IN 400 300 200 100 0

Iceland

ARCTIC OCEAN

Norwegian Sea

Norwegian Fiords

Faroe Islands

Shetland Islands

Orkney Islands

British Isles

Grampian Mountains

North Sea

Jutland

Vänern

Elbe

B a

English Channel

Severn

Thames

Rhine

Ardennes

E U R

Brittany

Seine

Black Forest

Bavarian Forest

Bay of Biscay

Loire

Massif Central

Lake Geneva

Mt Blanc 4,807 m

Alps

A p e n n i n e s

Adriatic Sea

Po

Cordillera Cantábrica

Pyrenees

Camargue

Rhône

Douro

Ebro

Corsica

Tagus

Iberian Peninsula

Balearic Islands

Sardinia

Sierra Nevada

Murcia 47°C

M e d i t e r r a n e a n

Sicily

Etna △ 3,263 m

Km
0    250    500

0    250    500
Miles

Malta

*Barents
Sea*

*Ural Mountains*

*Kola
Peninsula*

*White
Sea*

Ust' Shchugor
-58°C

*Lake
Onega*

*Lake
Ladoga*

*Volga*

*Bothnia*

*European Plain*

*Central
Russian
Upland*

*Don*

*Volga Uplands*

*Volga*

*Caspian
Depression
-28m*

*Caspian Sea*

*O P E*

*Dnieper*

*Vistula*

*Carpathian Mountains*

*Danube*

*Tisza*

*Great
Hungarian
Plain*

*Danube*

*Balkan Mountains*

*Caucasus*

Mount
El'brus
5,642m

*Black Sea*

*Aegean
Sea*

*Ionian
Sea*

*Peloponnese*

*Crete*

*Sea*

## ECOSYSTEMS

Temperate forests dominate the
continent, with steppe grasslands
to the east, boreal forest and tundra to
the north, and scrubby Mediterranean
vegetation in the south.

### KEY

- Temperate broadleaf forest
- Temperate coniferous forest
- Desert and scrub
- Temperate grassland
- Mediterranean woodland and scrub
- Montane grassland
- Boreal forest and taiga
- Tundra
- Ice

## GEOLOGY

The geological core of Europe is formed
by the ancient rocks beneath Russia,
Scandinavia, and the Caledonian mountains
of Scotland. Piled on top are the younger
rocks of the southern mountains.

### KEY

- Precambrian (pre-541 million years ago)
- Paleozoic (541–252 million years ago)
- Mesozoic (252–66 million years ago)
- Cenozoic (66 million years ago to present day)

B

A

### ELEVATION
M    FT

2,000 • 6,600

0 • 0

-2,000 • -6,600

Sicilian
mountains

Apennines

Adriatic
Sea

Mediterranean
Sea

Alps

North European Plain

North Sea

Norwegian
mountains

A ◀⋯⋯⋯⋯⋯ CROSS-SECTION ⋯⋯⋯⋯⋯▶ B

> **RISING PEAKS**
As the African Plate drives north, it has crushed southern Europe and thrown up the Alps and other high mountain ranges such as the Carpathians and Caucasus.

> **DEEP VALLEY**
A collision between tectonic plates 400 million years ago opened a gash across the Scottish Highlands called the Great Glen – which is now filled by Loch Ness.

# THE SHAPING OF EUROPE

Europe is a relatively small continent, yet has its own unique geology. Whereas most continents are formed from several stable sections of continental crust called cratons, Europe was forged as fragments formed and reformed around just one – the East European Craton.

## Europe rises
The East European Craton lies buried under sediments across the North European Plain that stretch from Holland to the Ural mountains. But in Sweden and Finland, the craton is exposed.

At first, the East European Craton formed the continent of Baltica. Then, 430 million years ago, Baltica collided with ancient landmasses geologists call Laurentia (now North America) and Avalonia (now the British Isles and part of North America) to create the supercontinent Euramerica. Along the collision zone, mountains were thrown up across Norway and Britain – the Scottish Highlands are remnants of this range. Much of Euramerica was a desert, and it is sometimes known as the Old Red Continent for the red sandstones that formed there.

In time, Euramerica crashed into the southern supercontinent of Gondwana to help form the giant supercontinent of Pangaea, piling up new mountains, such as the Vosges and the Harz, which have

now worn down. Pangaea eventually broke up to form the continents we know today, with North America splitting away from Europe as the Atlantic Ocean opened up. But the separated continent of Africa has for the last 40 million years been driving relentlessly north, forcing the southern edge of Europe up into a series of new mountain ranges from the Pyrenees to the Alps and the Caucasus.

## Deep freeze
For much of the last two million years, in extremely cold periods known as ice ages, vast ice-sheets covered northern Europe. Their grinding movement scoured large areas of lowland Europe flat, and left behind extensive deposits of debris, known as moraines, creating a bumpy landscape. In Scotland and the Lake District, glaciers carved deep, trough-like valleys. In Norway, they gouged

# 5.6 million years ago, the Mediterranean Basin was a desert

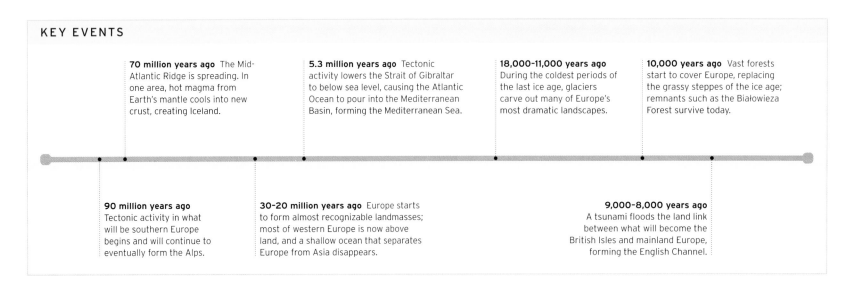

## KEY EVENTS

**70 million years ago** The Mid-Atlantic Ridge is spreading. In one area, hot magma from Earth's mantle cools into new crust, creating Iceland.

**5.3 million years ago** Tectonic activity lowers the Strait of Gibraltar to below sea level, causing the Atlantic Ocean to pour into the Mediterranean Basin, forming the Mediterranean Sea.

**18,000–11,000 years ago** During the coldest periods of the last ice age, glaciers carve out many of Europe's most dramatic landscapes.

**10,000 years ago** Vast forests start to cover Europe, replacing the grassy steppes of the ice age; remnants such as the Białowieza Forest survive today.

**90 million years ago** Tectonic activity in what will be southern Europe begins and will continue to eventually form the Alps.

**30–20 million years ago** Europe starts to form almost recognizable landmasses; most of western Europe is now above land, and a shallow ocean that separates Europe from Asia disappears.

**9,000–8,000 years ago** A tsunami floods the land link between what will become the British Isles and mainland Europe, forming the English Channel.

◁ **CARVING ICE**
The Jostedal ice-field in Norway, the largest in Europe, is a remnant of the ice-sheet that once carved out the deep valleys of the Norwegian fiords.

◁ **NARROW CHANNEL**
The Mediterranean Sea is linked to the Atlantic Ocean through the Straits of Gibraltar. The channel is so narrow that Morocco is visible from Spain.

▷ **NEW ISLAND**
Rising sea levels just 9,000 years ago cut the British Isles off from mainland Europe.

out valleys so deep that once the ice melted they were flooded by the sea to form the country's famous fiords.

## Sea levels fluctuate

During the ice age, Britain was joined to Europe by a long block of limestone – this acted as a dam against rising glacial meltwater. About 450,000 years ago, this natural dam burst, leaving behind the White Cliffs of Dover. Sea levels continued to rise and fall during each glacial and interglacial period, until 9,000–8,000 years ago, when the last ice-sheet melted and Britain was cut off from Europe for good.

The Mediterranean Sea has also come and gone. It was cut off from the Atlantic Ocean and most of it dried out to a salty desert. Then, the jostling of the African and the Eurasian plates caused the land underneath where the Straits of Gibraltar are now to suddenly fall. This let water flood through as a torrent from the Atlantic Ocean. At first the water pooled into the basin slowly over a period of a few thousand years. However, it is thought that 90 per cent of the Mediterranean Sea's water eventually came rushing into the basin in a matter of months to years.

### THE FORMATION OF THE MEDITERRANEAN SEA

About 5.6 million years ago, the Mediterranean Basin was cut off from the Atlantic Ocean and almost entirely dry. But 300,000 years later, the Straits of Gibraltar fell, and Atlantic waters flooded in and refilled the basin.

Atlantic Ocean

Mediterranean Basin

**5.6 MYA**

part of crust falls due to tectonic activity

water from Atlantic rushes in

**5.3 MYA**

erosion deepens Strait of Gibraltar

water from Mediterranean Sea mixes with water from Atlantic Ocean

**PRESENT DAY**

## BUILDING THE EUROPEAN CONTINENT

KEY ▬ Convergent boundary ▬ Divergent boundary

Mid-Atlantic Ridge spreads

land above sea level

Tethys Sea

Apennines form

African Plate subducts under Eurasian Plate

Europe linked to Asia by land

North Sea above sea level

African Plate continues to subduct

**94 MILLION YEARS AGO** The North Atlantic is spreading to push Europe and North America apart, while southern Europe is entirely inundated by the Tethys Sea between Europe and Africa.

**50-40 MILLION YEARS AGO** Africa drives north towards Europe. The Tethys Sea is squeezed out, and sediments pile up in southern Europe to form high mountain ranges such as the Apennines.

**18,000 YEARS AGO** As oceans recede, the Black, Baltic, and North seas dry out, and Europe has become linked to Asia by land. Europe is at its greatest extent.

△ SKYE LANDSCAPE
Skye, Scotland's second largest island, boasts some dramatic landscapes. The large rock pinnacle at the centre of this image is known as the Old Man of Storr.

# The Scottish Highlands

*A region of ancient mountains and extinct volcanoes, chiselled by glaciers and rainfall for many centuries, with a wild, rugged beauty*

NW. Europe

## HOW THE CALEDONIAN MOUNTAINS FORMED

During the Caledonian Orogeny, tectonic movements caused ancient continents to collide and push up mountains. Later, the combined landmass broke up, leaving the mountains scattered across the North Atlantic region.

Laurentia
Baltica
Avalonia

490–390 MILLION YEARS AGO

Greenland
Norway
Scotland
Ireland
Newfoundland

TODAY

The Scottish Highlands are part of the Caledonian Mountains – an ancient, much larger mountain chain that was formed in a series of events called the Caledonian Orogeny (see panel, left). They include the Northwest Highlands and the Grampians – which are separated by a 100-km- (60-mile-) long fault called the Great Glen Fault – as well as the mountainous islands of western Scotland. The Grampian Mountains are the highest in the British Isles, their tallest peak being Ben Nevis, at 1,345 m (4,413 ft).

### Volcanic remnants

Scattered among the highlands are remnants of past volcanic activity. Ben Nevis was once a large volcano that collapsed in on itself following an eruption 350 million years ago. Some highland subranges, such as the Cairngorms, are made of granites formed from solidification of magma underground, later exposed by erosion. The highlands have a characteristic rugged appearance. Forest still covers some areas, providing a home to wildlife such as the capercaillie, pine marten, and wildcat.

**Glen Coe**, a valley in the Highlands, **was** once a **supervolcano**

◁△ HIGHLAND HEATHER
Heather is an emblem of Scotland. In late summer, purple heather clothes the lower slopes of the Scottish Highlands, where highland cattle can often be seen grazing.

N. Europe

# Strokkur Geyser

*A famous Icelandic geyser, which erupts roughly every eight minutes*

Strokkur geyser is situated in an area of hot springs about 80 km (50 miles) east of Iceland's capital, Reykjavík. Nearby is another famous but less frequently erupting geyser, named Geysir, which was first described in 1294 and gave the world the word "geyser".

Strokkur first became active following an earthquake in 1789. The geyser erupts from a pool of water, and its typical eruption height is about 15–20 m (50–65 ft), but occasionally the fountain of hot water reaches 40 m (130 ft). Each blast lasts for just a few seconds and is accompanied by a dramatic explosion of sound.

△ **REGULAR CYCLE**
The geyser's eight-minute cycle owes its regularity to human intervention. In 1963, its conduit was cleaned to ensure regular eruptions.

◁ **READY TO BLOW**
When Strokkur erupts, first a dome of water bulges up as a steam bubble rises from below. A plume of boiling water then blasts violently skywards.

W. Europe

# The Massif Central

*A highland region of France with a superlative scenery of extinct volcanoes*

Covering about one-seventh of France, the Massif Central is separated from the Alps by the valley of the Rhône. Volcanism began here about 65 million years ago – possibly due to crustal thinning and upwelling of hot mantle rock – and is apparent today as volcanic remnants, known as the Chaîne des Puys, across a northern part of the region. The southern part of the massif features raised limestone plateaus cut through by deep canyons. The most famous of these is the magnificent Tarn Gorge, which is 53 km (33 miles) long and in places up to 600 m (2,000 ft) deep.

## INSIDE A CINDER CONE

Pyroclastic, or cinder, cones are relatively small, steep-sided volcanoes built mainly from loose volcanic cinders (glassy fragments of solidified lava). Some also contain volcanic ash. Most have just one main eruptive phase when they first appear. After erupting and growing for a few months to years, they go permanently quiet.

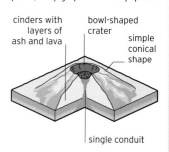

cinders with layers of ash and lava
bowl-shaped crater
simple conical shape
single conduit

◁ **THE CHAÎNE DES PUYS**
This line of small, extinct volcanoes, mostly in the form of cinder cones, is about 40 km (25 miles) long. Puy de Côme is shown in the foreground with a winter sprinkling of snow.

# One face of **Marmolada**, the **highest peak** in the Dolomites, is a near-vertical cliff **600 m (2,000 ft) high**

▽ **BOLD CLIFFS**
The topography of the Dolomites features massive rock walls, towers, and jagged pinnacles.

S. Europe

# The Dolomites

*A dramatic range in northeastern Italy, with 18 peaks rising above 3,000 m (10,000 ft)*

A subrange of the Alps, the Dolomites consist of multiple layers of sedimentary rock, thousands of metres thick, that originally formed as coral reefs in shallow seas hundreds of millions of years ago. Later, these rock layers were lifted up as a result of tectonic plate movements, then eroded by glaciers to form the dramatic landscape seen today. The mountains are renowned for the fossils of coral reef animals they contain.

## Pale mountains

The Dolomites are named after an 18th-century French mineralogist, Deodat de Dolomie, who was the first to describe the rock – dolomite – from which they are largely made. Dolomite is similar to limestone and typically white, cream, or light grey. Because of this, the Dolomites are also known as the Pale Mountains. Their highest peak is Marmolada at 3,343 m (10,968 ft). The spectacular scenery draws numerous climbers, skiers, paragliders, and hikers.

dolomite crystals

▷ **BUILDING BLOCK OF THE DOLOMITES**
Dolomite is the name given to a mineral, as well as a rock. In this specimen, cream-coloured dolomite crystals are seen mixed with crystals of quartz (pink) and a third, darker mineral.

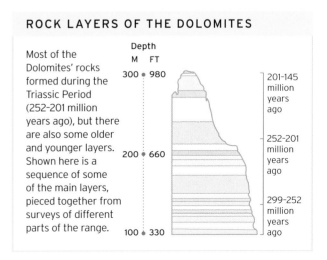

## ROCK LAYERS OF THE DOLOMITES

Most of the Dolomites' rocks formed during the Triassic Period (252–201 million years ago), but there are also some older and younger layers. Shown here is a sequence of some of the main layers, pieced together from surveys of different parts of the range.

Depth
M    FT
300 ● 980

200 ● 660

100 ● 330

201–145 million years ago

252–201 million years ago

299–252 million years ago

# The Pyrenees

*A high natural barrier between France and Spain that has played a significant role in the history of both countries*

W. Europe

The Pyrenees mainly formed in the last 55 million years (see panel, below). Over time, the uplifted land was eroded, resulting in a series of parallel mountain ranges, together with some flat-topped massifs, that in places reach over 3,000 m (10,000 ft) in height.

## Steep valleys and waterfalls

There are only a few places where the Pyrenees can be crossed through passes lower than 2,000 m (6,600 ft). During the last ice age, the Pyrenees experienced extensive glaciation. The mountains are characterized by jagged peaks that are steepest on the French side. They contain some of Europe's most spectacular waterfalls, the highest of which, at the Cirque de Gavarnie, plunges 422 m (1,385 ft). Major rivers, such as the Garonne of France and tributaries of the Ebro of Spain, flow out of the mountains. Wildlife in the region includes the rare lammergeier vulture, eagles, chamois, wild boars, and a small population of brown bears.

△ **SOURCE OF LEAD**
The Pyrenees contains deposits of some valuable minerals, including galena (an ore of lead). The crystals seen here have unusual shapes called truncated octahedrons.

## HOW THE PYRENEES FORMED

In the distant past, Iberia was an island being pushed away from Eurasia by a spreading mid-ocean ridge. But about 70 million years ago, a new convergent plate boundary formed between Iberia and Eurasia, which began to approach each other. This led to the collision that pushed up the Pyrenees.

Mid-Atlantic Ridge
spreading ridge

Eurasia

Iberia

**130 MILLION YEARS AGO**

new convergent boundary

Eurasia

Iberia

**70 MILLION YEARS AGO**

Pyrenees

France

Spain

**TODAY**

# The Alps

*The highest and most extensive of Europe's mountain ranges, containing many famous peaks*

C. Europe

Some 1,200km (750 miles) long and more than 200km (120 miles) wide at their broadest point, the Alps cover more than 207,000 square km (80,000 square miles) of western Europe. Pushed up over the last 100 million years by the coming together of the African and Eurasian plates, they form a curved belt, with more than 50 peaks rising to over 3,800m (12,500ft).

## Shaped by glaciers

The current landscape of the Alps is a result of glaciation over the past 2 million years. The glaciers that once covered these peaks sculpted out enormous amphitheatre-like cirques (mountain hollows), sharp arêtes (knife-like ridges), and majestic pyramidal peaks such as the Matterhorn on the Swiss–Italian border and Austria's highest mountain, Grossglockner. As the glaciers melted, they left behind some

U-shaped valleys, immense waterfalls pouring from hanging valleys, and elongated lakes of great depth. The longest remaining glacier is the Aletsch (see p.151).

## European playground

The Alps include famous mountains such as Mont Blanc on the French–Italian border, the Matterhorn, and the Eiger in Switzerland, whose north face has long been considered one of the world's climbing challenges. During summer, the region attracts mountaineers and people who enjoy hiking, mountain biking, and paragliding. Winter sports such as skiing and snowboarding are also popular.

◁ **LIVING AT ALTITUDE**
A type of large squirrel, the marmot is a common sight in high-altitude pastures during summer. Marmots hibernate for the rest of the year in burrows underground.

In 1991, a **5,000-year-old mummy**, since named **Ötzi**, was **found** in the **Ötztal Alps**

### SUBRANGES OF THE ALPS

The Alps contain many subranges, some of which are mapped below. The loftiest are the Pennine Alps, which contain 13 of the highest 20 alpine peaks; the Bernese Alps contain four; and the Mont Blanc Massif in the Graian Alps has three.

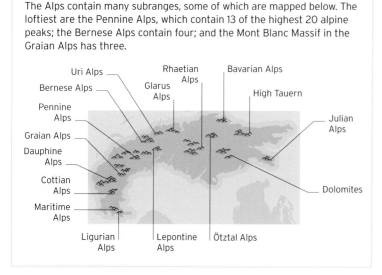

Uri Alps
Bernese Alps
Pennine Alps
Graian Alps
Dauphine Alps
Cottian Alps
Maritime Alps
Ligurian Alps
Glarus Alps
Rhaetian Alps
Lepontine Alps
Ötztal Alps
Bavarian Alps
High Tauern
Julian Alps
Dolomites

▽ **FLORAL MIGRATION**
As global temperatures have risen and glaciers retreated, flowering plants of all kinds have spread to much higher latitudes and altitudes in the Alps.

**ICONIC PEAK**
With its precipitous, glacier-sculpted walls, the Matterhorn is the quintessential alpine peak. Its summit, at 4,478 m (14,692 ft), was first conquered in 1865.

# Mount Etna

*The tallest and by far the largest active volcano in Europe,
with a long history of major, often spectacular eruptions*

S. Europe

Occupying 1,190 square km (459 square miles) of eastern
Sicily, Mount Etna is one of the world's largest, most
famous, and most active volcanoes. Towering above
Catania, it has one of the world's longest documented
records of volcanic activity, dating back to 1500 BCE.
Etna's origins are probably connected to it being close
to the boundary between the African and Eurasian
plates; it may also lie over a mantle hotspot.

## Complex structure

Etna is a stratovolcano, with a complex structure
that includes four separate summit craters – and
more than 300 smaller parasitic volcanic vents
and cones on its flanks. Its summit is at
3,330 m (10,925 ft), and lava flows cover
much of the surface of its flanks. Although
Etna has an overall conical shape, there is
a deep and prominent horseshoe-shaped
depression on its eastern side, known
as the Valle del Bove.

## Eruption types

Over the past several thousand years, Etna has been
almost continuously active, producing eruptions of
two main types. Spectacular explosive eruptions flare
up from one or more of its summit craters. These
can produce tall fountains of fiery lava, volcanic
bombs, cinder showers, and large ash clouds. Quieter
eruptions, with large-scale outpourings of lava, also
emerge from flank vents and fissures. Etna's most
destructive eruption occurred in March 1669. It
produced massive lava flows that destroyed most
of Catania's city walls. Despite the danger it
poses, most of the inhabitants of Sicily regard
Etna as a major asset.

◁ **LAVA LUMP**
The lava that erupts from Etna is mostly of
a composition and temperature that makes
it flow several kilometres before solidifying
into lumps like this.

In **1999**, Etna erupted **lava fountains 2 km
(1¼ miles) high** - the highest ever recorded

## HISTORIC LAVA FLOWS

This map shows areas
covered by Etna's lava
flows over the centuries,
going back to the 17th
century. It also shows
the extent of older
flows. Many eruptions
have come from Etna's
flanks rather than its
summit, and some flows
have invaded urban
areas. The pre-Etnian
sediments are more
than half a million
years old - Etna's
approximate age.

KEY

△ Summit of Mount Etna

    Historic summit lavas

Flank lava flows
- 21st century
- 20th century
- 19th century
- 18th century
- 17th century
- Pre-16th century
- Prehistoric lavas
- Pre-Etnean sediments
- Urban areas

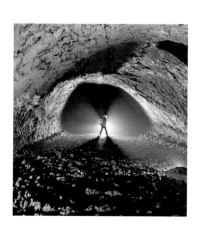

△ **FIRE TUNNEL**
This cave under Etna's southeastern
flank is a lava tube - a natural tunnel
within a solidified former lava flow.
Fiery magma once flowed through it.

**ASH COLUMN**
During an eruption in December 2015, Etna blasted ash to a height of 8 km (5 miles), where it flattened into a mushroom shape.

# Inside a
# STRATOVOLCANO

There are two main types of large volcano – shield volcanos and stratovolcanos. A shield volcano slopes gently, whereas a stratovolcano has steep sides and a conical shape built up from layers of eruptive products. The reason for these differences lies with the types of lava they emit. A shield volcano produces relatively runny lava that flows a long way before solidifying. A stratovolcano, such as Etna, also produces more viscous, water-rich lava that does not flow far. Instead, the lava tends to solidify around, and can sometimes block up, the volcano's main vent, or vents. As a result, much of the material erupted by a stratovolcano consists of fragmented solids – ash, pumice, and cinders – that are blasted into the air and later fall to the ground as the volcano clears its main vent. This helps explain not just the structure and shape of stratovolcanoes but also their long-term behaviour, with phases of violent activity interspersed with quiet periods, lasting from a few years to several thousand years.

## THE EVOLUTION OF A STRATOVOLCANO

Stratovolcanoes grow from the accumulation of their own eruptive products: solidified lava, ash, and cinders. The volcano initially grows rapidly, as eruptions add a lot of material relative to its existing size. A mature volcano gains height more slowly, partly due to erosion.

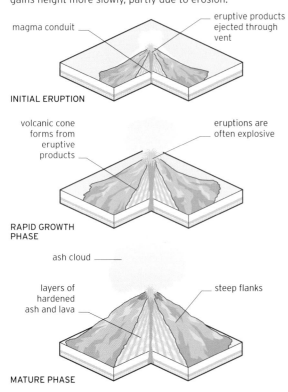

magma conduit — eruptive products ejected through vent

**INITIAL ERUPTION**

volcanic cone forms from eruptive products — eruptions are often explosive

**RAPID GROWTH PHASE**

ash cloud —

layers of hardened ash and lava — steep flanks

**MATURE PHASE**

**Ash cloud**
An ash cloud is formed by gas and magma that is explosively fragmented into tiny particles as it is blasted into the air.

▷ **FLANK CRATERS**
Etna's flanks are covered in hundreds of secondary cones, vents, and craters, many of them now inactive. This is one of a pair called the Silvestri craters, dating from an eruption in 1892. It is 110 m (360 ft) across.

**Solidified lava**
Etna's surface is covered in solidified lava of various ages.

directly underneath Etna are layers of ancient volcanoes that form the foundation for the present-day cone

▷ **LAVA FLOWS**
Some eruptions occur from fissures that open up quite low down on the volcano's flanks. These can produce huge flows of lava that destroy agricultural land.

**Lithospheric mantle**
Made of the coarse-grained igneous rock peridotite, the lithospheric mantle underlies the crust.

lightning occurs in ash clouds due to friction between the ash particles and normal atmosphere

secondary pipe or conduit

**Main pipe or conduit** Magma is carried up to the vents in the summit craters through the main pipe.

◁ **SUMMIT CRATERS**
A stratovolcano may have one or several summit craters. Etna has four. The two shown here are known as Voragine (meaning "the chasm") and Bucca Nuova ("new mouth").

▽ **EXPLOSIVE ERUPTIONS**
These can occur from summit craters or sometimes from new vents that open near the summit. They typically produce lava fountains, ash plumes, and lava flows.

**Crust**
Beneath Etna, the crust is about 30 km (20 miles) deep and consists mainly of faulted and fractured layers of sedimentary rock with intrusions of molten and solidified magma.

magma rises through fractures in crust

**Sill**
Horizontal sheets of magma between rock layers are called sills.

**Large magma chamber**
Beneath Etna is a cavity full of magma (molten rock) and dissolved gases.

**Dike**
Vertical channels of magma are called dikes.

◀ **MOUNT ETNA**
Etna has many features of a typical stratovolcano, although it is more active than most in terms of eruption frequency. It produces spectacular explosive eruptions from craters at or near its summit and quieter eruptions, with outpourings of lava, from flank vents and fissures.

# Stromboli

*A small volcanic island, known for thousands of years as the Lighthouse of the Mediterranean because of the regularity, reliability, and brilliance of its eruptions*

S. Europe

Stromboli is one of the few places on Earth that anyone can visit, on any day, with a high likelihood of being able to witness a volcanic eruption. Around every 20 minutes, one of its three summit vents blasts a fountain of lava fragments up to a height of 150 m (490 ft). These eruptions have been occurring for several thousand years and are so distinctive that geologists use the term Strombolian to describe similar eruptive activity at other volcanoes.

## Visiting Stromboli
Roughly 5 km (3 miles) in diameter and 924 m (3,032 ft) high, Stromboli lies north of Sicily in the Mediterranean. A hike to the summit is permitted only in the company of a local guide. The eruptions can also be viewed from a boat off the island's northwest coast.

▷ **GIANT FIREWORK**
This photograph shows a typical eruption at Stromboli. Hundreds of glowing lava particles blast into the air, then fall to the ground in gentle arcs.

**VOLCANIC ISLAND**

Steep paths from two shoreline villages lead to vantage points for viewing eruptions. During forcible explosions, streams of lava cascade down a depression on one side of the volcano called La Sciara del Fuoco.

La Sciara del Fuoco · Stromboli Village · Ginostra Village · crater · cliffs

# The Urals

*An upland divide between Europe and Asia and one of the most ancient and mineral-rich mountain ranges on Earth*

E. Europe

The Ural Mountains are about 2,500 km (1,600 miles) long and run north to south, from the Arctic Ocean to near the Aral Sea, encompassing a wide range of landscapes, from polar wasteland to semidesert. Located in Russia and Kazakhstan, they have traditionally been regarded as a boundary between Europe and Asia.

## Ancient mountains
Partly due to the long period of erosion since they were formed, the Urals are not exceptionally high. The highest peak is Mount Narodnaya, with an altitude of 1,895 m (6,217 ft). Much of the area is forested, although the northern Urals are strewn with glaciers and also feature alpine meadows and tundra. An extensive network of rivers and lakes covers the region, while karst topography, with many caves, is found on the western slopes. This mix makes the Urals one of the most beautiful areas in Russia. The richest and most varied fauna in the Urals are found in the forests and include brown bear, lynx, and wolverine. The mountains are particularly rich in natural resources, including timber, coal, metal ores, and a range of gemstones.

◁ **CRYSTALLINE TREASURE**
Amethysts, emeralds, and topaz are among the precious gemstones mined in the Urals. This uncut hexagonal crystal is an emerald.

## The Urals formed **300–250 million** years ago and are among the **world's oldest mountains**

# Vesuvius

*Europe's most dangerous volcano, famous for its large historic eruptions, including the massive and deadly eruption of 79 CE*

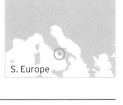

S. Europe

Vesuvius is a stratovolcano that lies just 8 km (5 miles) away from the Italian city of Naples, in the most densely populated volcanic area in the world. Its location, and capacity for especially violent eruptions, are what make it so dangerous. Vesuvius has a steep cone, with a height of 1,281 m (4,203 ft), and sits within the caldera of an older volcano called Mount Somma. Its lower slopes are dotted with villages and vineyards.

## Destructive past

Of Vesuvius's numerous large historic eruptions, the most infamous occurred in 79 CE. During the explosion, falls of volcanic ash and pyroclastic surges buried the towns of Pompeii and Herculaneum, killing about 2,100 people. Another enormous volcanic blast in 1631 killed more than 3,000 people, while an eruption in 1906 killed more than 200. There has been no eruption since 1944, suggesting that a large explosion is overdue. Plans have been drawn up for an evacuation in the event of warning signs, such as increased seismic activity.

△ **DECEPTIVE CALM**
Vesuvius's 400-m- (1,300-ft-) wide summit crater is outlined here. Part of the Bay of Naples is visible in the background, with the island of Capri in the far distance.

### THE EVOLUTION OF VESUVIUS

Before the eruption in 79 CE, Vesuvius was a single volcanic cone with a huge crater. Later, a new cone grew within this crater. Today, this new, much enlarged, cone is known as Vesuvius. Much of the older cone, now called Mount Somma, has eroded away.

ancient city of Naples — Bay of Naples

**BEFORE 79 CE ERUPTION**

new volcanic cone

**17TH-18TH CENTURY CE**

Mount Somma — Vesuvius — modern Naples

**TODAY**

△ **WINTER FOREST**
A large part of the Urals is covered in taiga, the coniferous forest that stretches across northern Eurasia. In winter, much of the forest is snow-covered.

◁ **ROCK GIANTS**
Known as the Manpupuner rocks, these stone pillars, more than 30 m (100 ft) high, jut out of a plateau in the western Urals. Legend has it that they were giants turned to stone by a shaman.

### ◁ BLUE CAVERN
Ice caves frequently form around the edges of the ice-cap. They are carved by meltwater streams beneath the glacier that have since dried up. Their locations tend to shift from year to year.

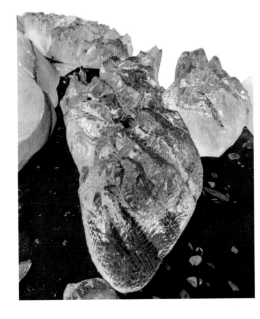

### ◁ BEACHED ICE
Blocks of ice are often seen on the beaches of black sand near the edge of Vatnajökull. These are remnants of icebergs that reached the sea but were driven back by waves.

### ▷ PANOPLY OF COLOURS
At sunrise, the undulating surface of Vatnajökull displays a glorious array of colours from soft pastel ochres and yellows to powder blue and bright turquoise.

N. Europe

# Vatnajökull

*A vast ice-cap covering eight per cent of Iceland's surface area and the largest glacier by volume in Europe*

One of several ice-caps in Iceland, Vatnajökull derives its name from the Icelandic words *vatna*, meaning "water", and *jökull*, meaning "glacier". The ice-cap completely covers the mountainous terrain of southern Iceland. Under the ice-cap are valleys, mountains, and plateaus.

## Shape and size

Roughly elliptical, the ice-cap forms a frozen dome, with its highest point lying at an altitude of over 2,000 m (6,600 ft). It covers a total area of about 8,100 square km (3,130 square miles). Around 30 outlet glaciers drain the ice from Vatnajökull, carrying it towards the sea. Some of the ice breaks off as icebergs into a large glacial lagoon, Jökulsárlón, and from there makes its way via channels to the sea. Owing to the light-bending effects of atmospheric refraction, the ice-cap is said to be visible sometimes from the top of the highest mountain in the Faroe Islands, which are 550 km (340 miles) away. This has been recorded as one of the world's longest sight lines.

## Subglacial volcanoes

Three active volcanoes, as well as some volcanic fissures, lie under Vatnajökull. The active volcanoes are named Bardarbunga and Öræfajökull (both stratovolcanoes), and Grímsvötn, which is a caldera volcano. When one of these volcanoes erupts, along with the usual effects and dangers of a volcanic eruption, there is the added risk of large amounts of ice melting and causing a catastrophic flood (see panel, below). Events of this sort, which are widely referred to by the Icelandic term *jökulhlaups* (meaning "glacier floods"), are most often associated with the Grímsvötn volcano, which has the highest eruption frequency of all Icelandic volcanoes. In November 1996, an eruption of Grímsvötn melted millions of tonnes of Vatnajökull's ice and triggered a flood that lasted two days. The floodplain between the glacier and the sea became littered with huge blocks of ice.

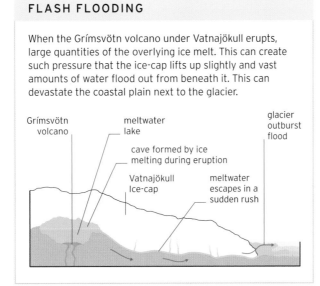

◁ **BORN SWIMMER**
One of six seal species found in Iceland, the common, or harbour, seal is often seen in a meltwater lagoon, Jökulsárlón, situated between Vatnajökull and the sea.

### FLASH FLOODING

When the Grímsvötn volcano under Vatnajökull erupts, large quantities of the overlying ice melt. This can create such pressure that the ice-cap lifts up slightly and vast amounts of water flood out from beneath it. This can devastate the coastal plain next to the glacier.

Grímsvötn volcano

meltwater lake

glacier outburst flood

cave formed by ice melting during eruption

Vatnajökull Ice-cap

meltwater escapes in a sudden rush

The **ice-cap** has a maximum thickness of **1,000 m (3,300 ft)** and is **400 m (1,300 ft) thick** on average

N. Europe

# The Monaco Glacier

*An immense glacier on Spitsbergen, the largest island of the Svalbard archipelago*

Spitsbergen, in northern Norway, is around 1,850 km (1,150 miles) from the North Pole, and about 80 per cent of it is covered in glaciers. The Monaco Glacier, situated on an area of the island called Haakon VII Land, is one of the largest. The glacier discharges copious amounts of ice into a long fiord, Liefdefiord, at its front end.

### A princely river of ice

Some 40 km (25 miles) long, the glacier is named after Prince Albert I of Monaco, a pioneering polar explorer of the early 20th century, who organized the first expeditions to explore and map the glacier. According to mapping studies, the glacier has retreated by more than 3 km (2 miles) in the last 50 years. Ringed seals, bearded seals, various seabird species, and polar bears are the main inhabitants of the neighbouring ice floes, coasts, and islands.

△ **SEA OF ICE**
Ice calving off the Monaco and other glaciers nearby is the source of imposing, ship-sized icebergs that float in the middle of Liefdefiord.

▷ **FROZEN WALL**
A flock of kittiwakes feed in front of the Monaco Glacier's enormous terminal wall, which is more than 50 m (165 ft) high and 4 km (2 miles) wide.

The **thundering sound** of ice breaking off the glacier can be heard for **tens of kilometres** around

# The Jostedalsbreen Ice-field

*The largest glacier in mainland Europe, a remnant of a vast ice-sheet that covered the whole of Norway until about 10,000 years ago*

N. Europe

The Jostedal Ice-field (*breen* is Norwegian for "glacier") lies in southwestern Norway and has an area of about 480 square km (185 square miles), with its highest point at 1,957 m (6,421 ft) above sea level. In places, its ice is more than 400 m (1,300 ft) thick. The ice-field survives primarily because of high regional snowfall rather than particularly cold temperatures.

## Locked-up water

The top, or central, part of the ice-field is a slightly undulating and largely featureless expanse of white ice and the source of numerous outlet glaciers. The total mass of ice corresponds to approximately 300 billion baths of water, or all of Norway's water use for 100 years. However, this mass is diminishing, as reflected by a high rate of melting at the snouts of Jostedalsbreen's outlet glaciers. Some of these, such as Nigardsbreen, contain spectacular blue ice caves.

△ **AZURE GROTTO**
Ice of such a deep blue colour as seen in this cave is very dense and strong, containing fewer air bubbles than white ice.

▽ **GLACIER TOUR**
A line of climbers, equipped with poles and crampons, make their way up Briksdalsbreen, one of the most accessible of Jostedalsbreen's outlet glaciers.

## JOSTEDALSBREEN'S OUTLET GLACIERS

The Jostedalsbreen Ice-field has a long, thin shape, with a maximum length of some 65 km (40 miles). About 50 outlet glaciers flow out of the ice-field and down to the valleys below. Four of the largest are marked below.

Briksdalsbreen

Lodalsbreen

Austerdalsbreen

Nigardsbreen

W. Europe

# The Mer de Glace

*A glacier on the northern, or French, side of Mont Blanc,
notable for the spectacular banding pattern on its surface*

The Mer de Glace is a valley glacier – a long glacier
with mountains on both sides – that extends down
from the Mont Blanc Massif towards the Chamonix
Valley in eastern France. It is some 11 km (7 miles) long.

The Mer de Glace, "Sea of Ice", gets its name
from the undulating, banded pattern seen on part
of the glacier below an ice-fall (a steeply descending
section) situated at an altitude of around 3,200 m
(10,500 ft). This surface pattern has been described as
waves frozen in the midst of a storm. The "waves" are
actually alternating bands of thicker, lighter ice and
thinner, darker ice, created due to seasonal factors
(see panel, right). The bands, known as ogives, are
curved – the result of the ice in the middle of the
glacier moving downhill slightly faster than the ice
at its sides.

▽ **SNAKING DOWNWARDS**
The Mer de Glace descends some 2,500 m (8,200 ft)
vertically from its ice accumulation area, which lies
below Mont Blanc, visible here at top left, to its snout.

## THE MER DE GLACE'S BANDING PATTERN

In summer, ice passing down
the ice-fall loses volume by
melting, and forms a trough
at the bottom, darkened by
the concentration of dust
inside it. In winter, thicker and
lighter-looking ice emerges
at the bottom.

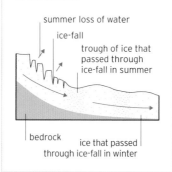

summer loss of water

ice-fall

trough of ice that
passed through
ice-fall in summer

bedrock

ice that passed
through ice-fall in winter

△ **ICY HIGHWAY**
Dark medial moraines (longitudinal stripes)
run down the glacier, reinforcing the appearance
of a winding highway of ice.

## TYPES OF MORAINE

Debris carried along by a glacier and later
deposited is called a moraine. Lateral
moraines occur at the sides of a glacier,
whereas medial moraines form from smaller
glaciers merging. Subglacial moraines are
found at the base, and englacial moraines
are trapped within the ice. A terminal
moraine (not shown here) is rock material
dropped at the melting snout of a glacier.

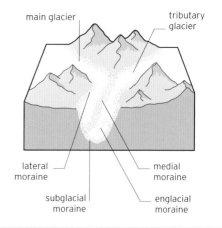

main glacier

tributary
glacier

lateral
moraine

medial
moraine

subglacial
moraine

englacial
moraine

# The Aletsch Glacier

*The longest and largest valley glacier in Europe, some 23 km (14 miles) long, which descends from a particularly scenic area of the Bernese Alps, in Switzerland*

C. Europe

---

The Aletsch Glacier forms from the merger of four smaller glaciers that originate in the southern flanks of the impressive Jungfrau and Monch mountains. The four glaciers converge at a flat ice plateau called Kondordiaplatz at an elevation of 2,750 m (9,022 ft), where the ice is up to 900 m (2,950 ft) deep.

### Movement and retreat

The glacier moves at speeds varying from about 200 m (660 ft) a year in the Konkordiaplatz to 10 m (33 ft) a year near its snout. The pattern of terminal moraines below the snout indicates that it has retreated about 5 km (3 miles) since 1860. Spectacular views down the glacier can be seen from Jungfraujoch, a saddle area in the mountains that is accessible via a steep-grade railway.

▷ **INSIDE THE ALETSCH**
Several ice caves have been carved out by meltwater streams. These are accessible from the glacier's edges and it is possible to enter them and view the brilliant blue interior.

## The Aletsch glacier **covers an area** of about **80 square km (31 square miles)**

# Litlanesfoss

*A stunning Icelandic waterfall cascading over
a remarkable outcrop of volcanic columns*

N. Europe

Situated on the Hengifossá River in eastern Iceland,
Litlanesfoss is a spectacular waterfall framed by a
series of tall and exceptionally regular hexagonal basalt
columns. Its total height of over 30 m (100 ft) includes
two separate falls. The upper, smaller tier sits at an
angle to the lower, near-vertical section, which falls
into a small, bright-blue plunge pool. Litlanesfoss
is one of several waterfalls dotted across the volcanic
landscape of this part of Iceland, including Hengifoss,
one of Iceland's tallest falls, just 1 km (½ mile) upstream.

## Ancient lava flow

The Litlanesfoss Falls effectively cascade down the
cross-section of a thick ancient lava flow. The wide
columns in the basalt rock are the result of the lava
cooling relatively slowly over time. As it cooled and
contracted, it formed joints, or fractures, that grew
perpendicular to the base and surface of the lava
flow. An array of straight and regular columnar jointing,
as seen at Litlanesfoss, is known as a colonnade and is
thought to have formed due to cooling from the base
of the flow upwards. Other well-known examples of
columnar jointing include the Giant's Causeway in
Ireland (see pp.164–65) and Fingal's Cave on the
Scottish island of Staffa (see p.166).

△ **HORSETAIL FALLS**
The lower, taller section of the
waterfall cascades down in what is
known as a horsetail formation, with
the water fanning out as it falls.

▽ **UPPER TIER**
The upper tier of Litlanesfoss is
much shorter and less steep. Water
pools at the bottom before it plunges
down the lower section.

## RADIAL DRAINAGE

The broadly circular Lake District comprises
a core of central mountains from which a
series of valleys, lakes, and waterways
radiates outwards. This pattern has been
produced by the uplifting of the underlying
dome of rock millions of years ago.

river flows downhill towards sea

rivers radiate
outwards from
high central point

Lake
Windermere

# The Lake District

*England's highest and arguably most picturesque region,
a landscape shaped by glaciers and loved by tourists*

NW. Europe

△ QUEEN OF THE LAKES
Seen here with a cloud of mist hanging over it, Derwentwater is thought by many to be the most beautiful of the Lake District's lakes and is referred to as the Queen of the Lakes.

The Lake District is a scenic area of mountains, moorlands, deep valleys, and lakes located within the county of Cumbria in northwest England. It contains most of the country's highest land, including Scafell Pike, its tallest peak, as well as England's longest and deepest lakes – Windermere and Wastwater, respectively. The region is named after the 16 lakes and many smaller water bodies, called tarns, which are found in higher areas. However, only one of the lakes, Bassenthwaite Lake, is actually called a lake; the others are known as waters or meres. Commonly defined by the boundary of the UK's largest national park, the Lake District covers an area of 2,362 square km (912 square miles).

## Glacial landscape

The underlying geological structure of the Lake District is a huge granite dome, on top of which sits a series of three broad bands of rock of different ages and origins. Variations in bedrock can be seen in the different features across the region, from volcanic outcrops to slate scree slopes.

However, much of the landscape of the Lake District has been shaped by episodes of glaciation over the last 2 million years. Indeed, the region is known for its classic examples of glacial landforms. The most obvious of these are the broad valleys with their U-shaped cross-sections, in the bottom of which often sit ribbon lakes.

△ COMMON PREDATOR
The blue damselfly is a common predator on many UK waterways and large lakes. Unlike dragonflies, damselflies fold their wings along their body when resting.

# The **Lake District National Park** is the **most visited** national park **in the UK**

# The Camargue

*One of Europe's major coastal wetlands, famous for its semi-wild horses and cattle, designated a wetland of international importance*

W. Europe

The Camargue is a vast wetland region located within the Rhône Delta – the largest river delta in Western Europe – on the south coast of France. It comprises more than 930 square km (360 square miles) of diverse habitats including salt marshes, sandbars, brackish lagoons, freshwater ponds, reedbeds, and low-lying islands. Many of these landforms are in a state of constant change, and the gradual accumulation of silt means the delta is steadily growing.

Historically, the Camargue has been prone to flooding, and a combination of dykes and canals has been built to help control the flow of water through the area. Large parts of the original wetlands at the periphery of the region have been drained for agricultural use, including the creation of rice paddies, which have been a feature of the region since the Middle Ages. The salt marshes in the southeast section of the Camargue are a major centre of commercial salt production.

## Famous wildlife

The Camargue's best-known animals include its semi-wild white horses and black cattle. However, its remarkable birdlife is one of the region's greatest attractions. More than 300 species are found in the Camargue, and it is an important stopover for birds migrating between Europe and Africa.

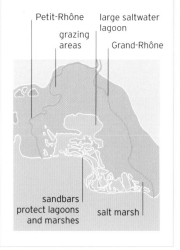

Petit-Rhône

grazing areas

large saltwater lagoon

Grand-Rhône

sandbars protect lagoons and marshes

salt marsh

◁ **REEDBED SPECIALIST**
Attracted to reedbeds, the bearded reedling is also called the bearded tit. In summer it eats insects and larvae, and in winter it eats reed seeds.

▽ **FLOCKING FLAMINGOS**
The Camargue is home to a large population of greater flamingos, which feed on brine shrimps that thrive in brackish waters.

## The **Camargue horse** is an **indigenous breed** that has lived here for **thousands of years**

# The Verdon Gorge

*A spectacular river-cut gorge in the South of France, referred to by some as the European Grand Canyon*

W. Europe

The Verdon Gorge is a deep river canyon in the limestone massifs of the alpine foothills straddling the Alpes-de-Haute-Provence and Var regions of southeastern France. The gorge lies on a 25-km (15-mile) stretch of the Verdon river, which has carved the gorge over the last several thousand years. The river's striking turquoise-green colour, produced by glacial rock flour suspended in the water, gives the river and gorge their name: *vert* means "green" in French. At the end of the canyon, the river flows into the artificial Lake Sainte-Croix.

## Limestone walls

The width of the gorge at river level ranges from 6 m (20 ft) to 100 m (330 ft), and on each side steep cliffs reach up to 700 m (2,300 ft) tall. The limestone rock in which the Verdon Gorge formed was laid down during the Triassic Period (252–201 million years ago) when the area was under an ocean. Subsequent episodes of tectonic uplift and rock fracturing, followed by a period of glaciation, prepared the landscape for the river to carve out the gorge as it is today.

◁ **LUSH VALLEY**
The bottom of the gorge, which has a mild and humid microclimate, is covered in lush vegetation that diminishes with elevation up the cliff face.

# Lake Geneva

*A large alpine lake on the French–Swiss border, once hit by a huge tsunami*

W. Europe

Known as Lac Léman in France, Lake Geneva is a crescent-shaped lake on the northern side of the Alps, stretching along the border between southeastern France and southwestern Switzerland. The largest of the alpine lakes, it is 73 km (45 miles) long, up to 14 km (9 miles) wide, and has a shoreline of just over 200 km (120 miles).

▷ **BUSY SHORES**
The shores of Lake Geneva are home to more than 1 million people, including the residents of Lausanne and Geneva.

## Lake of two halves

Lake Geneva lies on the course of the River Rhône, which enters through a delta at its eastern end and leaves at the western end through the city of Geneva. The lake is geographically divided into two basins separated by the Strait of Promenthoux. The Large Lake, to the east, is the widest section, and reaches a maximum depth of 310 m (1,020 ft). The Small Lake, to the west, is much narrower and shallower. Lake Geneva is notable for its rhythmical fluctuations in water level, known as seiches, caused by strong winds and rapid changes in atmospheric pressure.

### THE LAKE GENEVA TSUNAMI

Historical accounts and evidence from the lakebed suggest that Lake Geneva was struck by a large tsunami in 563 CE. Thought to have been caused by a rock fall near the Rhône Delta, a wave up to 16 m (52 ft) high travelled the length of the lake, devastating the shorelines in its wake. Moving at speeds of up to 70 kph (45 mph), the wave would have reached Geneva in just over an hour.

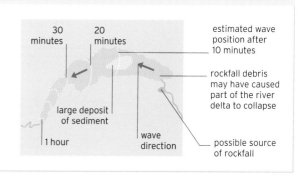

30 minutes
20 minutes

estimated wave position after 10 minutes

rockfall debris may have caused part of the river delta to collapse

large deposit of sediment

1 hour

wave direction

possible source of rockfall

C. Europe

# The Danube

*One of Europe's great waterways, linking the west of the continent to the east, crossing and forming many borders*

Europe's second longest river, after the Volga, the Danube forms at the confluence of the Brege and Brigach rivers in the Black Forest, in southwest Germany. Over a long journey eastwards to its mouth at the Black Sea, the river flows a total of 2,860 km (1,770 miles). The Danube can lay claim to being the most international river in the world. It passes through, or partly borders, 10 countries – more than any other river – and its huge drainage basin includes parts of nine others.

## Three courses

The course of the Danube is often divided into three major sections. The upper course runs from its source eastwards across southern Germany and northern Austria to the Devin Gate, a natural gorge to the east of Vienna. At the start of its middle section, the river slows down and deposits large amounts of gravel and sand. Turning south, it flows through Budapest and across the Great Hungarian Plain to become a border between Romania

and Serbia, where its banks narrow into a series of towering limestone cliffs known as the Iron Gate. The Danube's lower course becomes wider and shallower as it starts its journey across a broad, flat plain. As the river approaches the Black Sea, it winds north around the Dobruja Hills before turning east, splitting into three main distributaries and forming Europe's second largest delta after the Volga.

## Vital artery

The Danube has played a central role in the social and economic history of Central and Eastern Europe. At some times, it has formed the border between territories; at others, it has been a vital link allowing the flow of commerce between nations, enabling the countries it crosses to thrive.

◁ **BRIGHT FROG**
Pool frogs live in ponds, lakes, and slow-moving rivers over much of Europe. Those found in Central and Southern Europe are often a brighter green than their western cousins.

## The Danube is the **only major European river** that flows **west to east** from Central to Eastern Europe

### HOW RIVERS TRANSPORT SEDIMENT

The amount of material that a river is able to transport depends on the energy of its waters and the size of the material. While dissolved materials are carried in solution, the smallest particles, such as clay and silt, are suspended in the water column when its currents are strong enough. Larger items of sediment that are not held in suspension are called its bedload.

When a river has enough energy, it can transport smaller bedload by pushing or bouncing it along the riverbed. Rivers with high energy levels and strong currents - during a flash flood, for example - can move boulders by rolling them downstream.

lighter material held in suspension by turbulent eddies, with lightest particles nearer to surface

direction of flow

dissolved material carried in solution

bedrock

bedload moved by bouncing (saltation)

heaviest material moved by rolling (traction)

△ **CHANGING HABITATS**
Over its long course, the Danube passes through a variety of habitats, including deciduous woodland, coniferous forests, semiarid plains, and marshland.

▷ **PELICAN HAVEN**
Home to 70 per cent of the world's white pelican population, the Danube Delta is a key sanctuary for more than 300 bird species.

◁ **THE DANUBE DELTA**
The river's delta sprawls across an area of more than 4,000 square km (1,500 square miles). It is one of Europe's largest wetland areas, with one of the world's largest reedbeds.

C. Europe

# The Biebrza Marshes

*A large Central European wetland area with a wide variety of habitats supporting flourishing populations of birds and mammals*

Centred on the Biebrza river valley in the far northeast of Poland, the Biebrza Marshes are a diverse wetland complex covering an area of more than 1,000 square km (390 square miles). The region features a mosaic of habitats including extensive marshes and reedbeds (see panel, right), grasslands, river channels, lakes, low islands, and wooded higher ground. The area also includes one of the largest and best-preserved peat bogs in Central Europe. In many sections of the Biebrza Marshes, classic examples of ecological succession (see p.60) can be seen, as marshes grade first into raised peat bogs and then gradually into wet woodland.

The varied habitats of the Biebrza Marshes support a wide range of wildlife, and it has been recognized as a site of global significance. The area is one of the most important breeding sites for wetland birds in Central Europe, and more than 270 bird species have been recorded. Mammals include wetland specialists such as elk, beavers, and muskrats, as well as a few packs of wolves.

◁ **WATER LOVER**
Known as moose in North America, elk usually live close to water and are excellent swimmers. While they also graze trees and bushes, up to half of their diet consists of aquatic plants.

# The Hortobágy Wetlands

*A vast flooded grassland area in the Hungarian steppe region, regarded as one of the world's most important wetland reserves*

C. Europe

Located on the upper reaches of the River Tisza, in an area known as Puszta on the Great Hungarian Plain, the Hortobágy Wetlands is a steppe – a plain of grasslands and shrubs – that experiences regular flooding. This has turned an otherwise semiarid area into a wetland ecoregion of marshes, streams, and reedbed-lined lakes and lagoons. Part of the largest continuous natural grassland in Europe, the Hortobágy Wetlands is a key section of a national park that covers an area of 800 square km (310 square miles).

One of the best birdwatching sites in Europe, the Hortobágy Wetlands has a year-round wealth of birds. More than 340 bird species have been sighted here, and almost half have nest sites in the area.

In **autumn**, up to **70,000 common cranes rest** and **feed** at Hortobágy

▷ **STEALTHY PREDATOR**
Common in European wetland areas, the great egret wades through shallow water feeding mainly on fish, frogs, and small mammals, which it spears with its long, sharp bill.

## REEDBED STRUCTURE

This cross-section of a reedbed shows the succession from young reeds in open water to tightly packed old reeds on higher land. Old reeds give the dense cover needed by nesting birds, while young reeds provide shelter for small fish.

dense, older reeds

younger reeds

flooded reedbed

◁ **RICH PLANT LIFE**
The rich alluvial soils produced by regular flooding in the marshes support a wide variety of plants, including many rare species.

# Slovak Karst Caves

*An exceptionally large group of Eastern European caves with a great diversity of formations, a system of major geological and archaeological significance*

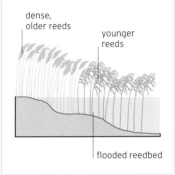

C. Europe

The Slovak Karst cave system consists of more than 1,000 caves within an area of about 550 square km (210 square miles) along the border between southeastern Slovakia and northeastern Hungary. These diverse and complex caves formed as limestone and dolomite were dissolved by water.

## Cave formations

Throughout the system, the caves are richly decorated by an array of structures and formations. The 25-km- (15-mile-) long Baradla-Domica cave complex, one of the most studied, has an elaborate display of stalagmites and stalactites of various shapes and colours, and a stream running through its main section. Like some of the other caves in the region, it shows evidence of ancient human occupation. The Slovak Karst system also includes many ice caves containing year-round ice formations. Perhaps the best known of these, the Dobšinská Ice Cave, was discovered in 1870 and is thought to be about 250,000 years old.

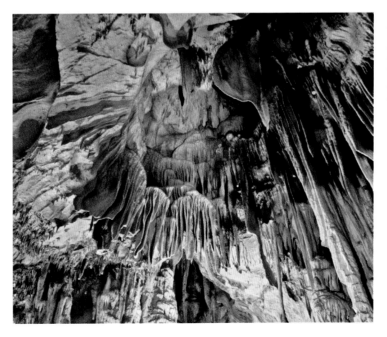

◁ **COMPLEX CAVE**
The ceiling of the Jasovská Cave is decorated with a variety of stalactites and curtain-like calcite flowstones.

# Lake Ladoga

*The largest lake in Europe, with a huge drainage basin made up of thousands of smaller lakes and rivers*

NE. Europe

▷ AQUATIC BLOOM
The amphibious bistort, an aquatic plant in the knotweed family, is found in the shallows of Lake Ladoga and along the River Neva.

Located in northwestern Russia, near the border with southeastern Finland, Lake Ladoga is Europe's largest lake. Its total area is about 17,700 square km (6,800 square miles); from north to south it is 219 km (136 miles) long, and it is 138 km (86 miles) across at its widest point. The deepest part of the lake is near the tall, rocky cliffs of its northern shore, where it reaches 230 m (750 ft) just west of Valaam Island, one of the largest of more than 650 islands in the lake. The southern part of the lake is shallower and has a lower shoreline.

## Lake and sea

Lake Ladoga formed in a graben, a depression between two fault lines, which has since been reshaped by glaciers. During the last ice age, it was part of the Baltic Ice Lake, which eventually formed the Baltic Sea when the ice retreated. Lake Ladoga is now separated from the Baltic Sea to the southwest by the Karelian Isthmus. The Neva, the only river that flows from the lake, leaves at its southwestern corner, passes through St Petersburg, and drains into the Gulf of Finland (part of the Baltic Sea).

There are more than **50,000 lakes** and **3,500 rivers** in Lake Ladoga's drainage basin

# The Volga

*The longest river in Europe and the core of its biggest river system, regarded as Russia's national waterway*

SE. Europe

The Volga is Europe's longest river, at the heart of the continent's largest river system. Rising in the Valdai Hills, northwest of Moscow, it flows 3,690 km (2,293 miles) southeastwards through western Russia before emptying into the Caspian Sea. The lush Volga Delta, which spans 160 km (100 miles), is the largest river estuary in Europe. With a catchment area extending eastwards to the Ural Mountains, the Volga drains most of European Russia. Along its course, it collects more than 200 tributaries, and its watershed contains 11 major cities. The river is traditionally referred to by Russians as Mother Volga.

△ THICK BILL
The largest species of tern, the Caspian tern is a common sight in the Volga Delta. This species has a distinctly chunkier bill than other terns.

▷ ICE FLOES
The Volga is affected by ice over much of its length for about 100 days each year.

△ FROZEN INLET
The edges and inlets of the lake start to freeze in December, and by the end of February the entire surface of the lake is frozen over.

## EUROPE'S LONGEST RIVERS

Four of the five longest rivers in Europe pass through Russia at some part of their course, and the Volga and the Don are contained within its borders. The Danube crosses 10 different countries on its course.

| | KM 0 | 1,000 | 2,000 | 3,000 | 4,000 |
|---|---|---|---|---|---|
| Don | | | | | |
| Dnieper | | | | | |
| Ural | | | | | |
| Danube | | | | | |
| Volga | | | | | |

| MILES 0 | 620 | 1,200 | 1,900 | 2,500 |
|---|---|---|---|---|

Length

△ GREEN DELTA
The Volga brings life to an otherwise arid landscape as it flows into the saline Caspian Sea, the world's largest inland body of water.

NE. Europe

# Norwegian Fiords

*The most complex coastline in the world, a maze of long, deep marine channels, many overlooked by high cliffs*

A fiord is an elongated ocean inlet, originally created by the action of glaciers over thousands or millions of years. While fiords are found in many countries, those of Norway are particularly magnificent, numerous (around 1,000), and in many cases exceptionally long.

A fiord-dominated landscape runs as a strip all around Norway's coast. This has a unique appearance, with a tangle of islands and peninsulas interspersed between the fiords themselves. Except in the area where they meet the sea, fiords are much deeper than typical coast-plain estuaries. Sognefiord, for example,

reaches a maximum depth of 1,300 m (4,300 ft). Many Norwegian fiords have vertical cliffs on one or both sides, rising to a height of up to 1,000 m (3,300 ft).

## Fiord life

Life in Norway's fiordland centres on picturesque fishing villages, which are connected by an extensive system of ferries. Wildlife includes marine mammals such as seals and porpoises, sea eagles, and millions of other sea birds, including colourful puffins. The harbour porpoise – one of the smallest marine mammals – is commonly seen here and feeds on small schooling fish.

## Fiords **lengthen Norway's coast** from 3,000 km (1,900 miles) to over **30,000 km (19,000 miles)**

### HOW FIORDS FORM

During the last glacial period, huge glaciers extending down from a vast ice-sheet scoured deep U-shaped valleys. As the glaciers melted and the sea level rose, seawater flooded into these valleys, forming fiords.

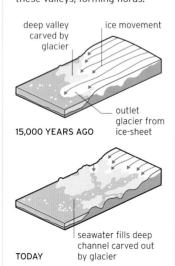

deep valley carved by glacier

ice movement

outlet glacier from ice-sheet

**15,000 YEARS AGO**

seawater fills deep channel carved out by glacier

**TODAY**

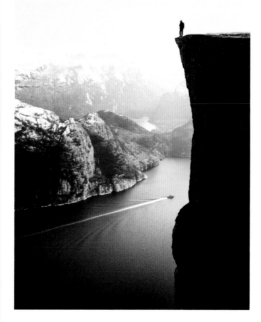

△ **SHEER DROP**
Soaring 600 m (2,000 ft) above Lysefiord, Pulpit Rock is one of Norway's most visited landmarks. It is popular for activities such as cliff jumping.

△ **TURQUOISE PASSAGE**
The waters of Hjeltefiord meander between islands in Hordaland County, southwestern Norway. Hjeltefiord is around 40 km (25 miles) long.

▷ **STUNNING ENTRANCE**
Traditional fishing huts can be seen near the mouth of Kierkfiord in the Lofoten Islands. A short distance from the mainland, the Lofotens are blessed with exquisite scenery.

NW. Europe

# The Giant's Causeway

*An array of huge columnar rocks, famed worldwide for its spectacular geometric stepping-stone appearance*

The Giant's Causeway is a group of thousands of interlocking columns of basalt rock at the foot of a cliff on the coast of Northern Ireland. The tops of the columns form a staircase that leads from the cliff foot up to a mound-shaped area, and then disappear under the sea.

## Fiery origins

Legend has it that the Causeway was built by a giant, Fionn MacCoul, as a means of crossing the sea to fight a Scottish rival. It actually originated from lava poured out by volcanic fissures in the area roughly 55 million years ago. The lava cooled and solidified to form thick layers of rock, some of which fractured into columns beneath the surface. After millions of years of erosion by glaciers, and later the sea, these structures were revealed at the surface. Apart from the causeway itself, the surrounding area is excellent for studying sea birds and also for botanical studies, since some rare plants thrive here.

◁ FINE-GRAINED ROCK
The Giant's Causeway is composed of basalt, a dark rock that forms when a particular type of runny lava cools and solidifies rapidly.

## Some of the **40,000 basalt columns** that form the causeway are more than **11 m (36 ft)** high

◁ △ STEPPING-STONE COAST
These columns were once a mass of cooling lava, which solidified, shrank, and fractured to form geometric patterns similar to those sometimes seen in drying mud.

▷ ERODED CLIFFS
Behind the Giant's Causeway are cliffs eroded out of the plateau of northern Antrim. The basalt columns jut out of these cliff faces, forming jagged steps that creep into the sea.

NW. Europe

# Fingal's Cave

*An atmospheric sea cave on the uninhabited Scottish island of Staffa, known for its wonderful acoustics*

Staffa is one of the Inner Hebridean islands off the west coast of Scotland, close to the much larger island of Mull. Its cave, named after the hero of an epic poem, is about 20 m (65 ft) high and extends about 60 m (200 ft) into the island.

In summer, local boating companies organize regular boat trips to Staffa. In calm conditions, visitors can explore Fingal's Cave, including its interior, via a walkway. The cave's size and arched roof, and the echoes of lapping waves, give it a special acoustic quality. The German composer Felix Mendelssohn, after visiting the cave in 1829, was so moved that he wrote an overture, "The Hebrides", in remembrance. This piece of music is now known as the "Fingal's Cave overture".

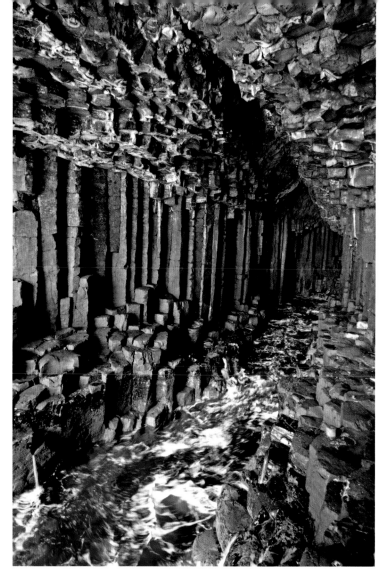

◁ **HEXAGONAL FRAMING**
The cave's arched entrance is framed by an array of hexagonal basalt columns. The sea covers the cave floor, with waves regularly washing and splashing in and out.

# The Cliffs of Moher

*An imposing procession of high, banded grey cliffs on the west coast of Ireland, home to tens of thousands of seabirds during the annual nesting season*

NW. Europe

The Cliffs of Moher extend along part of the coast of County Clare, Ireland, at the edge of a region called the Burren, which is dominated by karst topography (partially dissolved limestone). They take their name from an ancient fort called Moher, or Mothar, that once stood on Hag's Head, their southernmost point.

## Ancient heights

The cliffs consist of layers of the sedimentary rocks shale and sandstone, which originally formed more than 300 million years ago. In parts, they rise more than 200 m (660 ft) above the Atlantic Ocean, and their tops offer fantastic views of various offshore islands and mountain ranges of western Ireland. These include the three Aran Islands in Galway Bay and the Maumturks mountain range in Connemara. Between April and July, more than 30,000 pairs of seabirds – including guillemots, razorbills, puffins, and shags – nest here, while numerous other species, including ravens, choughs, and peregrine falcons, can be seen all year round.

△ **SUMMER VISITOR**
From May to July, thousands of pairs of puffins, with their brightly coloured beaks, nest on the cliffs.

◁ **BANDED GIANTS**
The cliffs wind in and out over a stretch of more than 8 km (5 miles). The banding pattern is due to colour variation in the sedimentary rock layers.

# The Cliffs of Dover

*One of England's most famous natural landmarks, a parade of tall,*
*brilliant white chalk cliffs, facing in the direction of France*

NW.
Europe

▽ **WHITER THAN WHITE**
The cliffs owe their gleaming
whiteness to the almost 100 per
cent purity of the chalk.

The Cliffs of Dover, which form part of the coast
of the English county of Kent, run along the
northwestern side of the narrowest part of
the English Channel. They are made from chalk
formed during the Cretaceous period, between
100 and 70 million years ago. At that time, a large
part of what is now northwestern Europe was
underwater. The shells of tiny marine organisms
accumulated on the sea floor and became compressed
into a layer of chalk hundreds of metres thick. Later,
after sea levels fell, this chalk formed a land bridge
between Britain and mainland Europe. Around
8,500 years ago, this bridge was swept away in
a catastrophic flood, leaving behind the cliffs
of Dover and similar white cliffs
at Cap Blanc Nez in France.

▷ **BRILLIANT BLUE**
The Adonis Blue butterfly
favours chalk downland, such as
the area of Kent behind the cliffs,
and is found there in modest numbers.

**Fossils** found in the **cliffs** include
**sharks' teeth, sponges,** and **corals**

# The Dune du Pilat

*The tallest sand dune in Europe, located close to Arcachon Bay*
*in southwestern France, with magnificent views from the top*

W. Europe

### EVOLUTION OF THE DUNE

Analysis of sand samples from
the dune and the underlying soil
has shown how the dune has
grown over several centuries.
Wind power is currently moving
it a few metres per year towards
the forest to its east.

Containing 60 million cubic metres (more than
2 billion cubic feet) of sand and running parallel to
the shoreline, the Dune du Pilat rises gently to a height
of 110 m (360 ft) above the west coast of France. It is
nearly 3 km (2 miles) long and 500 m (1,600 ft) wide.

The word Pilat comes from the Gascon word *Pilhar,*
which means "heap" or "mound". It takes about
30 minutes for a fit person to climb to the dune's top,
where the wind can be very strong. It is a great vantage
point in autumn for viewing flocks of migratory birds.

5th-17th
century:
20-40 m
(65-130 ft) high

18th-19th
century:
70-80 m
(230-260 ft)
high

2,000-
500 BCE:
3-4 m
(10-12 ft) high

8th-3rd
century BCE:
only slightly
above sea
level

today:
110 m
(360 ft)
high

◁ **SHIFTING SAND**
The dune is slowly migrating away from
the sea. Unless this movement can be
stopped, the sand will eventually engulf
a small forest on its land side.

▷ **COASTAL TREASURE**
The species of ammonoid fossilized
in this rock specimen lived 195 million
years ago. Called Promicroceras,
its fossils are very common in
parts of the Jurassic Coast.

limestone
groundmass

## WHAT ARE AMMONITES?

The rock strata of the Jurassic
Coast are especially rich in ammonites
– spiral-shaped fossils of the shells of
extinct sea creatures called ammonoids.
These predatory molluscs, alive
200–70 million years ago, were
swimming organisms whose soft
parts somewhat resembled those
of present-day squids.

living ammonoid has long
tentacles for catching prey

bouyant shell helped with
swimming and defence

ammonite fossil
embedded in rock

The **rock layers** of the
Jurassic Coast cover
**187 million years of**
**Earth's history**

fossils are 2.5–5 cm
(1–2 in) in diameter

densely packed
fossils show that
animals lived in
great numbers

NW. Europe

# The Jurassic Coast

*A stretch of English coastline famed for its fossils of ancient, extinct animals
and plants, as well as some striking natural landforms*

Approximately 154 km (96 miles) long, the Jurassic
Coast occupies part of two English counties, Dorset
and Devon. The coast has attracted worldwide
attention mainly because of the cliffs, composed
of layers of sedimentary rock, that run along
most of its length.

### A walk through Earth's past

Overall, the rock layers of the Jurassic Coast
contain a near-complete record of Earth's Triassic,
Jurassic, and Cretaceous periods. At different stages
during this vast timespan, the area where the coastline
now extends has been a desert, a tropical sea,
an ancient forest, and a lush swamp. All these
times and environments have been recorded
in different rock layers. The area's significance
in palaeontology only began to be realized
in the early 19th century. In 1811, a local
fossil collector, Mary Anning, and her
brother discovered the complete fossil

of a large, extinct marine reptile called an ichthyosaur.
Subsequently, she made more discoveries, including
fossils of a pterosaur (an extinct flying reptile).
This helped trigger an explosion of interest in
the area's fossil legacy, which has continued to the
present. Some of the more recent discoveries here
include fossils of extinct crustaceans, insects, and
amphibians, and a fossil forest of ancient
conifers and tree-ferns.

In addition to its interest to palaeontologists, the
Jurassic Coast contains some superb examples of various
types of natural landforms, including sea arches, rock
pinnacles, and an enormously long shingle bank.

marcasite crystal

chalk groundmass

**◁ SHINY CLUSTER**
Nodules of the gleaming mineral
marcasite, a form of iron sulphide,
are commonly found in rocks on
the Jurassic Coast.

**△ SECLUDED BANK**
This remarkable shingle structure is
Chesil Beach. It is 30 km (20 miles)
long, and has the sea on one side
and a lagoon on the other.

**◁ NATURAL ARCH**
Durdle Door is a limestone arch
formed by wave erosion. The name
Durdle comes from an Anglo-Saxon
word *thirl* meaning "bore" or "drill".

# The Algarve Coast

*A coast famous for its breathtaking cliffs, golden beaches, limestone caves and rock formations, scalloped bays, and sandy islands*

SE. Europe

### △ CAVE OF THE EYE
This cave is close to Benagil Beach, near Lagoa, and can be visited by boat or by swimming. The enormous "eye" in its roof is 16 m (52 ft) across.

The Algarve is the southernmost region of mainland Portugal, and its coast has two sections. The first of these, some 160 km (100 miles) long, is south-facing and extends westwards from Portugal's border with Spain out to the southwestern tip of the Iberian Peninsula, at Cape St Vincent. The second section is west-facing and extends for 50 km (30 miles) northwards from Cape St Vincent.

## Honey-coloured cliffs
Bathed by the warm North Atlantic Current (an extension of the Gulf Stream), the Algarve Coast is notable for its picturesque, honey-coloured limestone cliffs, small bays and coves, sheltered beaches of fine sand, and turquoise- or emerald-coloured seas. Features typically formed by marine erosion can be seen along many sections. These include some large caves at the feet of cliffs, blowholes, and holes that have been excavated from headlands to create sea arches.

Parts of the shoreline are dotted with rocky sea stacks (isolated pillars of rock set off from headlands), some of which appear to be precariously balanced on the shoreline. Although limestone is a primary component of the landscape, other rocks, including sandstone and shale, are visible along scattered stretches of the coast. The region's warm climate and strikingly beautiful scenery have made this coast a popular holiday destination.

### ▷ ERODED STACKS
At Praia da Rocha, near Portimao, heavily eroded limestone sea stacks embellish the shoreline. Their coloration is typical of the whole of the Algarve Coast.

### ◁ GOLD RUSH
In Roman times, gold was mined across the Algarve region. Some new gold deposits have recently been discovered, so mining may restart.

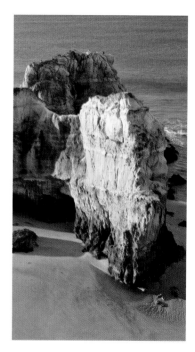

# The Dalmatian Coast

*A fractured but highly scenic coast made up of hundreds of islands and straits, set in the azure waters of the Adriatic*

S. Europe

The Dalmatian Coast extends down part of the eastern side of the Adriatic Sea (an arm of the Mediterranean Sea) and comprises a large section of the coastline of Croatia. From the seaport of Dubrovnik in the south, the coast meanders in a northwesterly direction for a distance of roughly 375 km (230 miles).

## A drowned coast

The Dalmatian Coast is a classic of its type – a particular kind of drowned coast, formed from the submergence of what was once a landscape of many parallel mountain ridges close to a former coastline. As a result of the drowning, the higher parts of many of the old ridges are now long, narrow islands running parallel to the present-day coastline. The island of Dugi Otok (Croatian for "Long Island"), for example, is 44.5 km (27¾ miles) long, but only 4.8 km (3.0 miles) wide. Other parts of the former landscape are now long promontories, running parallel to the overall direction of the coastline, or have split up into innumerable small islands.

◁ **DOLPHIN-DOTTED WATERS**
Bottlenose dolphins are the only dolphin species commonly seen around the Dalmatian Coast. These sleek swimmers can reach speeds of more than 30 kph (20 mph).

### HOW THE DALMATIAN COAST FORMED

Around 10,000 years ago, the region contained mountains and valleys parallel to the coast. Over time, the rising sea level drowned the area, producing the coastal features seen today.

narrow streams run through valleys

**BEFORE SEA-LEVEL RISE OF PAST 10,000 YEARS**

long islands and promontories are left after sea has flooded in

**AFTER SEA-LEVEL RISE OF PAST 10,000 YEARS**

## The Dalmatian Coast is one of the **densest archipelagos** in the world

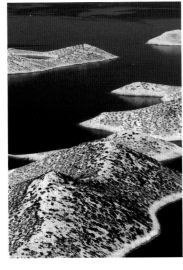

△ **PARCHED ISLANDS**
This string of islands forms part of the Kornati Archipelago in northern Dalmatia. Their appearance is typical of many small islands of the region.

▷ **PRISTINE WATERS**
The clear waters and scenic beauty of Stara Baska in the southern part of Krk make it a popular tourist destination.

NW. Europe

# The Hallerbos

*A centuries-old Belgian woodland, where one wild flower species produces a seasonal transformation*

Every spring, the Belgian woodland called the Hallerbos undergoes a spectacular colour change. From mid-April to May, thousands of bluebells burst into bloom, covering its 5 square km (2 square miles) of forest floor in an intense blue-violet. The change is so striking that it has given rise to the woodland's other name: the Blue Forest.

### Old and new

Usually, native bluebells grow in old woodland, but most trees in the Hallerbos date from a replanting effort conducted between 1930 and 1950. Its few ancient beeches and oaks are the only ones to survive World War I, when occupying forces devastated what had been a centuries-old forest. The first mention of the Hallerbos dates from 686 CE, and there are references throughout the medieval period; for this reason, official literature classes it as "an ancient forest with young trees".

△ **DISTINCTIVE DROOP**
Bell-shaped flowers of native bluebells produce flowering stalks up to 50 cm (20 in) high, which sag at the top.

▽ **TRANSIENT CARPET**
Hallerbos bluebells display their most intense colour just as beech leaves unfurl. Once the leaves open fully, less light hits the flowers beneath.

▷ **TOWERING FOLIAGE**
Spruce, including the towering Norway spruce, shares the Bavarian Forest with beech and fir on hillsides, and alder, beech, and willow in valleys.

▽ **TRAPPED AIR**
Fog is common in the Bavarian Forest and contributes to the area's precipitation. In autumn and winter, damp valley floors also produce cold air lakes - areas of trapped frigid, foggy air.

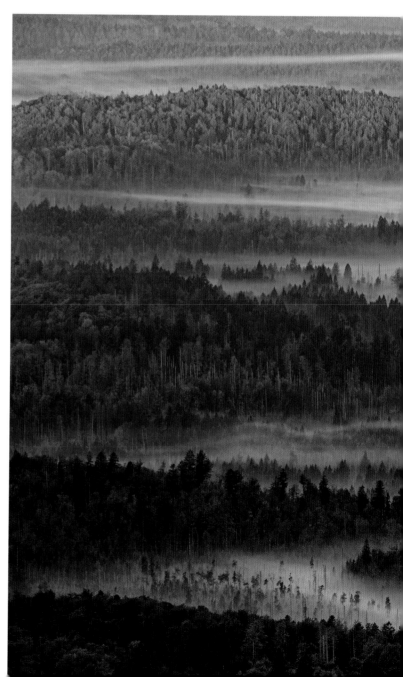

# The Bavarian Forest

*A mixed mountain forest, dominated by spruce, fir, and beech, where plants and animals must contend with acidic soil, high rainfall, and months of cold winter weather*

C. Europe

Southeastern Germany's Bavarian Forest covers the high region between the Danube river valley and the Bohemian Forest along Bavaria's border with the Czech Republic. Together, these two forests form Europe's largest area of continuous woodland. Bavarian Forest terrain consists mainly of tree-covered granite and gneiss hills and mountain domes, interspersed with valleys and elevated plains. Native spruce is the dominant tree species, mixed with fir and beech; all must cope with the area's highly acidic, nutrient-poor soils. According to a local saying, the climate is "three-quarters winter and one-quarter cold", which aptly describes the annual average temperature range of 3–7.5°C (37–46°F). Depending on altitude, yearly precipitation is 1,100–2,500 mm (43–98 in), 30–40 per cent of which falls as snow.

△ **SEED SEEKER**
The Eurasian red squirrel makes the most of the Bavarian Forest's plentiful supply of spruce seeds. It emerges just after sunrise to begin daily foraging.

## Around **10,000 animal species** are believed to live in the Bavarian Forest

# The Białowieża Forest

*The only old-growth forest in lowland Europe and a refuge for the largest population of free-ranging European bison*

E. Europe

Covering more than 1,500 square km (580 square miles), the Białowieża Forest is shared between Poland and Belarus. It is the only large remnant of the primeval forest that once covered northeastern Europe after the last ice age and is still home to some of the continent's tallest trees; old-growth spruce reaches more than 50 m (165 ft) here. Part of the Central European mixed-forest ecoregion that extends

◁ **STRIPED COATS**
Wild boar young, known as boarlets or piglets, have striped coats to blend in with their dappled woodland surroundings.

from eastern Germany to northeastern Romania, Białowieża contains tree species from both northern (boreal) and lowland mixed deciduous forests. The latter include English oaks – some of which are thought to be 150–500 years old – as well as the small-leaved lime and hornbeam. Mixed broadleaf and conifer stands are also interspersed with pocket ecosystems, such as the increasingly rare boreal spruce bog forest.

## Dead wood, new life

A high percentage of standing and fallen deadwood supports thousands of fungi, insects, and bird species, including the protected three-toed woodpecker. Wild boar, elk, wolf, and lynx are common forest residents, and Białowieża is an important refuge for the continent's largest mammal, the European bison or wisent, with the largest free-ranging herd in the world.

### TREES AND FUNGI

Many trees and fungi share a mutually beneficial relationship called a mycorrhiza. A fungus sends out hundreds of thin, root-like threads called hyphae that grow around a tree's rootlets, absorbing sugars from them. In return, the tree uses the hyphae network to absorb more nutrients from the soil.

hyphae network | tree roots

**SWAMPY SANCTUARY**
The Białowieża Forest contains several different habitats, including sections of spruce bog forest that are home to many rare plant species.

# Kola Peninsula Tundra

*A treeless landscape of harsh beauty with permafrost, clear lakes, and rivers, and a vital home for Arctic plant and animal species*

N. Europe

Northern Russia's Kola Peninsula is located almost completely within the Arctic Circle, between the Barents and White Seas. Tundra covers nearly all of its 100,000 square km (40,000 square miles), but approximately 58,800 square km (22,700 square miles) are classed as critically endangered coastal Arctic tundra. The landscape is covered with moss, lichen, grasses, and wild flowers, with shrubs such as dwarf Arctic birch and cloudberry; the frozen subsoil known as permafrost (see panel, right) prevents trees from growing. Kola winters are long and extreme; frost often occurs from August through to the following June, and strong winds blow up to 120 days a year. Despite its harsh environment, the Kola's tundra, lakes, and river systems are home to 200 species of birds and 32 mammal species, including migrating herds of reindeer. However, Soviet nuclear testing and mining have already harmed what was once a pristine environment.

▷ **FRUITFUL SURVIVOR**
Cloudberries can survive temperatures as low as -40°C (-40°F). Their sweet fruit provides food for birds and mammals in autumn.

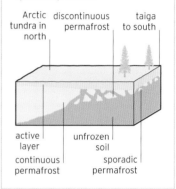

## PERMAFROST

Permafrost is ground that has been frozen for two years or more. It usually forms above a layer of unfrozen soil and beneath a layer, called the active layer, that freezes and thaws each year. The permafrost itself may be continuous or discontinuous, or form sporadic pockets.

Arctic tundra in north | discontinuous permafrost | taiga to south

active layer | unfrozen soil
continuous permafrost | sporadic permafrost

◁ **LIFE IN SMALL DOSES**
Tundra plants remain short to avoid freezing winds. Their shallow roots cannot penetrate permafrost.

# The Western Steppe

*A flat, windswept plain stretching thousands of kilometres and forming part of the largest temperate grassland on the planet*

E. Europe

The Western Steppe lies chiefly in Ukraine, Russia, and Kazakhstan. With the Altai Mountains as its eastern boundary, it forms half of the vast Eurasian Steppe, the largest temperate grassland on Earth. Steppes get enough precipitation for grass growth but not enough for trees. The Western Steppe is 4,000 km (2,500 miles) long and is often called a "sea of grass", although where rivers and streams cut through the otherwise unbroken landscape, some trees do grow. Bordered to the north by taiga, and to the south by desert, the Western Steppe is relatively mild compared to eastern grasslands, but is still subjected to howling winds and temperatures ranging from 27°C (80°F) during the day to below freezing at night.

◁ **GRASSLAND FOOD CHAIN**
The grasses and wild flowers of the region abound with insects such as grasshoppers and beetles, which in turn attract many birds, including the western yellow wagtail.

**High dunes**
Captured in visible light and infrared, these sand dunes are in Africa's ancient Namib Desert. Winds blowing off the Atlantic Ocean produce some of the world's highest dunes here, some reaching heights of 300 m (1,000 ft).

# Africa

# LAND OF THE RIFT

## Africa

Africa is a gigantic continent; only Asia is larger. It stretches nearly 8,000 km (5,000 miles) from its most northerly point, Ras ben Sakka in Tunisia, to its most southerly point, Cape Agulhas in South Africa. It straddles the equator, which divides the continent in half, and consists of two large lobes – with the northern lobe extending far out west into the Atlantic Ocean, and the southern lobe running south between the Atlantic and Indian oceans. The northern half is dominated by the world's largest desert – the Sahara. It is such an overwhelming geographical feature that Africa is often divided into the

Sahara and sub-Saharan Africa, the latter dominated by the western river basins filled with lush, tropical forests, and the drier, bushy grasslands or savanna on the eastern plateaus.

Despite its enormous size, Africa has no high mountain ranges, except for the Atlas in the extreme northwest. Much of the east and south are dominated by huge, high plateaus. In the northeast, the plateaus are split apart by the deep trench of the East African Rift valley, dotted by ribbons of huge lakes. The rift was created by tectonic forces beneath the crust cleaving the African Plate apart.

### KEY DATA

▲ **Highest point** Mount Kilimanjaro, Tanzania: 5,895 m (19,341 ft)
▼ **Lowest point** Lake Assal, Djibouti: -156 m (-512 ft)
● **Hottest record** Kebili, Tunisia: 55°C (131°F)
● **Coldest record** Ifrane, Morocco: -24°C (-11°F)

### CLIMATE

Most of Africa lies within the warm tropics. Except in the forest areas of the west, rainfall is scarce with just one wet season in the savanna and almost none in the Sahara.

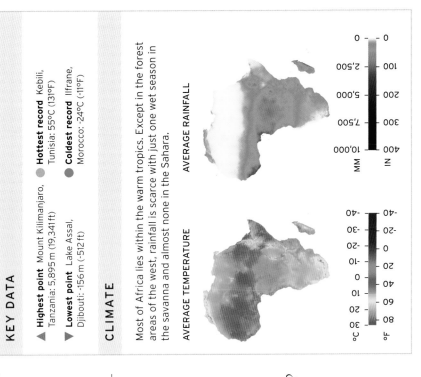

AVERAGE TEMPERATURE

| °C | | °F |
|----|----|----|
| 30 | | 80 |
| 20 | | 70 |
| 10 | | 60 |
| 0 | | 50 |
| -10 | | 40 |
| -20 | | 30 |
| -30 | | 20 |
| -40 | | -40 |

AVERAGE RAINFALL

| MM | | IN |
|----|----|----|
| 10,000 | | 400 |
| 7,500 | | 300 |
| 5,000 | | 200 |
| 2,500 | | 100 |
| 0 | | 0 |

Gulf of Guinea

INDIAN OCEAN

ATLANTIC OCEAN

Mozambique Channel

Madagascar

Comoro Islands

Zanzibar

Mount Kilimanjaro 5,895m

Lake Natron

Lake Rudolf

Great Rift Valley

Lake Victoria

Serengeti Plain

Lake Tanganyika

Lake Nyasa

Zambezi

Victoria Falls

Lake Kariba

Limpopo

Congo Basin

Bié Plateau

Okavango Delta

Kalahari Desert

Drakensberg

Great Karoo

Fynbos

Cape of Good Hope

Namib Desert

Crystal Mountains

Congo

São Tomé

Bioko

## ECOSYSTEMS

There are three broad ecosystems: the desert stretching across the north, the tropical forest of the centre, and savanna to the west, east, and south.

**KEY**

- Tropical broadleaf forest
- Tropical dry broadleaf forest
- Mediterranean woodland and scrub
- Tropical and subtropical grassland
- Wetland
- Desert and scrub
- Montane grassland

**KEY**

- Precambrian (pre-541 million years ago)
- Paleozoic (541–252 million years ago)
- Mesozoic (252–66 million years ago)
- Cenozoic (66 million years ago to present day)

**ELEVATION**

| M | FT |
|---|---|
| 4,000 | 13,000 |
| 2,000 | 6,600 |
| 0 | 0 |

Km 0 500 1,000

Miles 0 500 1,000

## GEOLOGY

Africa's geology is built around several slabs, or cratons, of granite and gneiss that create the high plateaus, surrounded by vast flatlands of more recent sedimentary rocks.

**CROSS-SECTION**

A

B

Great Rift Valley

slope of Mount Kilimanjaro

Serengeti Plain

Lake Victoria

Great Rift Valley

Congo Basin

Crystal Mountains

▷ **CRATON REVEALED**
In the Drakensberg, South Africa, the Blyde River has cut down to the ancient rocks of the Kaapvaal Craton.

# THE SHAPING OF AFRICA

Africa is formed around five stable sections of Earth's crust called cratons. They were the heart of the giant southern continent Gondwana during the age of the dinosaurs. About 65 million years ago, the land around these cratons broke away to create Africa.

## Ancient heart

The five African cratons were formed over a billion years ago when magma welled up from Earth's interior and solidified. In places, they have since been altered by heat and pressure, and in others, they have been covered by a veneer of sediments. The rocks they are made of are very tough – they have endured as other rocks came and went, and continents and oceans shifted around them. The cores of the cratons are ancient and date back to one of the oldest periods in Earth's history, the Archean Era, over 2.5 billion years ago. The oldest rocks are the granite, gneiss, and greenstone terrains of parts of South Africa, Zimbabwe, and Tanzania. The Kaapvaal Craton is noted for its distinctive komatiite rocks that formed when Earth was very hot in its earliest days. The area is also famous for its deposits of gold, nickel, and uranium.

## Mountain fringes

For long periods in Earth's history, the African cratons were separate continents. A billion years ago, the craton to the west was attached to a landmass that would eventually hold the Amazon Basin within an ancient supercontinent, while the central and southernmost cratons were islands. Gradually, they were all glued together by mobile belts of rock and became part of the Pangaea supercontinent 250 million years ago. Meanwhile, the mobile belts were crunched between the cratons, folding up rocks to form mountain chains such as the Atlas and the Cape Fold mountains, some of which survive in South Africa (others in South America, Australia, and Antarctica).

**BUILDING BLOCKS**
Africa is built around five major cratons of ancient rock, with the most ancient of all being the Kaapvaal Craton to the south. The Arabian-Nubian Craton is being rifted apart.

Arabian-Nubian Craton

craton

Kaapvaal Craton

## KEY EVENTS

**140 million years ago** The landmass that will become Africa splits from what will become South America due to the opening of the Mid-Atlantic Rift, creating a new ocean – the Atlantic.

**30 million years ago** The Arabian Peninsula pushes away from Africa as magma forces through Earth's crust and cools, creating new land that is denser than the older, surrounding land. As a result of this higher density, it sinks – creating the Afar Depression.

**1.8 million years ago** Mount Kilimanjaro forms due to volcanic activity linked to the opening of the Great Rift Valley.

**160 million years ago** Madagascar (along with the Seychelles and India) splits from Africa, and then about 84 million years ago it splits again from the Seychelles and India – both events are due to rifting as the Indian Ocean opens up.

**30 million years ago** Many faults open between various plates in eastern Africa, and the land between three neighbouring plates "drops" to form the Great Rift Valley.

**2.5 million years ago** North Africa becomes increasingly dry, eventually forming the Sahara Desert. Changes in the tilt of Earth's axis lead to brief periods of increased humidity and plant growth.

**400,000 years ago** Tilting of Earth's crust creates a basin. Water collects within it and Lake Victoria forms. During the ice age it drains away multiple times as the rate of rainfall rises and falls.

◁ **COLLIDING CONTINENTS**
The High Atlas mountains to the north of Ouarzazate were formed less than 50 million years ago when the African and Eurasian Plates started colliding.

▷ **PARCHED LAND**
Almost all of northern Africa is covered by the huge Sahara Desert. Half of the Sahara receives less than 2.5 cm (1 in) of rain per year – the rest gets less than 10 cm (4 in).

▷ **FIERY CRACKS**
The Erta Ale volcano lies between the splitting sides of the African and Arabian plates.

▷ **ACTIVE VOLCANO**
Oldoinyo Lengai in Tanzania is one of many erupting volcanoes along the widening crack in the African Plate that forms the Great Rift Valley. Uniquely, it erupts natrocarbonatite lava.

In time, volcanic activity forced cracks to open in Pangaea, and as a result Africa split from Europe and South America and started its own journey. Madagascar had split from Gondwana about 20 million years earlier.

## A splitting continent

Around 25–22 million years ago, a big crack began to split east Africa in two as the tectonic plates began to pull either side apart and plumes of hot magma thrust their way up from underneath. The widening crack became one of the world's most dramatic geographical features – the Great Rift Valley – which runs over 6,000 km (3,700 miles) from the Dead Sea in Jordan to Mozambique. North of Ethiopia, the rift is splitting the Arabian-Nubian Craton apart and has already gaped wide enough to allow the Indian Ocean to flood in and create both the Red Sea and the Gulf of Aden. In about 10 million years' time, the two sides of the Great Rift Valley will have pulled far enough apart to eventually open up a new ocean right through the middle of East Africa.

## Africa dries out

Africa's location straddling the tropics has created three huge deserts – the Sahara in the north, along the Tropic of Cancer, and the Namib and Kalahari in the south along the Tropic of Capricorn. The Namib is the world's oldest desert and formed 80 million years ago. The Sahara is the world's largest hot desert, yet developed less than 2.5 million years ago – much of it was green until a dramatic climatic shift just 5,000 years ago scorched the north of Africa dry.

# Africa was still attached to South America when Madagascar split away

## BUILDING THE AFRICAN CONTINENT

KEY ▬ Convergent boundary ▬ Divergent boundary

South America joined to Africa

Gondwana begins to rift apart

southern Africa is under water

**237 MILLION YEARS AGO** The ancient cratons that form the core of Africa today are locked deep within the supercontinent of Pangaea. Africa is sandwiched between South America and Asia.

an ocean separates Africa from Eurasia

plates diverge, Atlantic widens

Madagascar attached to India

**94 MILLION YEARS AGO** Africa splits away from South America, and as South America drifts west, the South Atlantic opens up between them. Meanwhile, an ocean separates Africa from Eurasia.

lower sea levels means Africa's land exceeds modern coastlines

**18,000 YEARS AGO** Africa moves north to collide with Eurasia again, squeezing the ocean separating the continents out of existence, but it retains its separate identity as a continent.

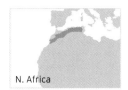

N. Africa

# The Atlas Mountains

*A barrier between the Mediterranean and Sahara, stretching from Morocco to Tunisia*

Extending for around 2,500 km (1,600 miles), the Atlas Mountains consist of several distinct subranges, which are interspersed with plateaus and gorges. The subranges include the Anti-Atlas and High Atlas in Morocco and the Tell Atlas in Algeria.

## Product of collision

The bulk of the Atlas Mountains were pushed up by a collision between the African and Eurasian plates, with the main uplift taking place 30–20 million years ago. However, certain parts were created during an earlier period of mountain building, from around 250 million years ago. Today, the appearance of the Atlas varies between forested slopes in the north and much more arid southerly subranges. The region contains some of the world's largest and most diverse mineral resources.

◁ **MINERAL WEALTH**
The Tell Atlas range of Algeria is mined in several locations for iron ore as well as gold and phosphate. The chunk of iron ore shown here is of a type called haematite.

Although in **subtropical Africa**, the **tallest peaks** of the Atlas were **covered in glaciers** during **the last ice age**

△ **MOUNT TOUKBAL**
This small village, seen here in winter, is at the base of Mount Toukbal, the highest peak in the Atlas. Although there is snow on the peaks, the lower slopes are hot and arid.

◁ **INTRICATE FOLDING**
The Anti-Atlas is geologically complex. This satellite image, taken with an infrared camera sensitive to differences in surface composition, reveals intricate folding among the rock outcrops.

E. Africa

# The Afar Depression

*One of the hottest and driest places on Earth, a region of volcanoes, hot springs, and salt lakes*

Also called the Afar Triangle due to its shape, the Afar Depression is a low-lying region touching Ethiopia, Eritrea, and Djibouti. It is an area where Earth is thought to be rifting apart with thinning and stretching of the crust above an upwelling plume of hot mantle rocks. The presence of the magma underground is apparent in the form of many volcanoes at the surface, including a massive shield volcano (a broad volcano with shallow slopes), Erta Ale, which has a permanent lava lake. Partly because much of it lies below sea level, the Afar Depression is one of the hottest places on Earth.

### ▽ COLOURFUL CRATERS
These hot springs, with salt and sulphur deposits, lie 45 m (150 ft) below sea level at Dallol in the northern part of the Afar region.

### THE AFAR TRIPLE JUNCTION

The Afar Triple Junction is a spot on Earth's surface where the Arabian Plate and two parts of the African Plate – the Nubian and Somalian Plates – are slowly separating. Projecting from this site are three arms, or rifts (down-faulted sections of crust). Two are occupied by the Red Sea and the Gulf of Aden. The northern part of the third arm is the Afar Depression.

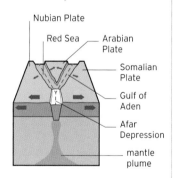

Nubian Plate

Red Sea — Arabian Plate

Somalian Plate

Gulf of Aden

Afar Depression

mantle plume

### ◁ NORTHERN SURVIVOR
The only surviving primate in Africa north of the Sahara Desert, the Barbary macaque has adapted to living in the rocky cliffs and high-altitude oak and cedar forests of the Atlas range.

E. Africa

# The Great Rift Valley

*A system of faults in Earth's crust that extends for 6,000 km (3,700 miles) through part of southwestern Asia and then down the length of East Africa*

The Great Rift Valley marks a region where plumes of hot material from Earth's mantle are welling up and causing the crust to stretch and split apart along a series of fissures, or rifts. These rifts start in southern Lebanon and pass southwards, via the Red Sea, to the Afar Depression (see p.183) and finally down through East Africa, ending on the coast of Mozambique. The average width of the Great Rift Valley is around 50 km (30 miles). In many parts, sections of crust have sunk down, leaving escarpments on either side, with steep walls that are

▷ **SOCIAL ANIMALS**
The Great Rift is home to many endemic species of African wildlife, including olive baboons. These live in groups of 15 to 150 individuals, with a complex social structure.

typically around 900 m (2,950 ft) high. The African section of the Great Rift, known as the East African Rift System, is gradually splitting the African tectonic plate in two.

## Volcanoes and lakes

Dotted throughout the rift system are numerous volcanoes, formed where magma has welled up through fissures in the crust and erupted at the surface. Several, such as Ol Doinyo Lengai in Tanzania, have been active in the past 50 years. In the East African Rift System, many lakes – some very deep – fill the valley floors. Ancient fossils of various extinct relatives of modern humans have been discovered in different parts of the region.

# The Great Rift Valley has **formed over** approximately the **last 35 million years**

## EASTERN AND WESTERN RIFT VALLEYS

The oldest and best defined rift occurs in the Afar region of Ethiopia. This is usually referred to as the Ethiopian Rift. In East Africa, the Great Rift splits into two branches called the Western (or Albertine) and the Eastern (or Gregory) rifts. Each contains active volcanoes and a string of lakes. Some of these are highly alkaline and are termed soda lakes; others are very deep.

Ethiopian Rift

Eastern Rift

Western Rift

Afar region

**KEY**

■ Fault line

▲ Major active volcanoes

■ Tectonic plate boundary

▲ Major dormant volcanoes

△ **SALINE LAKE**
Kenya's Lake Bogoria is a saline and alkaline lake that is home to one of Earth's largest populations of lesser flamingos. There are also geysers and hot springs along the banks.

▷ **UNIQUE LANDSCAPE**
The crater of Ol Doinyo Lengai is filled with steep-sided cones and a unique dark lava that turns white as it comes into contact with moisture in the air and solidifies.

# The formation of a
# RIFT VALLEY

Continental rifting is a process in which part of one of Earth's tectonic plates, made up of the crust and upper mantle (together known as the lithosphere), becomes stretched and thinned to the point where fissures and faults appear. Blocks of crust collapse downwards, leaving a series of depressions, called a rift valley, at the surface. The fissuring also allows magma (hot melted rock from the mantle) to intrude into the crust and create volcanic activity at the surface.

## Rift valleys in East Africa

An excellent example of continental rifting can be seen in East Africa. The cause here is thought to be a mantle plume, or possibly plumes, under the region (see panel, below). The East African Rift system has two arms, east and west (see p.184), on either side of an area containing Lake Victoria. Rifting and volcanism started in a part of the eastern branch to the northeast of Lake Victoria around 35 million years ago, probably due to a mantle plume, then spread north and south. But as the rifting process encountered a stable core of lithosphere, around and to the south of Lake Victoria, which it could not go through, it is theorized to have diverged around it, leading to the two rift arms. Alternatively, it could be that separate mantle plumes exist, or once existed, under the eastern and western rift arms.

▷ **DEEP AND HOT**
The lava lake of Nyiragongo volcano is sometimes 600 m (2,000 ft) deep. It connects to a magma reservoir in the crust beneath it, which has intruded there due to the rifting process.

Lake Edward

Lake Kivu

▶ **EAST AFRICAN RIFT SYSTEM**
In East Africa, two arms of a rift system are separated by a broad, stable plateau. Each arm is a curved depression in the landscape and the site of several remarkable natural features, many of them having a volcanic origin.

▷ **RIFT FAULTING**
In a rift, some blocks of crust drop down, along faults created by stretching forces. In the segment shown here, a long major fault has developed on one side. Along a whole rift, major faults may switch between sides, giving an overall sinuous appearance.

**Lake Tanganyika**
With a maximum depth of 1,470 m (4,820 ft), Tanganyika is the world's second deepest freshwater lake.

**Lake sediments**
Due to the lake's age, of up to 12 million years, its bottom sediments are several kilometres deep.

**Western Rift arm**
This is marked by a string of four large lakes, which fill the lower parts of the rift's depressions.

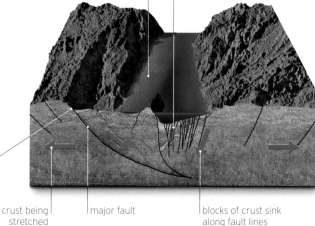

steep escarpment on shoulder of rift valley

crust being stretched

major fault

blocks of crust sink along fault lines

## HOW THE EAST AFRICAN RIFT FORMED

The system of rift valleys in East Africa formed when a mantle plume, or possibly plumes, caused upward warping of the crust, then stretching and thinning as parts of the African Plate on either side began to move apart. Tensional faults appeared, and blocks of crust dropped down, producing depressions, which became partly filled by lakes, bordered by escarpments or mountains.

stretching and thinning of crust

crust

lithospheric mantle

BEFORE

mantle plume

Lake Victoria

Mount Nyiragongo

Ol Doinyo Lengai

dropped-down blocks

NOW

**Continental crust**
This is laterally stretched and faulted in a rift region, with some crustal blocks collapsing downwards. In places, magma intrudes into the crust.

**Lithosperic mantle**
The uppermost layer of mantle, with overlying crust, comprises the lithosphere, which is divided into plates.

**Asthenosphere**
This relatively deformable layer of Earth's upper mantle is warmer than the overlying lithosphere.

**Speculative mantle plume**
Volcanism in the Western Rift arm indicates that in some places magma has definitely intruded into the crust.

**◁ GREAT ELEVATION**
Mount Stanley in the Rwenzori Mountains is Africa's third highest mountain. The Rwenzoris originated on one side of the western rift arm as a gigantic horst – a crustal block that lifted up when land on either side subsided.

**◁ SCARLET GLOW**
The strange lava, called natrocarbonatite, exuded by the Ol Doinyo Lengai volcano glows an unusual scarlet colour at night. The volcano developed as a result of rifting, but the reason it produces this unique lava is not yet known.

Lake Albert

Lake Kyoga

Lake Bogoria

Mount Kenya is an extinct volcano and Africa's second highest mountain

Lake Natron

**Eastern Rift arm**
This contains relatively small, shallow lakes and is bounded by some spectacular escarpments.

**Lake Victoria**
Africa's largest lake by area, Lake Victoria occupies a shallow depression in the central plateau.

**Central plateau**
The area between the two rift arms is 1,300 km (810 miles) wide, underlain by a stable ancient block of continental lithosphere.

**Ngorongoro Crater**
This large extinct volcanic caldera is famed for its wildlife, especially lions and other large mammals.

**Mantle plume**
A mantle plume is strongly suspected to be present under the Eastern Rift arm and has caused stretching and faulting.

# There are over **50 volcanoes** in the **East African rift system**

**▷ SKY HIGH**
Kilimanjaro – a dormant volcano – is Africa's highest mountain at 5,895 m (19,341 ft). Its origins are connected to the development of the Eastern Rift arm.

# The Brandberg Massif

*An isolated, dome-shaped mountain, or inselberg, that towers over the Sun-scorched gravel plains of the northwestern Namib Desert*

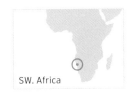

SW. Africa

The Brandberg formed around 130 million years ago when a large glob of magma from deep in Earth's crust pushed up through the surrounding rock layers, before solidifying. Later, the granite mass was exposed by erosion of the surrounding rocks.

## The burning mountain

The Brandberg is the highest mountain in Namibia, with its most elevated point at an altitude of 2,573 m (8,442 ft). The indigenous San people call it the "burning mountain" as it appears to glow red in the light of the setting Sun. With its considerable height and breadth, the Brandberg influences the local climate, drawing more rain to its flanks than the desert below. The rain slowly seeps out through springs. Unique animal and plant species thrive in its high-altitude environment, and prehistoric cave paintings decorate rock faces hidden within ravines around the base of the mountain.

### ◁ RING OF ROCKS

The Brandberg is around 23 km (14 miles) long and 20 km (12 miles) wide, and has a rugged, undulating surface. A ring of dark, steep-sided rocks encircle the granite mass.

S. Africa

# Table Mountain

*A long, flat-topped mountain that forms a dramatic backdrop to the city of Cape Town*

Approximately 3 km (2 miles) long and edged by impressive cliffs, Table Mountain is one of the most iconic landmarks in Africa. With its high point at 1,086 m (3,563 ft), it lies at the northern end of a sandstone mountain range that terminates about 50 km (30 miles) to the south at the Cape of Good Hope. Table Mountain once had much thicker, softer layers of rock above it, but these eroded away to produce the flat top. This is often enveloped in a "table cloth" of clouds, formed when onshore winds are directed up the mountain's slopes into colder air, where the moisture they carry condenses.

### △ SUNSET VIEW

Table Mountain is seen here covered by its "table cloth" of white clouds just before sunset. Below and to its left are the lights of Cape Town and to the right lies the suburb of Camps Bay.

### ◁ ICONIC FLORA

Also known as the Pride of Table Mountain, the spectacular Red Disa orchid typically grows near streams and waterfalls. It is the emblem of South Africa's Cape Province.

## INSIDE TABLE MOUNTAIN

The mountain's upper part is a 600-m- (2,000-ft-) thick slab of hard sandstone, around 450 million years old. This lies above several layers of sedimentary rock, except at one end, where there is a large granite intrusion.

Table Mountain group (sandstone)

Graafwater formation (sandstone, siltstone, and shale)

Malmesbury group (sandstone and slate)

Peninsula granite

# The Drakensberg

*A gigantic, step-like, eroded landform that separates part of southern Africa's narrow coastal plain from its vast interior of savanna and desert*

S. Africa

△ SAVAGE ARENA
In this part of the uKhahlamba-Drakensberg Park, the escarpment is shaped like an amphitheatre. The pinnacles on the right are known as the Dragon's Teeth.

Meaning "Dragon's Mountain" when translated from Afrikaans, the Drakensberg is the eastern part of a long, winding escarpment, the Great Escarpment, which forms the edge of a high plateau of central southern Africa. The Drakensberg is South Africa's highest mountain range and runs for more than 1,000 km (620 miles) across eastern South Africa, with some peaks exceeding 3,000 m (10,000 ft) in height.

## A giant, eroded step

From its base, the escarpment appears like a giant step. This is continuously worn down by gravity and the flow of water. The landscape is diverse, with steep cliffs, pinnacles, and caves. Rivers rising in the plateau behind the escarpment have carved some impressive gorges. The local fauna and flora includes vultures, rock hyrax, and 300 species of indigenous plants.

### INSIDE THE DRAKENSBERG

The escarpment's steep sides consist of a thick layer of basalt (an igneous rock formed by cooling of lava), created as a result of ancient volcanic outpourings. Lower down are layers of older sedimentary rocks. Together, the basalt and sedimentary strata form part of a sequence of rocks found in a large part of Africa known as the Karoo Supergroup.

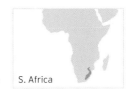

sedimentary rocks | basalt

Karoo Supergroup

150 million years ago (MYA)

190 MYA
200 MYA
210 MYA
220 MYA

240 MYA

▽ FREE FALL
The Drakensberg boasts some of the world's tallest waterfalls, and other beautiful cascades, such as the Berlin Falls, shown here.

◁ GROUND-DWELLER
The indigenous Drakensberg rockjumper is a ground-nesting bird, which is often seen perching on rocks.

In places, the **escarpment** is **1,500 m (4,900 ft) high** from base to top

## △ DESERT LIFELINE

Passing through arid landscapes over most of its course, the Nile provides a vital water supply for drinking and agricultural use. It is also an important transport route used by millions of people each day.

## THE WORLD'S LONGEST RIVERS

The accurate measurement of a river's length can be problematic. In some cases, it is difficult to identify a river's source or mouth, and over its course a river may split into many different channels. Although figures for the lengths of the longest rivers vary, the measurements below are the most widely recognized.

△ VERSATILE PLANT

A species of aquatic sedge, papyrus forms tall, reedlike beds in the Nile's shallow waters. Since ancient Egyptian times, papyrus has been used to make various items, from scrolls to boats.

# The Nile

*The longest river on Earth, running from central Africa to the Mediterranean, and a lifeline for millions of people on its route*

NE. Africa

The longest river in the world, the Nile flows more than 6,650 km (4,130 miles), from its source near the equator, through East and northeast Africa to its mouth on the Mediterranean Sea. The Nile basin drains more than 3 million square km (1 million square miles) – about 10 per cent of Africa's land area.

## White and Blue

The Nile has two major tributaries. Considered the river's primary stream, the White Nile flows from the north of Lake Victoria (see p.196), in Uganda, and descends into the Great Rift Valley (see pp.184–87) over the Murchison Falls. After crossing Lake Albert, it passes onto the huge wetland plain of the Sudd. At Khartoum, Sudan, it is met by the Blue Nile, which has its source at Lake Tana in the Ethiopian Highlands. From here, the Nile runs over a series of cataracts, or rapids, before passing through Egypt's historic valleys. North of Cairo, the river splits into two branches – the Rosetta to the west and the Damietta to the east – forming the Nile Delta as it enters the Mediterranean.

▽ **MISTY FALLS**
The Blue Nile Falls, in Ethiopia, is known locally as Tis Abay, meaning "great smoke". Spanning up to 400 m (1,300 ft) in the rainy season, its waters plunge 45 m (150 ft) to the bottom.

## Welcome floods

The Nile Delta, one of the most populated and highly cultivated parts of Egypt, was originally formed of sediment brought down by the Blue Nile from the Ethiopian Plateau. Indeed, the flooding of the Nile has been an important annual cycle, delivering not only water but also fertile silt that replenishes the soil across the floodplain. However, since the completion of the Aswan High Dam in 1970, the flow of the Nile has been significantly moderated, and traditional farming practices have been greatly affected.

△ **SOLITARY FORAGER**
The shoebill is a large, storklike bird that typically forages alone for fish, which it snaps up with its huge, hook-tipped bill.

It takes **three months** for the **water leaving Lake Victoria** to reach the **Mediterranean**

# Lake Retba

*A West African lake with a bright pink hue during the dry season and very salty water*

W. Africa

Also called Lac Rose, meaning "Pink Lake" in French, Lake Retba is a small, shallow lake in Senegal, West Africa, best known for its vivid colour and the high salt content of its water. The colour, which is deeper during the dry season (November to June), is created by an algae species called *Dunaliella salina* that produces a red pigment to help absorb sunlight for energy (photosynthesis). The salinity of Lake Retba's water is comparable to that of the Dead Sea (see p.236), and this has led to adaptations in the lake's fish, with many growing to only a quarter of the usual adult size.

▷ **SALTY SHORES**
Scores of salt collectors harvest the mineral from the lakebed by hand and export it across West Africa. They cover their skin in shea butter to protect it from the salty conditions.

# The Congo

*One of Earth's mightiest rivers, draining much of West Africa's vast tropical rainforest*

W. Africa

Africa's second longest river after the Nile (see pp.190–91), the Congo carries more water to the sea than any other river on Earth except the Amazon (see pp.106–09). It is also the deepest river in the world, with sections reaching more than 200 m (660 ft). From its source in the highlands of Zambia, it flows in a giant anticlockwise arc, draining a vast circular depression that covers most of the Democratic Republic of the Congo, before discharging into the Atlantic Ocean. Its route, which crosses the equator twice, covers more than 4,700 km (2,900 miles) and includes many waterfalls, rapids, broad waterways, and lakes.

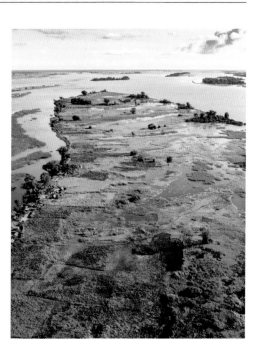

▷ **MARSHY BANKS**
For most of its course, the Congo flows through tropical rainforest, where extensive marshes line its banks and run between tributaries.

**ENDEMIC FISH**
The tropical waters of Lake Malawi are thought to be home to as many as 1,000 species of cichlids, most of which are endemic to the lake.

## LAKE MALAWI'S LAYERS

Lake Malawi has a definite stratification, with layers of water having different nutrient and oxygen contents that generally do not mix. However, from May to August, southern parts of the lake experience strong winds and cooling that create a large upwelling of water. This circulation sends nutrients from the mid-depths up the water column to mix with shallow, more oxygenated water, creating ideal conditions for fish to thrive.

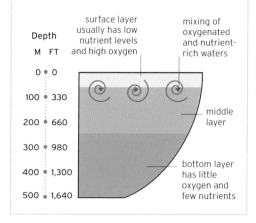

surface layer usually has low nutrient levels and high oxygen

mixing of oxygenated and nutrient-rich waters

| Depth | |
|---|---|
| M | FT |
| 0 | 0 |
| 100 | 330 |
| 200 | 660 |
| 300 | 980 |
| 400 | 1,300 |
| 500 | 1,640 |

middle layer

bottom layer has little oxygen and few nutrients

# Lake Malawi

*A large, deep lake at the southern end of Africa's Great Rift Valley with a remarkably diverse population of fish*

E. Africa

Also known as Nyasa and Niassa, Lake Malawi is the southernmost and third largest of the East African Great Lakes, lying between Malawi, Tanzania, and Mozambique. It is about 580 km (360 miles) long, up to 75 km (47 miles) wide, and reaches a depth of more than 700 m (2,300 ft). The lake is fed by the Ruhuhu River, from the north, and several smaller rivers. From the southern end flows its only outlet, the Shire River, a tributary of the Zambezi. Lake Malawi runs north-to-south, occupying a single basin in part of the southern end of the Great Rift Valley (see pp.184–87). Above its north and east shores rise steep mountain slopes, some covered by thick forest. The much flatter and shallower shores of the southern end of the lake feature many white, sandy beaches.

▷ **WINGED ARCHITECT**
The male village weaver builds a dense nest out of grass and leaves. Leaving an opening at the bottom, it suspends the nest from a tree branch.

## Multitudes of fish

Lake Malawi's waters and shores provide a variety of habitats and are home to many animals including crocodiles, hippopotamuses, and the African fish eagle. The lake is best known for containing as many as 3,000 species of fish – more than any other lake on Earth.

△ **LAKE OF STARS**
The explorer David Livingstone called Lake Malawi the Lake of Stars, due to the dazzling reflections off its still waters. However, it is also known for the violent gales that blow through.

S. Africa

# Victoria Falls

*An immense African waterfall with a unique geological history, plunging into a deep, narrow chasm*

Part of the mighty Zambezi River, Victoria Falls crosses the border between Zimbabwe and Zambia in southern Africa. Challenged only by South America's Iguaçu Falls (see pp.110–11) for the title of Earth's largest waterfall, it spans more than 1,700 m (5,600 ft) and reaches a height of 108 m (354 ft) – twice that of Niagara Falls (see p.52). It is often described as having the largest sheet of falling water in the world.

last 100,000 years, the river has eroded the sandstone to leave a series of narrow gorges, the latest of which the river plunges over (see panel, below). The zig-zagging of the gorges, which traces the falls' movement upstream, is due to a series of joints running across the ends of the gorges that have "captured" the river and changed its direction. The narrow chasm formation of the falls produces a spray that can be seen 50 km (30 miles) away.

### Unique formation

The Zambezi runs across a basalt plateau that has large cracks, roughly perpendicular to the river's flow, filled with softer sandstone. Over the

◁ **FEARSOME PREDATOR**
The Nile crocodile, the largest crocodile species in Africa, is the most fearsome predator found along the Zambezi.

## The falls' local name, **Mosi-oa-Tunya**, translates as "**the smoke that thunders**"

◁ **ON THE EDGE**
The falls form part of two national parks, both with large populations of elephants. Occasionally, one will venture out onto the falls to feed on trees on the islands.

△ **FALLING IN SECTIONS**
The islands in the river on the edge of the falls separate the flow into several recognized sections (from left): the Devil's Cataract, Main Falls, Rainbow Falls, and the Eastern Cataract.

### THE VICTORIA FALLS GORGES

The steep, narrow gorges below Victoria Falls mark their changing positions over time. The Devil's Cataract, on the far west side, is the point at which a new waterfall is in the early stages of cutting back.

Zambezi River

position where next full-width falls will form

Devil's Cataract

Victoria Falls

2nd gorge

3rd gorge

4th gorge

5th gorge

E. Africa

# Lake Victoria

*Africa's largest lake and the second largest in the world, the major source of the Nile*

Lake Victoria sits in a shallow depression on a plateau between two arms of the Great Rift Valley (see pp.184–87) in East Africa and is bordered by Uganda, Kenya, and Tanzania. It is the largest of the African Great Lakes and, with an area of more than 68,800 square km (26,500 square miles), it is the second largest freshwater lake in the world by area after Lake Superior (see pp.50–51). However, it is shallow for its size, with an average depth of only 40 m (130 ft).

Lake Victoria is the source of the White Nile (see pp.190–91), its only outlet, which is known as the Victoria Nile at its start. The Kagera is the largest river entering the lake, from the west, but the lake receives most of its water from rainfall.

## ▷ LONG TOES
The African jacana is a tropical wading bird with long toes and claws that enable it to walk across floating vegetation, such as lily pads, in shallow lakes.

## △ INTRICATE SHORELINE
The lengthy shores of Lake Victoria, seen here from space, provide access to fresh water for millions of people, including those in several large cities such as Uganda's capital, Kampala.

## ▽ AQUATIC ANTELOPE
An antelope found in the wetlands of southern Africa, the red lechwe has long, widely splayed hooves that help it to run on soft ground and a water-repellent coat that aids swift movement through water.

# The Okavango Delta

*A vast inland delta in southern Africa, an oasis within the Kalahari Desert that attracts millions of animals on annual migrations*

S. Africa

The Okavango Delta is a gently sloping inland delta in northwestern Botswana. It is fed by the Okavango River, which flows southeastwards from its source in Angola through the Kalahari Desert but never reaches the ocean. Instead, its water drains into a depression in the Kalahari Basin, where it spreads out to create an oasis comprising permanent and seasonally flooded marshlands, grassland plains, islands, and waterways. The annual flooding of the delta by water that fell as rain in the Angolan highlands, more than 1,000 km (620 miles) away, swells it to three times its usual size. Coinciding with Botswana's dry season, the flooding attracts a huge influx of animals, including herbivores such as buffaloes and elephants, creating one of the greatest concentrations of wildlife in Africa. Over 97 per cent of the water that enters the delta evaporates or eventually seeps into the sand of the Kalahari Desert. What is left drains into Lake Ngami to the west.

◁ **BOLD MARKINGS**
Marbled reed frogs display a large range of colours and markings with a combination of stripes, spots, and dashes.

◁ **FLAT DELTA**
The Okavango Delta is mostly flat, with less than 2 m (6 ft) variation in height across the area. The pattern of its waterways changes every year as channels become blocked and the slow-moving water finds a new route.

## ENDORHEIC BASINS

The Okavango Delta is part of an endorheic basin – a drainage system that has no outlet to the sea. These basins usually converge on a lake or swamp. Most endorheic basins are located in arid areas, and most water that enters evaporates.

Okavango River

Okavango Delta

NW. Africa

# Los Gigantes Cliffs

*A line of breathtakingly high cliffs on the western coast of Tenerife in the Canary Islands*

Los Gigantes, or "the Giants", rise up to 800 m (2,620 ft) in places and are made of the igneous rock basalt. The cliffs continue under the sea to the seabed some 30 m (100 ft) deep, which harbours a rich variety of marine life.

Tenerife is built from three volcanoes that coalesced about 3 million years ago into a larger volcano. The Los Gigantes Cliffs are the result of the erosion of a huge mass of solidified lava.

▷ **GEMS IN THE CLIFFS**
Crystals of the olive-green mineral peridot are occasionally found in rocks and beaches around Tenerife.

## LOCATION OF LOS GIGANTES

The Los Gigantes Cliffs are located where multiple layers of basalt - solidified lava erupted by one of Tenerife's original volcanoes - have been eroded away by the sea. The mass of basalt makes up an area in the western part of Tenerife called the Teno Mountains.

Los Gigantes Cliffs
Teno Mountains
Teide Volcano
Las Cañadas Caldera
Island of Tenerife
ancient lava flows, millions of years old

△ **THE END OF THE WORLD**
In classical times, 3,000 years ago, mariners thought these cliffs marked the end of the dry world, beyond which there was only ocean.

## ZONES OF A FRINGING CORAL REEF

A fringing reef has several zones. Facing the sea, the fore-reef is populated by different corals at different depths. At its top is the reef crest, which takes the brunt of wave action. On the shoreward side is the inner reef (or reef flat), which may become partly exposed at low tide.

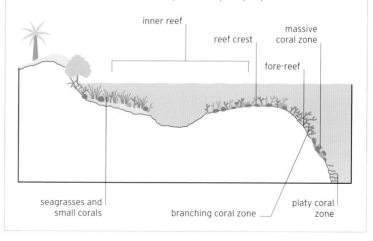

inner reef
reef crest
massive coral zone
fore-reef
seagrasses and small corals
branching coral zone
platy coral zone

# The Red Sea Coast

*The margins of the world's most northerly tropical sea with magnificent reefs and the site of a new ocean in the making*

NE.
Africa

(see p.185)

The Red Sea is an inlet of the Indian Ocean and has coasts on seven countries – Egypt, Israel, Jordan, Saudi Arabia, Sudan, Eritrea, and Yemen. It stretches for 2,250 km (1,400 miles) over a tectonic plate boundary where the Arabian Plate is gradually moving away from the African Plate – a process that will eventually create a new ocean (see p.185).

## Red Sea reefs

Reflecting the fact that it sits over an area where Earth's crust is rifting apart, the centre of the Red Sea plunges to a maximum depth of 2,211 m (7,254 ft). However, most of

its coast is rimmed by shallow submarine shelves and extensive fringing coral reefs (see panel, opposite). These are some of the most biologically well-endowed reefs, home to over 1,000 invertebrate species, 1,200 fish species, and several hundred types of hard and soft coral. With this rich marine life and high water clarity, the Red Sea Coast is a highly popular destination for scuba divers. Unfortunately, intense diving tourism, along with coastal development, has damaged the reefs in some areas.

△ RIOT OF COLOUR
Tiny green chromis fish surround scarlet and magenta soft corals - colonies of numerous minuscule animals called polyps - in a typical scene from a Red Sea reef.

▽ VIVID CORAL
This mushroom coral is one of over 300 different species of hard coral found on Red Sea reefs.

C. Africa

# The Congo Rainforest

*The second largest tropical rainforest on Earth, home to more than 11,000 plant species*

One of the world's most important remaining wilderness areas, the Congo Rainforest covers around 2 million square km (770,000 square miles) of central Africa's Congo River Basin, stretching across six countries.

## Last refuge

Second in size to the Amazon, this warm, wet region is hugely biodiverse, with more than 600 tree species and 10,000 animal species. Vegetation is so thick in parts of the forest that it has become a last refuge for some of the most endangered species on Earth. Rare forest elephants, okapis, bonobos, and both eastern and western gorillas live in the Congo, as do hundreds of plant species found nowhere else.

▷ RED FLAGS
The tropical dogwood, or red flag bush, is often found beside streams in the rainforest.

## THE WORLD'S TROPICAL RAINFORESTS

Tropical rainforests occur around Earth's equatorial regions, between the latitude lines of the Tropic of Cancer (23° North) and the Tropic of Capricorn (23° South). The largest coincide with South America's Amazon and Africa's Congo river basins, while Southeast Asia and India also have significant rainforest coverage. Northeast Australia has a small area of rainforest on its eastern side, stretching from coastal hills down to the sea.

△ VERSATILE MIMIC
The African grey parrot feeds on nuts, seeds, and fruit. It mimics other bird and mammal calls as well as human speech.

E. Africa

# Madagascan Rainforest

*A narrow strip of rainforest, where isolation has led to an extraordinary number of unusual native species*

Madagascar's rainforest runs along the island's eastern coastal region. Separated from the west by a chain of mountains – the Tsaratanana, Ankaratra, and Ivakoany massifs – and bordered to the east by the Indian Ocean, this area receives more than 2,540 mm (100 in) of rainfall a year. Constant warmth and humidity have created an ideal habitat for tropical plants to flourish, including around 1,000 orchid species, as well as unique animals such as lemurs. Madagascar's evolutionary isolation has also led to an extremely high number of native species: 80–90 per cent of its rainforest plants and animals are found nowhere else on Earth.

△ **HIGH DINING**
Critically endangered mountain gorillas (a subspecies of eastern gorilla) live high in the Congo, at elevations of 2,440-3,960 m (8,000-13,000 ft). Their diet consists of plants such as wild celery and bamboo.

◁ **DENSE FOREST**
Rainforest vegetation in the Congo is so dense that many sections of the forest have never been explored by humans. In some areas, only 1 per cent of sunlight reaches the ground.

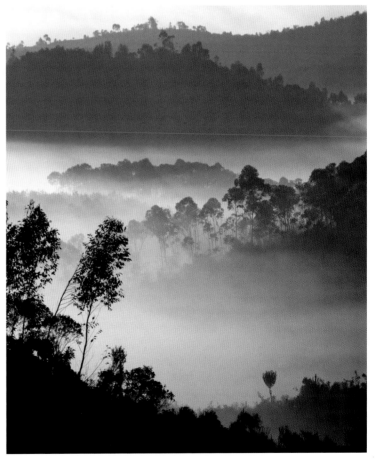

△ **DAWN MIST**
Clouds of moisture often linger over Madagascar's rainforest. Around 12,000 plant species are native to the island, which has led to parts of the rainforest receiving UNESCO World Heritage status.

E. Africa

# Madagascan Dry Forest

*A distinctive, diverse, and rapidly disappearing ecosystem, with an extremely high percentage of native plant and animal species found nowhere else on Earth*

Madagascar's tropical dry forest covers about 151,000 square km (58,300 square miles), mainly in small, fragmented blocks along the island's western and northwestern coasts. Characterized by a dry season that lasts up to eight months a year, it is a habitat full of unique, sometimes bizarre-looking plants that have evolved to cope with an extreme lack of water. Seven species of baobab tree grow here – six native to Madagascar – alongside shorter species that have also adapted to dry conditions, such as moringas and spiky succulents of the *Pachypodium* genus. The animals are also highly specialized, so much so that many, such as numerous lemurs, chameleons, and the island's largest predator, the cat-like fossa, are found nowhere else. Slash-and-burn agriculture has already destroyed 97 per cent of the forest and threatens the remainder.

◁ **CLINGING ON**
Numerous species of chameleons cling on to existence in what remains of Madagascar's dry tropical forest. The panther chameleon hunts by day in low trees or bushes.

## ADAPTING TO DRY CONDITIONS

Drought-survival methods vary with tree species. Baobabs store water in their trunks. Their roots are shallow but widespread, absorbing moisture near the surface. Acacias have widespread surface roots as well as a taproot that extracts deeper groundwater.

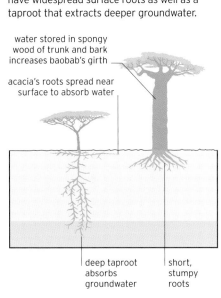

water stored in spongy wood of trunk and bark increases baobab's girth

acacia's roots spread near surface to absorb water

deep taproot absorbs groundwater

short, stumpy roots

△ **VERTICALLY MOBILE**
The golden-crowned sifaka's long, powerful back legs enable it to leap upright from tree to tree in search of seeds, fruit, and leaves.

◁ **DROUGHT SURVIVOR**
Madagascar's boadaka tree stores water in its trunk, like a cactus. Its trunk and branches can also photosynthesize, which means it can shed leaves during periods of drought.

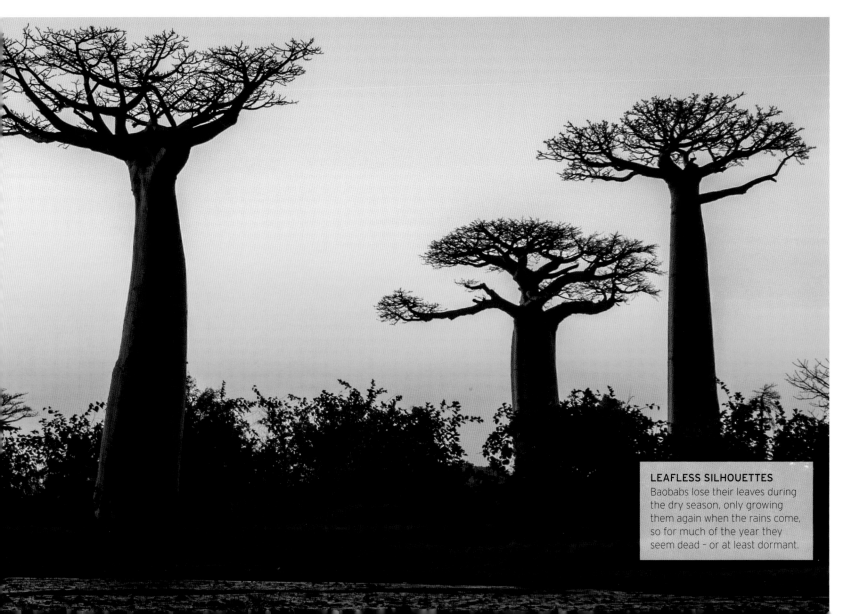

**LEAFLESS SILHOUETTES**
Baobabs lose their leaves during the dry season, only growing them again when the rains come, so for much of the year they seem dead - or at least dormant.

# Ethiopian Montane Grasslands

*A region of rare high-altitude grasslands, situated amid some of Africa's tallest mountains and a refuge for many endangered species*

NE. Africa

Northeastern Africa's montane grasslands are set mainly amid the eastern and western highlands of Ethiopia, with parts extending into the Eritrean highlands. Encompassing the Simien Mountains in the northwest, the Bale Mountains in the southeast, and part of the Great Rift Valley (see pp.184–87) between the two, this is a striking region, where jagged peaks surround high plateaus of grass and shrubland, as well as pockets of woodland, heather moorland, and alpine steppe. The most fertile montane grasslands occur at elevations of 1,800–3,000 m (5,900–10,000 ft). These attract not only a host of endemic reptiles, birds, and small mammals, but also a high human population, whose agricultural activities and removal of the area's natural resources have resulted in a loss of about 97 per cent of its original vegetation. Even so, these grasslands are a vital habitat for some of the continent's most endangered creatures, including the world's only population of Walia ibex and the Ethiopian wolf – Africa's sole wolf species.

◁ **GRASS FORAGER**
Geladas live almost exclusively on grasses from high-altitude meadows. They are also called bleeding-heart baboons due to their red chest patches.

**84 small mammal species**, many **endemic**, are found throughout the **Ethiopian highlands**

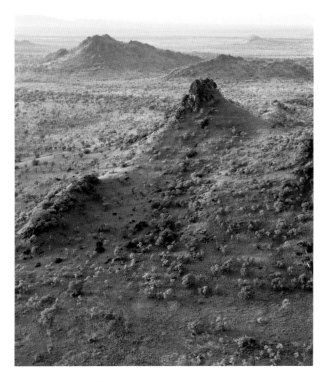

△ **GREEN EXPANSE**
South Sudan's Boma National Park turns a lush green after the annual rains, heralding the arrival of more than a million savanna grazing animals such as the tiang and white-eared kob - two antelope species that have converged here for centuries.

# The Sudanian Savannas

*A vast expanse of dry woodland and grassland, and a region of great botanical biodiversity*

W. and C. Africa

The band of mixed grasslands and trees known as the Sudanian Savannas stretches across western and central Africa from the Atlantic Ocean eastwards to the Ethiopian highlands. To the north is the Sahel, the transition zone between the savannas and the Sahara Desert, while to the south lie the forests of Guinea and Congo.

## West meets east

The ecoregion as a whole covers an estimated 917,600 square km (354,000 square miles) but divides geographically into two main sections, the West and East Sudanian savannas, which are separated by a much higher plateau in Cameroon. Large tracts of hot, dry woodland comprising acacia and other deciduous trees characterize this landscape. There is an understorey of shrubs, such as bushwillow, and grasses including the bamboo-like Napier grass, with an average height of up to 3.5 m (11 ft), as well as shorter species such as thatching grass, which grows up to 80 cm (31 in). More animals are found in the western savanna, while the eastern section contains a much higher degree of botanical biodiversity, including around 1,000 species of endemic plants. The entire savanna ecoregion forms an important habitat for elephants, wild dogs, cheetahs, and lions, as well as migrating birds.

## CLIMATE

Ethiopian highland climate is classed as tropical monsoon, but due to the region's elevation, temperatures are cooler here than would be expected this close to the equator. Rainfall increases from April to October, with the southwestern section receiving the most, while the annual average for the rest of the region is around 1,600 mm (63 in).

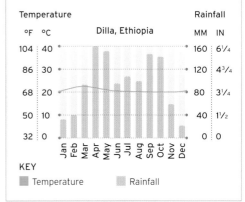

Temperature — Dilla, Ethiopia — Rainfall

KEY

◾ Temperature ◾ Rainfall

◁ JAGGED PEAKS

The rugged basalt cliffs and gorges of Ethiopia's Simien Mountains were created 30-20 million years ago by volcanic eruptions.

# The Cape Floral Region

*A biological hotspot, one of the most botanically diverse regions in the world, containing thousands of species found nowhere else*

S. Africa

Situated on the southwestern tip of South Africa, the Cape Floral Region is one of the country's eight World Heritage sites. This spectacular area covers just 10,947 square km (4,227 square miles) from the Cape Peninsula to the Eastern Cape – around 0.04 per cent of Africa's landmass – yet it contains up to 20 per cent of the entire continent's plant species. Of the Cape Floral Region's almost 9,000 plant species, a high percentage belong to the rare fynbos shrubland that makes up part of the area. Fynbos vegetation, which includes plants such as proteas and heathers, is found nowhere else on Earth outside South Africa; it thrives in the Cape's Mediterranean climate.

However, because many species are found only in localized, extremely small areas, this is a highly threatened, fragile environment, where just ploughing a field or building a house can eradicate an entire species.

▽▷ BLOOM TIME

The term fynbos originates from a Dutch word meaning "fine-leaved plants". Fynbos plants include hundreds of protea species, such as the king protea (below), which bloom at varying times of year.

E. Africa

# The Serengeti

*The famous immense African grassland, the site of one of the most spectacular annual migrations on Earth*

Covering 30,000 square km (11,600 square miles) across Kenya and Tanzania, the Serengeti lies near the equator and consists mainly of grassland in the southeast, open woodlands of acacia trees in the north, and a mix of grassland and acacias in the western regions. Ash from nearby volcanoes has added to the fertility of the region's rolling plains, nourishing its short, intermediate, and long grasses.

## Mass migrations

"Serengeti" is derived from a Masai word meaning "endless plain", but even its vast grassland sections are occasionally broken up by kopjes – weathered granite outcrops – that not only provide much-needed shade for the savannas' wildlife species but also hold rainwater in rocky depressions, providing small water holes for animals. Hordes of wildlife, particularly herds of herbivores, are part of the fabric of the Serengeti. Each year, more than a million wildebeest and hundreds of thousands of zebras and gazelles come together to form a vast "super herd" in an amazing migration of 1,000 km (620 miles), following the seasonal rains that bring fresh grass and vital water supplies to this arid area of Africa.

◁ **STANDING TALL**
The tallest living land animals on Earth, giraffes grow to heights of 4.5–6 m (15–20 ft) – useful for reaching acacia leaves.

△ **SYMBOL OF THE SERENGETI**
Acacias dominate parts of the Serengeti. Among the few trees found in the region, acacias provide food and vital shade for many animal species.

△ **NATURE'S RECYCLERS**
Dung beetles provide waste removal in the Serengeti by rolling and burying dung to use as food and as a place to lay eggs.

◁ **WATER BREAK**
Water holes provide a lifeline to migrating herbivores, such as zebra – but also for the carnivores that prey on them.

## TYPES OF SAVANNA

There are several types of African savanna vegetation, mainly influenced by the amount of rainfall, which varies with distance from the equator. Rain in the Serengeti falls in two main seasons: from March to May and November to December. In the more wooded northwestern Serengeti, the average annual rainfall is 1,050 mm (41 in), while in the grass-dominated open savannas of the southeast, it is just 550 mm (22 in).

rainforest (rain all the year) | parkland or closed savanna | savanna (summer rain; winter drought) | semi-desert scrub or open savanna | desert (annual drought)

The **largest numbers** of **grazing and browsing animals** in the **African continent** are found in the Serengeti

N. Africa

# The Sahara Desert

*The world's largest hot desert, shaped by relentless winds and shifting sand dunes that are constantly on the move*

Stretching from the Atlantic Ocean in the west to the Red Sea in the east, the Sahara covers almost the whole of North Africa. At 9.4 million square km (3.6 million square miles), it is slightly larger than the US, and surveys indicate that this vast, almost uninterrupted arid region is still expanding. Since 1962, the desert has grown by 647,500 square km (250,000 square miles).

## More than sand

Though often associated with red sand dunes, most of the Sahara consists of barren, rocky plateaus called hamadas. These contain little sand, most having been swept away by the winds that continually transform the desert into a harsh landscape full of rocks, boulders, and gravel. In other areas, sand is piled up in dune

fields. The largest of these are called ergs – vast areas spanning more than 125 square km (50 square miles). Many contain shifting dunes, some as high as 150 m (490 ft). Because ergs are highly active regions of loose sand, they are notoriously difficult to cross.

Several mountain ranges lie in the Sahara and many contain volcanoes. The desert also has gravel fields and salt flats, as well as dry river- and lakebeds. Despite such harsh terrain and little to no rainfall – half the Sahara gets less than 25 mm (1 in) a year, the rest possibly up to 100 mm (4 in) – around 500 plant and 70 animal species are found here.

◁ **DESERT FOX**
The smallest of all foxes, the desert-dwelling fennec fox is the size of a house cat. It has massive ears, which keep it cool by releasing excess body heat.

## Each year, **thousands of tonnes** of Saharan phosphorus dust reach the **Amazon rainforest**

△ **BULL'S-EYE**
Known as the Eye of the Sahara, the huge Richat Structure was thought to have been formed by a meteor strike. Scientists now believe it formed when igneous rocks were uplifted and later eroded.

◁ **SAND WAVES**
The red dunes of the Great Western Erg, near Taghit oasis, Algeria, resemble ocean waves. High winds here cause deadly sandstorms that can last for days.

◁ **ROCKY PEAKS**
At Tassili du Hoggar in southern Algeria, the central Saharan winds have sculpted a large sandstone plateau into jagged peaks, rock arches, and spires that resemble the ruins of a forgotten city.

# The White Desert

*A desert of chalk, shaped by wind-driven sand into an eerie landscape of stone monoliths*

NE. Africa

Often compared to a lunar landscape, the White Desert is located northeast of the oasis town of Farafra, Egypt. This gleaming terrain was originally an ancient seabed.

## From chalk to chickens

When the prehistoric sea that once covered this vast area dried up, the bodies of millions of marine creatures calcified, forming an enormous plateau of chalk and chalky limestone. Millions of years of wind erosion carved the upland into pinnacles, spires, boulders, and other shapes – some so bizarre that they have been likened to giant mushrooms, ice-cream cones, and even an oversized chicken. Occasionally, chalky inselbergs (isolated, tower-like hills jutting out of the flat plain) are also found among the smaller formations. The sand surrounding the shapes glitters with quartz crystals, adding to the White Desert's overall frosty appearance, particularly when viewed under moonlight.

▽ **DESERT SCULPTURE**
Wind-blown sand has carved these striking stone formations that seem to balance dangerously on severely eroded bases.

---

NW. Africa

# The Adrar Plateau

*Mauritania's high, hot plateau, with a unique collection of desert landforms, and its own geological "eye"*

The Adrar Plateau can be considered a miniature version of the Sahara Desert (see pp.208–209). With huge shifting sand dunes, rugged gorges, palm-filled oases, stone deserts, and rocky cliffs rising to about 240 m (790 ft) above their surroundings, this northwest African region has all the features normally associated with the Sahara as a whole. The centre of the plateau, which lies in central Mauritania, is arid, with little to no vegetation, but enough water concentrates around the base of the surrounding uplands to allow the growth of some crops.

## Desert eye

West of the Adrar, the plateau merges into the Richat Structure – a gigantic "bull's-eye" first noted from space. This geological dome, also known as the Eye of the Sahara (see p.209), is 40 km (25 miles) in diameter.

△ **CUSTOM BILL**
The greater hoopoe-lark uses its long, curved bill to probe for termites and other insects in the desert sands.

▽ **TABLETOP MOUNTAINS**
The dark tabletop mountains, or mesas, of the Adrar Plateau rise above almost barren sands.

# The Kalahari Desert

*Not a true desert but a massive sandy savanna with adequate rainfall and a covering of vegetation in parts, yet no water on the surface or in the soil*

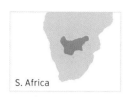
S. Africa

Despite summer temperatures that top 40°C (104°F) and seemingly endless sand sheets, the Kalahari is not, technically, a desert. Parts of this semiarid region, which covers most of Botswana and some of Namibia and South Africa, receive more rainfall annually than the 250mm (10in) that is traditionally used to classify desert areas. While rainfall in the Kalahari's southwestern half falls within this range, the northeast gets almost double that figure – yet no surface water remains here. Any rain drains immediately through the deep sands, creating a soil completely devoid of moisture. Despite this and the huge dune-chains that characterize its western half, the Kalahari contains considerable plant life. Deep-rooted acacia trees, shrubs, and some grasses grow in the centre, while the north boasts forested areas. More animals are found in the north, but in the arid south, grazing species such as kudu and gemsbok track down enough vegetation to survive.

## RAINFALL

Total rainfall is only one factor in classifying deserts. Although the Kalahari and Karoo receive over twice the precipitation of true deserts, their rainfall is seasonal, which means that there may be no rain in some areas for 6–8 months.

| Rainfall | | Kalahari | Karoo | Sahara | Namib |
|---|---|---|---|---|---|
| MM | IN | | | | |
| 500 | 20 | | | | |
| 400 | 16 | | | | |
| 300 | 12 | | | | |
| 200 | 8 | | | | |
| 100 | 4 | | | | |
| 0 | 0 | | | | |

KEY   Minimum rainfall   Maximum rainfall

◁ **RUST RED**
A herd of gemsbok stands out against the red Kalahari sand, which gets its intense colour from iron oxide.

# The Karoo

*A semiarid cape zone, containing many of the world's succulent plant species*

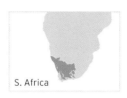
S. Africa

Covering about a third of South Africa, the Karoo is one of the driest regions of the country. Noted for its vast open spaces interspersed with Karoo koppies (flat-topped hills), rugged mountains, and fossil deposits, this desert also contains a fertile zone along the coast that is home to one-third of the world's succulent plant species. In the west, spring rains transform the landscape, covering it with carpets of vivid orange and pink flowers.

The region is divided into sub-Karoos – the more desert-like Upper Karoo, the lower-lying plateau basin called the Great (Groot) Karoo, and the relatively fertile Little (Klein) Karoo.

▷ **BRANCHING SUCCULENT**
The quiver tree is a tree-sized succulent, with a stem that grows to a diameter of 1m (3ft). Desert bushmen hollow out its tubular branches to make quivers for arrows, giving it its name.

SW. Africa

# The Namib Desert

*A dramatic coastal desert, where terrain ranges from bare bedrock to high, mobile dunes that meet the sea*

Stretching from the Atlantic to a high, inland plateau, the narrow Namib Desert extends for about 1,300 km (810 miles) along the coast of southwest Africa. Considered to be one of the world's oldest deserts, it is bordered to the north by the Kaokoveld Desert, which stretches into Angola, and to the south by South Africa's Karoo. In most places, the narrow Namib measures less than 160 km (100 miles) wide. Despite this, it is often viewed as having three sections: the Atlantic-influenced coastal strip; the Outer Namib, which takes up what remains of the western half; and the Inner Namib, comprising the desert's eastern half. Commonly called the Skeleton Coast, the coastal Namib receives almost no rainfall and instead relies on regular sea fogs for moisture. From sea level, the terrain rises, reaching about 900 m (2,950 ft) by the time the desert meets the base of the Great Escarpment to the east. In the Inner Namib, average annual rainfall is only around 50 mm (2 in).

## Living the dry life

Dryness is such a feature of the Namib that it is believed to have been in an arid state for at least 55 million years. This makes it all the more surprising to find that the desert is home to animals ranging from adders and geckos to zebras and elephants. Parts of the Namib are also rich in plant life, including the ground-hugging, ragged-leaved tree tumbo (*Welwitschia mirabilis*), which captures moisture from sea fog and can live for more than 1,000 years.

▷ **SEA OF SAND**
At the heart of the desert lies the Namib Sand Sea, spanning an area of 31,000 square km (12,000 square miles). It is composed of two dune systems – an ancient, semi-stable one topped by a younger, active layer.

▽ **ADAPTED FEET**
The web-footed gecko's feet enable it to walk easily on top of desert sand, as well as bury itself beneath the sand to avoid the daytime heat.

In the **southern Namib**, some **dunes** are up to **32 km (20 miles) long** and **240 m (790 ft) high**

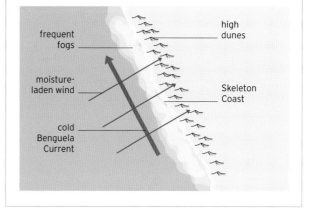

## SKELETON COAST FOG

As the Atlantic's cold Benguela Current runs north along the coast of Namibia, it cools the moist air above the ocean. When this meets the hot desert air, dense fog banks form, rolling up to 100 km (60 miles) into the central Namib until they are burned off by the Sun.

frequent fogs

high dunes

moisture-laden wind

Skeleton Coast

cold Benguela Current

△ FOG HARVESTER
The darkling beetle survives in the Namib by using its own body to harvest water vapour. Microgrooves on its forewings collect water droplets, which are channelled to its mouth when it raises its back legs.

◁ A PLACE OF BONES
Dunes absorb myriad shipwrecks and bones that litter the Skeleton Coast. Here, the thick fogs that caused sailors to run aground bring precious moisture inland.

**Red pillars**
Sandstone pillars rise up to 200 m (660 ft) at Zhangjiajie in southern China, close to the Tianzi Mountains. The pillars have been shaped by water erosion and weathering by ice and plant life.

# Asia

# TUNDRA TO TROPICS
## Asia

Asia is the world's largest continent, taking up almost one-third of Earth's land surface. A large part of the north is the West Siberian Plain, the largest area of flat land in the world. It is dotted with huge swamps and blanketed by dark Siberian forests, ending in the far north with bleak, open tundra – which can often be colder than the Arctic in winter. Further south are the extensive steppe grasslands of Central Asia, merging into the Gobi desert. Asia's other great desert is the sand sea of the Rub' al Khali, scorching across the Arabian Peninsula.

In the heart of Asia is the Tibetan Plateau. The world's highest mountains soar around it: to the north, the Altai and Tian Shan; and to the south, the Himalayas, the highest of all, rising to the peak of Everest. Beyond the mountains, to the south and east, long stretches of land extend out into the Indian and Pacific oceans. They are cut by great rivers, filled by rain and melting snow from the mountains – the Indus and Ganges in Pakistan and India, the Mekong in Southeast Asia, and the Yellow and Yangtze in China.

### GEOLOGY

Asia is divided between the huge, stable area of continental rock under Siberia and the more varied, unstable south. In between lie the mountains of Tibet, thrown up as India pushes north.

### KEY
- ◼ Precambrian (pre-541 million years ago)
- ◼ Paleozoic (541–252 million years ago)
- ◼ Mesozoic (252–66 million years ago)
- ◼ Cenozoic (66 million years ago to present day)

### ELEVATION
M FT

5,000 ● 16,000

0 ● 0

-3,000 ● -10,000

Km
0    400    800
0    400    800
Miles

West Siberian Plain

Ural Mountains

Kirghiz Steppe

Kazakh Uplands

Lake Balkhash

Aral Sea

Anatolia

Caucasus

Caspian Sea

Karakum

Kyzyl Kum

Tian Shan

Taklamakan Desert

Zagros Mountains

Euphrates

Tigris

Tirat Tsvi 54°C

Dead Sea shore, –400m

Red Sea

Iranian Plateau

Arabian Peninsula

Persian Gulf

Hindu Kush

Plateau of Tibet

Himalaya

Indus

Thar Desert

Gulf of Oman

Arabian Highlands

Rub' al Khali

Arabian Sea

Ganges

Deccan

Western Ghats

INDIAN OCEAN

Bay of

Sri Lanka

**CROSS-SECTION**

A ◀ West Siberian Plain   Kazakh Uplands   Lake Balkhash   Tian Shan   Takla Makan Desert   Himalayas   Ganges floodplain   Deccan Traps   Indian Ocean   Sri Lanka ▶ B

## KEY DATA

▲ **Highest point** Mount Everest, China/Nepal: 8,848 m (29,029 ft)

▼ **Lowest point** Dead Sea shore, Israel/Jordan: -400 m (-1,312 ft)

● **Hottest record** Tirat Tsvi, Israel: 54°C (129°F)

● **Coldest record** Oimaykon, Russia: -68°C (-90°F)

## CLIMATE

Asia's size makes its climate extreme. The centre veers between perishingly cold winters and blistering summers. Land near the southern coasts is nurtured by monsoon rains and warm air.

AVERAGE TEMPERATURE

AVERAGE RAINFALL

°C 30 20 10 0 -10 -20 -30 -40

°F 80 60 40 20 0 -20 -40

MM 10,000 7,500 5,000 2,500 0

IN 400 300 200 100 0

## ECOSYSTEMS

The north of the continent is covered by coniferous forests, or taiga, while the centre is grassland or high desert. The south is more varied, with large but dwindling areas of tropical rainforest.

### KEY

- Tropical broadleaf forest
- Tropical dry broadleaf forest
- Tropical coniferous forest
- Temperate broadleaf forest
- Temperate coniferous forest
- Tropical and subtropical grassland
- Desert and scrub
- Temperate grassland
- Wetland
- Montane grassland
- Boreal forest and taiga
- Tundra

▷ **WORN HILLS**
The Urals are one of the world's oldest mountain ranges, although they have been worn down from their original heights by millions of years of weathering and erosion.

▷ **MOLTEN REMNANTS**
The rock left over from the basalt lava eruptions that formed the Deccan Traps 66 million years ago have been exposed by the erosion of less resilient rocks around them.

◁ **ANCIENT LAND**
The ancient rock beneath the Kazakh steppe was once an entirely separate island continent geologists call Kazakhstania. Now, it is entirely trapped deep inside Asia.

# THE SHAPING OF ASIA

Asia is bigger than any other continent and contains the world's highest mountain and largest mountain range. Yet it is also a young continent, formed from a complicated assembly of fragments.

## The giant continent

All the continents have joined each other and split apart again and again – except Asia. It has stayed largely intact through Earth's history and is a recent creation, an amalgam of many pieces. The vast slab of ancient rock beneath Siberia, known as the Siberian, or Angaran, Platform, forms the continent's core, but it was joined by the other parts relatively recently. Even when the Siberian Platform was part of the supercontinent Pangaea 200 million years ago, south and east Asia were an assortment of separate island continents.

Between 100 and 50 million years ago, these fragments began to come together rapidly. First, north and south China and Southeast Asia fused onto Eurasia's southeast corner. Then, far to the south, India's extraordinary journey began. About 80 million years ago, India was attached to southern Africa, but then it broke away and hurtled north across an ancient ocean and smashed into the southern edge of Eurasia 60–40 million years ago, helping to cement all the other fragments in place. Asia as we know it was born.

## Volcanic India

India's Deccan Plateau is dominated by one of the largest volcanic features in the world, the Deccan Traps. The traps are what geologists call a large igneous province and are made of layers of volcanic basalt over 2,000 m (6,600 ft) thick. Even now, they spread over an area of 500,000 square km (193,000 square miles), but before being worn away by erosion they covered half of India.

Unlike some lavas, basalt is very runny and can flood onto the surface in huge quantities. The eruptions began 66.5 million years ago, when India had only recently become detached from Africa. One theory suggests that at that time India was above a hotspot in Earth's interior where the island of Réunion is now. At a hotspot, a fountain of hot magma

### CREATION OF THE DECCAN TRAPS
Basalt lava erupted via a fissure in Earth's crust above a mantle plume, a fountain of hot rock rising in the mantle. The lava hardened and formed the Deccan Traps.

fissure
erupting flood of basalt lava
crust
lithosphere mantle
mantle plume
asthenosphere

## KEY EVENTS

**130 million years ago**
A rainforest forms in an area that will become Malaysia. A portion of this survives in Taman Negara National Park today.

**60-40 million years ago**
The Indian Plate crashes into the southern part of the Eurasian Plate, causing the crust to crumple into the Himalayas.

**5.5 million years ago** An ancient sea that had spread from eastern Europe to western Asia dries out as land lifts and sea levels drop, leaving behind the Caspian, Aral, and Black seas and Lake Urmia.

**252-250 million years ago** Intense volcanic activity due to a mantle plume causes eruptions over a large area that will become eastern Siberia, creating a huge plain of volcanic rock called the Siberian Traps that still exists today.

**66 million years ago** Huge volcanic eruptions on a landmass that will become India occur. Lava spreads and solidifies into the Deccan Traps.

**50-40 million years ago** The rise of the Himalayas blocks rainfall heading north of the new mountain range, causing the area to dry out and become the Gobi Desert.

**11,000 years ago** Sumatra, Java, and Borneo become islands as sea levels rise.

◁ ROOF OF
THE WORLD
The vast extent
of the Tibetan
Plateau, with its
fringing ranges of
lofty, snow-capped
Himalayan mountains,
can be seen from space.

called a mantle plume rises and melts its way through the crust.
The giant eruption may have been a contributing factor towards
the extinction of the dinosaurs – the main culprit being the meteor
strike – and scientists once believed that they might have been
killed off by a cooling of the world's climate caused by gases
released by this eruption.

▷ SOURCE OF
THE GANGES
The rivers flowing down from
the Himalayas, including the
Ganges, sweep down 25 per
cent of all the world's sediment,
to create wide, fertile plains.

### Mountains rising

Following its separation from Africa, India moved north at a rate
of about 20 cm (8 in) a year, faster than any other known plate
movement. As it rammed into the southern edge of Asia, it exerted
a huge impact.

In collisions between plates, one plate is usually subducted
beneath the other. But these two plates were so closely matched in
density that neither would sink, and so the crust between them
crumpled up into the highest and largest mountain range the world

has ever seen – the Himalayas. In just 50 million years, the rocks
have been thrust up more than 9 km (6 miles) to create jagged
peaks – including Everest, the world's highest mountain at 8,848 m
(29,029 ft). In fact, the impact has not stopped the Indian Plate from
moving – it still creeps forwards at 3 cm (1¼ in) a year, and as a
result the Himalayas rise 1 cm (½ in) every year. Geologists think
the only reason they are not higher still is that, as it piles up, the
rock also flows outwards – in effect, the Himalayas are similar to
the bow wave in front of a moving boat.

# Violent volcanic eruptions in India 66.5 million years ago probably contributed to the extinction of the dinosaurs

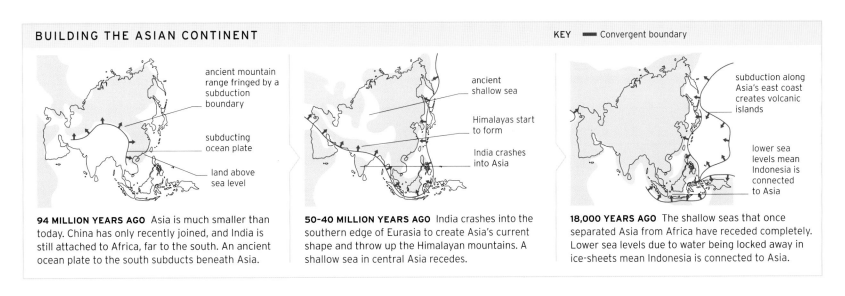

BUILDING THE ASIAN CONTINENT

KEY ▬ Convergent boundary

ancient mountain
range fringed by a
subduction
boundary

subducting
ocean plate

land above
sea level

ancient
shallow sea

Himalayas start
to form

India crashes
into Asia

subduction along
Asia's east coast
creates volcanic
islands

lower sea
levels mean
Indonesia is
connected
to Asia

**94 MILLION YEARS AGO** Asia is much smaller than
today. China has only recently joined, and India is
still attached to Africa, far to the south. An ancient
ocean plate to the south subducts beneath Asia.

**50-40 MILLION YEARS AGO** India crashes into the
southern edge of Eurasia to create Asia's current
shape and throw up the Himalayan mountains. A
shallow sea in central Asia recedes.

**18,000 YEARS AGO** The shallow seas that once
separated Asia from Africa have receded completely.
Lower sea levels due to water being locked away in
ice-sheets mean Indonesia is connected to Asia.

# Pamukkale Hot Springs

*Staircase-like terraces of hot springs in southwest Turkey*

W. Asia

Pamukkale, which means "cotton castle" in Turkish, evokes a sense of awe due to its white landscape and turquoise pools – arranged in a series of steps, or terraces – that carry water from hot springs.

## Travertine steps

Pamukkale's terraces are composed of travertine, a bright white substance made from calcium carbonate deposited by water from the hot springs. There are 17 springs with temperatures ranging from 35 to 97°C (95 to 207°F). For thousands of years, water that has percolated deep into Earth's crust and been heated there has re-emerged near the top of a 100-m (330-ft) high hill. As it flows downhill, it passes through a series of pools lined by travertine. The mineral-rich pools are reputed to have health-giving properties.

▷ **TERRACED POOLS**
This area of stepped pools at Pamukkale is 65 m (215 ft) long and 27 m (90 ft) wide. It has around 20 levels, and its pools contain water at about 36°C (97°F).

# The Zagros Mountains

*An imposing natural barrier running through the Middle East, which displays evidence of the ongoing effects of a tectonic plate collision*

W. Asia

The Zagros is the largest mountain range in Iran, also extending into parts of Turkey and Iraq. Some sections rise above 4,000 m (13,000 ft) and are permanently snow-capped, with the highest peak being Ghash Mastan at 4,435 m (14,550 ft). Wild, and mostly semiarid, the Zagros started forming around 30–25 million years ago when two tectonic plates – the Arabian Plate and the Eurasian Plate – collided, resulting in the folding of underlying rock layers. A sizeable amount of deformation is still occurring today as a result of this collision, and the region is subject to frequent earthquakes.

◁ **RIPPLED EARTH**
Its many parallel ridges give the Zagros a wave-like appearance.

▷ **NIGHT FEEDER**
A nocturnal predator and scavenger, the striped hyena occurs in small numbers throughout the Zagros. It features prominently in Middle Eastern folklore.

## HOW A TRAVERTINE TERRACE FORMS

Travertine terraces form when a hot spring containing calcium carbonate erupts on a hill. The flowing water deposits travertine on the slope below the opening, while a build-up of the mineral in the opening eventually chokes it up. A new opening is then forced to develop a little higher up. In this way, the site of the spring gradually shifts uphill, leading to a travertine staircase.

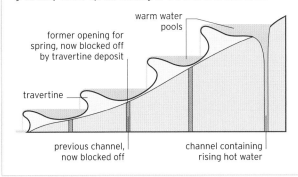

warm water pools

former opening for spring, now blocked off by travertine deposit

travertine

previous channel, now blocked off

channel containing rising hot water

## The **springs flow** at about **400 litres (106 gallons) per second**

---

# The Caucasus Mountains

*A wall of mountains stretching across the meeting point of Asia and Europe*

W. Asia

The formidable Caucasus mountain range is situated mainly in the Asian countries of Georgia, Armenia, and Azerbaijan but also partly in European Russia. It consists of two main subranges: the Greater and Lesser Caucasus. The highest peak is Mount Elbrus (in the Russian section) at 5,642 m (18,510 ft). The mountains feature a remarkable array of vegetation zones, ranging from forests and alpine meadows to highland semideserts, and also the world's deepest-known cave, the Krubera Cave in Georgia.

▷ **MAGNETIC MINERAL**
Dashkesan, a region of the Lesser Caucasus, is a source of the iron-rich mineral magnetite, the most magnetic of all minerals.

▷ **SUNRISE VIEW**
The Caucasus are steep, craggy, and relatively young mountains. Many of the peaks in the range carry glaciers – there are more than 2,000 glaciers in the entire range.

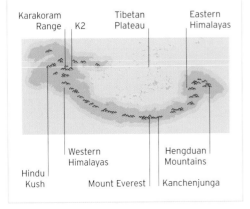
# The Himalayas

*The world's highest mountain range, with more than 50 peaks reaching a height of 7,200 m (24,000 ft) or more*

C. Asia

Running across northern Pakistan and India to Nepal and Bhutan, with some parts in China, the Himalayas are not only Earth's highest mountain range but also relatively young, having formed within the last 50 million years.

## Abode of snow

Some 2,300 km (1,400 miles) long and 250–350 km (160–220 miles) wide, the Himalayas form a barrier between the Tibetan Plateau to the north and the plains of the Indian subcontinent to the south. The name of the range comes from the Sanskrit word *Himalaya*, which means "abode of snow". Although extremely inhospitable, the highest peaks of the Himalayas present an irresistible magnet to mountaineers seeking the greatest possible challenges.

### △ HIGH-ALTITUDE HUNTER

A few thousand snow leopards are believed to live in the Himalayas. In summer, their range extends as high as 6,000 m (20,000 ft).

Parts of the Himalayas are **still rising** at a rate of about **4 mm** (³/₁₆ **in) a year**

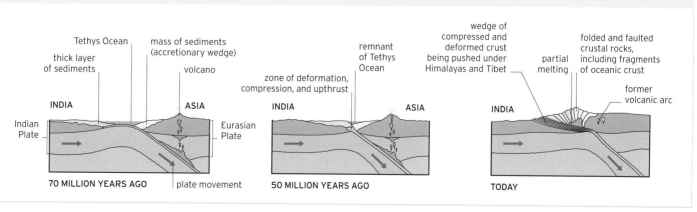

△ **FIRST LIGHT**
At dawn every day, the high snow-clad peaks light up while the valleys remain in darkness.

◁ **FLORAL CARPET**
In July and August, some lower slopes are covered in brightly coloured flowers, such as these Himalayan fleeceflowers.

## HOW THE HIMALAYAS FORMED

The Himalayas were pushed up by tectonic plate movements that caused India, previously an island, to collide with Asia. In the process, crustal rocks were crumpled and thrust up by as much as 8 km (5 miles).

thick layer of sediments

Tethys Ocean

mass of sediments (accretionary wedge)

volcano

INDIA

ASIA

Indian Plate

Eurasian Plate

**70 MILLION YEARS AGO** | plate movement

zone of deformation, compression, and upthrust

remnant of Tethys Ocean

INDIA

ASIA

**50 MILLION YEARS AGO**

wedge of compressed and deformed crust being pushed under Himalayas and Tibet

partial melting

folded and faulted crustal rocks, including fragments of oceanic crust

former volcanic arc

INDIA

**TODAY**

# The Altai Mountains

*A beautiful range of high mountains situated almost in the middle of Asia*

C. Asia

Straddling an area where Russia, China, Mongolia, and Kazakhstan meet, the Altai stretch for 2,000 km (1,200 miles) across central Asia. Much of the range exceeds 3,000 m (10,000 ft) in height, and there are numerous glaciated peaks over 4,000 m (13,000 ft) in the Mongolian sections. The highest peak, Mount Belukha (4,506 m/14,783 ft), lies in the north at the headwaters of the Ob–Irtysh river system in Russian Siberia. In the east, the Altai merge into the high plateau of the Gobi desert (see pp.266–67) and the cold steppes of Mongolia. Partly because they include a variety of habitats – such as tundra, forest, and alpine vegetation – the Altai are home to a wide range of wildlife, including Siberian ibex and several species of deer, wolves, lynx, and brown bear.

◁ **HEAVY FRAME**
Resembling a heavily built goat, the Siberian ibex is typically found on steep rocky slopes above the tree line. It feeds on small shrubs, grass, and sedges.

▽ **GOLDEN MOUNTAINS**
*Altai* means "gold peak" in Mongolian. The snow-dusted peaks of the Altai mountains are seen here with yellow steppe grass in the foreground.

# The Tian Shan Mountains

*One of the world's longest and highest mountain ranges, boasting more than 60 peaks that exceed 6,000 m (20,000 ft) in height*

C. Asia

Also known as the Tien Shan, the Tian Shan consists of a series of subranges that extend through Kyrgyzstan, a part of Kazakhstan, and western China. Their overall length is about 2,800 km (1,700 miles), which is longer than the Himalayas. The peaks above 4,000 m (13,000 ft) – including the highest, Victory Peak at 7,439 m (24,406 ft) – are capped by glaciers. The Chinese words *Tian Shan* mean "heavenly mountains". The range has been listed as a World Heritage Site – partly because of its unique and rich biodiversity, as well as its diverse landscapes that contrast with the vast adjacent deserts.

▷ **BLUE MOUNTAINS**
A part of the Borohoros, one of the Tian Shan's subranges, appears vivid blue due to atmospheric scattering of blue-wavelength light.

C. Asia

# The Karakoram Range

*The greatest concentration of high mountains in the world*

Spanning the northern regions of Pakistan and India, with parts in China and Tajikistan, the Karakoram lies to the northwest of the Himalayas and contains nearly all of the world's 100 highest peaks that are not in the Himalayas (see pp.222–23). These include the world's second highest mountain, K2, which rises to a height of 8,611 m (28,251 ft).

## Climbing challenges

The Karakoram Range is about 500 km (310 miles) in length and is heavily glaciated. Some individual glaciers, such as the Baltoro (see p.233), are among the world's longest glaciers outside the polar regions. The Karakoram is sparsely populated, and visitors are almost exclusively intent on scaling one or more of its peaks, some of which are extremely difficult to climb. They include K2 and the Trango Towers, a group of granite columns situated close to the Baltoro Glacier.

## About **1 in 5 people** who have **tried to climb K2** have **died** in the attempt

△ **NAMELESS TOWER**
Called the Nameless Tower, the granite peak at the centre here is part of the Trango Towers. Its near-vertical sides rise some 900 m (2,950 ft) above the ridgeline, and the spire reaches up to 6,239 m (20,469 ft).

▷ **SAVAGE MOUNTAIN**
Seen here with a serac (pointed block of glacial ice) in the foreground, K2, also called Mount Godwin-Austen, is known as the Savage Mountain because of the extreme difficulty of its ascent.

**HIGH PEAKS OF THE KARAKORAM**

1  2  3  4  5

1  K2 8,611 m (28,251 ft)
2  Gasherbrum I 8,080 m (26,509 ft)
3  Broad Peak 8,051 m (26,414 ft)
4  Gasherbrum II 8,035 m (26,362 ft)
5  Gasherbrum III 7,952 m (26,089 ft)

S. Asia

# Mount Everest

*Earth's highest mountain, which sits in the Eastern Himalayas*

Also known in Nepal as Sagamartha (meaning "sky head") and in China as Chomolungma (meaning "Holy Mother"), Mount Everest soars to 8,848 m (29,029 ft) above sea level.

## Faces and ridges

Everest is a highly inhospitable place. It is dangerous to climb: fatalities are mainly a result of avalanches, hypothermia, and altitude sickness. The mountain is roughly the shape of a three-sided pyramid, with a southwest face, a north face, and an east face, each very steep. At the intersections of these faces are three ridges – southeast, northeast, and west – leading to the summit. The first confirmed ascent took place in 1953, by Tenzing Norgay and Edmund Hillary, via a route using the southeast ridge. Successful ascents have now been made by at least 18 different routes of varying difficulty using various combinations of ridges and faces.

## THE GEOLOGY OF EVEREST

Everest's summit pyramid is composed of limestone that formed on the seafloor some 470 million years ago, during Earth's Ordovician period. Beneath that are other layers of sedimentary rock, and deeper still, metamorphic rocks such as gneiss, with intrusions of igneous rocks.

leucogranite

Ordovician limestone

sedimentary rocks such as shale and limestone

gneiss

igneous rocks intrusion metamorphic rocks

▷ **AVALANCHE**
Avalanches are frequent on Everest. In April 2015, an earthquake triggered a deadly avalanche that swept into the southern base camp and killed 22 people.

northeast ridge

north face

**Changtse**
This peak at 7,543 m (24,747 ft) connects to Everest's summit via the north col.

north col

**Rongbuk Glacier**

**Khumbutse**
Lying on the border between Nepal and China, this peak reaches 6,636 m (21,772 ft).

▶ **EVEREST MASSIF**
Viewed here from the west, the Everest massif includes Everest itself and the nearby peaks Lhotse and Nuptse. The ridges between the peaks form a horseshoe shape, enclosing a basin.

▷ **BASE CAMP**
There are two main base camps on Everest on different sides of the mountain. These campsites are used by climbers before and after ascents and by trekkers. The south camp (seen here), in Nepal, is close to the Khumbu Icefall at 5,364 m (17,598 ft).

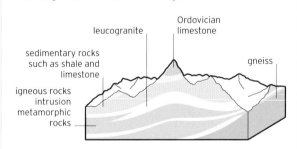

There is **66 per cent less oxygen** at **Everest's summit** than at sea level

**Everest summit**
Everest rose by about 3 mm (⅛ in) per year up to 2015 but may have dropped slightly during an earthquake in that year.

west ridge

southeast ridge

southwest face

south col

**Lhotse**
At 8,516 m (27,940 ft), Lhotse is the world's fourth highest mountain. Its western flank – the Lhotse face – is a steep wall of blue glacial ice.

Lhotse face

▷ **WEATHER CONDITIONS**
The weather is notoriously changeable on the Everest massif, including on Nuptse (pictured here). Temperatures at different times and places on the mountain can vary from -62°C (-80°F) to 38°C (100°F), and hurricane-force winds are not uncommon.

**Nuptse**
This peak to the southwest of Everest, has its summit at 7,861 m (25,791 ft).

**Western Cwm**
At the foot of the Lhotse face is a broad valley basin containing hidden crevasses.

Nuptse southwest face

**Crust**
Under the Everest massif, the crust extends as deep as 80 km (50 miles) and is made of many different rock types.

Lower Khumbu Glacier

△ **KHUMBU ICEFALL**
A dangerous part of the most popular route up Everest (which later passes via the South Col and southeast ridge), this broken-up and crevassed part of the Khumbu Glacier descends from around 6,000 m (20,000 ft) to 5,400 m (17,700 ft).

# Zhangye Danxia

*An extraordinary landscape of brightly coloured sedimentary rocks and sculpted landforms in northern China*

E. Asia

△ **RAINBOW RIDGES**
The rock strata of the Zhangye Danxia are predominantly reddish in colour, but also include some orange, yellow, pale blue, and green-hued layers.

Stretching across 500 square km (200 square miles) in Gansu Province, China, the Zhangye Danxia Landform Geopark is known for the unusual colours and shapes of its sedimentary rocks. With its undulating, mainly red-hued ridges, parts of it resemble a sea of fire with rolling waves. Danxia is a broad term used to describe the types of geological formations seen in this area, and Zhangye is a city close to the park.

## Source of colour

From around 80 million years ago, sedimentary rocks began forming in the area through the deposition of sediments in lakes. Variations in the mineral composition of the sediments account for the range of rock colours seen today. Starting from about 20 million years ago, the whole area was uplifted and the sedimentary layers tilted as a result of tectonic stresses. Subsequently, the processes of weathering and erosion have resulted in the spectacular sights visible today.

▷ **ROCK FORTRESS**
To the west of the Zhangye Geopark lies the Binggou Danxia. It features sandstone outcrops in shapes that resemble manmade fortresses.

The Danxia's coloured rock layers took **tens of millions** of years to form

E. Asia

# Mount Fuji

*An iconic, beautifully symmetrical volcano that can be seen from Tokyo on a clear day*

Rising to a height of 3,776 m (12,388 ft), and located 90 km (55 miles) southwest of Tokyo, Mount Fuji is the highest mountain in Japan. Mostly formed 11,000–8,000 years ago, its attractive cone is one of the country's most famous symbols, often celebrated in art and literature, as well as visited by sightseers and climbers. Although it is classified as an active volcano, Fuji's last eruption – which covered a huge area to its east in ash – was in December 1707. Since then, there have been no further signs of volcanic activity.

### ▷ BLOSSOM SEASON
The spring flowering of cherry trees is admired so much in the Mount Fuji area that there is an official cherry blossom viewing season, called *Hanami*.

### △ CELEBRATED CONE
Fuji's cone has a base around 20 km (12 miles) wide and is almost perfectly symmetrical. It is snow-capped from October to June.

### INSIDE MOUNT FUJI

Fuji is a stratovolcano - a steep-sided volcano composed of layers of solidified lava alternating with ash and other ejected material. The successive layers have built the volcano into a tall cone.

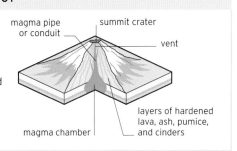

magma pipe or conduit — summit crater — vent — layers of hardened lava, ash, pumice, and cinders — magma chamber

# Kliuchevskoi

*The highest active volcano outside the Americas, situated on Russia's Kamchatka Peninsula and regarded as sacred by the native population*

N. Asia

### ▽ LAVA RIVER
In this photo taken in 2015, a lava flow runs down Kliuchevskoi while plumes of steam, ash, and other volcanic gases escape into the atmosphere.

Also known as Klyuchevskaya Sopka, Kliuchevskoi is one of the most active as well as the highest of a dense chain of more than 160 volcanoes that run down the Kamchatka Peninsula. The volcanoes are the result of the Pacific and North American plates subducting (sliding under) the small Okhotsk Plate, on which Kamchatka sits. This process provoked the formation of magma at depth, which moved upwards to create the volcanoes.

## Dangerous beauty
Originally formed around 6,000 years ago, Kliuchevskoi has been erupting almost continually since at least the late 17th century, including significant eruptions in 2007, 2010, 2012, 2013, and 2015. Due to its frequent eruptions, the peak has not been climbed often. A beautiful, symmetrical, snow-capped volcano, Kliuchevskoi is regarded by some of Kamchatka's native people as the place where the world was created.

# The Tengger Volcanic Complex

*An unearthly looking group of volcanic cones inside an ancient caldera, all set within lush tropical vegetation*

SE. Asia

The Tengger Volcanic Complex is part of a national park on the Indonesian island of Java. The complex lies within a large caldera, the remains of an enormous volcano that disintegrated in a cataclysmic eruption over 45,000 years ago. Over the past few thousand years, several new volcanic cones have grown from its floor. Of these, the most easily recognized – as its top has been blown off, leaving a wide summit crater – is Mount Bromo.

## Recent eruptions

Mount Bromo is the youngest, most active, and most frequently visited of the volcanoes. It has erupted repeatedly since 1590. Outside the caldera is a larger volcano, Semeru, which also regularly belches out huge clouds of steam and smoke. At 3,676 m (12,060 ft), Semeru is Java's highest mountain.

### ▽ DAWN PERSPECTIVE
Here, part of the Tengger Caldera's rim is visible, with a hamlet - a base for exploring the complex - perched on it. On the right is the mist-covered caldera.

### △ TENGGER PANORAMA
In this view, the volcanic cones comprising the Tengger Complex can be seen at the centre, with smoke emanating from Mount Bromo. In the distance is Mount Semeru.

## TENGGER AND SEMERU

The Tengger Caldera is roughly square in shape. Its flat, sand-filled floor is called the Sand Sea. Within the caldera are five overlapping stratovolcanoes. A few kilometres to the south is the large active stratovolcano Semeru.

Tengger Volcanic Complex

Mount Bromo

rim of Tengger Caldera

sand sea

Mount Semeru

The **caldera** within which **the Tengger complex** sits is **16 km (10 miles)** wide

# Mount Pinatubo

*A brooding volcano in the Philippines with a history of extraordinarily violent eruptions*

SE. Asia

△ **SUMMIT LAKE**
Since 1992, Pinatubo's 2.5-km
(1½-mile) wide caldera has been filled
with a turquoise lake. With a depth of
600 m (2,000 ft), it is the deepest lake
in the Philippines.

The volcano Pinatubo on the island of Luzon in the Philippines does not erupt frequently, but when it does, it explodes with extreme ferocity. It has produced some exceptionally large eruptions at intervals of anything from 500 to 8,000 years.

## Pinatubo in 1991 and today

The most recent of Pinatubo's eruptions occurred in June 1991, when a series of explosions sent a vast quantity of rock and ash into the atmosphere and produced pyroclastic flows (avalanches of hot gas and ash) that incinerated land up to 17 km (11 miles) away. More than 800 people died, and thousands more were displaced. Part of the top of the volcano collapsed, leaving a summit caldera that is now filled with a tranquil lake. Ash darkened the sky for many days, and fine particles of dust and tiny droplets of sulphuric acid spread all around Earth. These droplets blocked out sunlight, lowering global temperatures by 0.5°C (1°F) for a year.

Even today, some of the effects of the eruption can be seen in the form of vast, deep fields of ash all around the volcano. Pinatubo is closely monitored for any further activity.

### THE 1991 ERUPTION OF PINATUBO

Pinatubo's 1991 eruption is the largest, most violent volcanic event of the past 100 years in terms of the volume of ash, lava, pumice, and other materials ejected. Here, it is compared to four other massive eruptions since 1950.

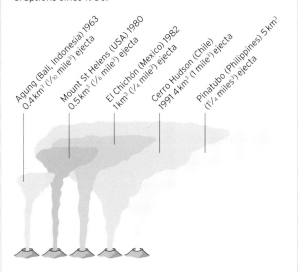

Agung (Bali, Indonesia) 1963
0.4 km³ (¹⁄₁₀ mile³) ejecta

Mount St Helens (USA) 1980
0.5 km³ (¹⁄₆ mile³) ejecta

El Chichón (Mexico) 1982
1 km³ (¹⁄₄ mile³) ejecta

Cerro Hudson (Chile)
1991 4 km³ (1 mile³) ejecta

Pinatubo (Philippines) 5 km³
(1¹⁄₄ miles³) ejecta

△ **MUSHROOM CLOUD**
During Pinatubo's 1991 eruption, a
cloud of volcanic ash was launched
34 km (21 miles) into the air, forming
a mushroom shape. This is far higher
than most commercial airliners fly.

# The Fedchenko Glacier

*The longest glacier in the world outside of the polar regions, flowing through valleys in the Pamir Mountains of Tajikistan*

C. Asia

The Fedchenko is an exceptionally long, narrow glacier, whose length has been measured from satellite imagery at 77 km (48 miles) – slightly longer than some other very long non-polar glaciers, such as the Siachen Glacier and the Baltoro Glacier (see opposite) in the Karakoram mountain range. The Fedchenko Glacier covers a total area of about 700 square km (270 square miles) and has dozens of tributaries. It is located in such a remote part of the world that it remained undiscovered until 1878 and was not fully explored until 1928. Subsequently, it was named after Alexei Fedchenko, a Russian explorer well known for his travels in Central Asia.

▷ **NARROW COURSE**
Here, the lower part of the glacier snakes diagonally across the landscape from bottom left to top right. A few tributaries can also be seen.

# The Yulong Glacier

*One of the best known glaciers in China, located on Jade Dragon Snow Mountain*

E. Asia

The Yulong Glacier is one of 19 small glaciers that adorn the flanks of a massif (a compact group of mountains) in Yunan Province, southwestern China. This massif is known as Mount Yulong or Jade Dragon Snow Mountain. The Yulong Glacier is a cirque glacier (one occupying a depression high up on a mountain) with an extra tongue of ice, called a hanging glacier, below it. It is one of the most accessible glaciers in China since it lies near the top of a cable car route. Unfortunately, the glacier is shrinking, and at its current rate of ice loss it is expected to disappear in about 50 years' time.

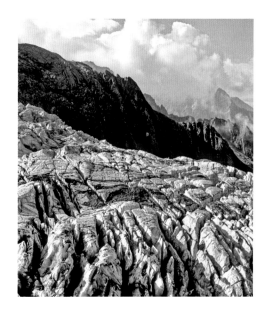

▷ **CRACKED SURFACE**
The steeply descending "hanging" part of the Yulong Glacier, situated at an altitude of about 3,650 m (12,000 ft), has a heavily crevassed, broken up surface.

△ **WHITE FLURRY**
A powder snow avalanche falls from one of the peaks in the Gasherbrum group onto the Baltoro Glacier, which is in the foreground.

C. Asia

# The Baltoro Glacier

*A major glacier running through part of the Karakoram mountain range, with glorious views of many of the world's highest peaks*

At 63 km (39 miles) long, the Baltoro Glacier runs roughly west to east through Pakistan-administered Kashmir. It forms part of the route for mountaineers on their way to climb either K2, the world's second highest mountain, or a collection of the world's highest peaks known as the Gasherbrum group.

## Peak views

The surface of the glacier is rough and uneven, broken by crevasses and covered by seracs and rock debris. Where the Baltoro comes together with one of its largest tributaries, the Godwin-Austen Glacier, is a broad area known as Concordia. This spot offers mountaineers magnificent views, in different directions, of K2 and three other peaks rising over 8,000 m (26,000 ft).

◁ **ICE CONES**
Huge pinnacles and cones of ice called seracs – some of them as large as houses – can be seen all over the glacier.

### HOW CREVASSES AND SERACS FORM

Crevasses are caused by tensional stretching of a glacier as it passes over humped or unevenly sloping bedrock. They can be transverse or longitudinal but sometimes the two types combine, producing pillars or pinnacles of ice – seracs – where they intersect.

direction of stretching
crevasse at right angle to flow
crevasse along direction of flow
serac between crevasses
flow direction

**TRANSVERSE CREVASSES**  **LONGITUDINAL CREVASSES**  **SERACS**

**Six** of the world's **17 highest peaks can be viewed** on a trek up the glacier

C. Asia

# The Biafo Glacier

*A long, near-straight glacier in the Karakoram Mountains of Pakistan*

The Biafo Glacier is a 63-km- (39-mile-) long glacier that runs northwest to southeast in a western region of the Karakoram (see p.225) and is heavily covered with rock debris on its lower parts. It is possible to walk along the entire length of the glacier, up to its highest point, at 5,128 m (16,824 ft). At its top end, the Biafo meets another highway of ice, the Hispar Glacier. Nearby is one of the largest basins of snow outside polar regions, known as Snow Lake. Occasionally, evidence of wildlife can also be seen in the vicinity of the glacier, including ibex and markhors (types of mountain goats) or, much more rarely, snow leopards.

### SOURCES OF DEBRIS IN GLACIERS

Rocks that have come loose from nearby mountains are the main source of debris in or on a glacier. Dust is also continuously blown onto the glacier surface, and rocks are "plucked" from the bedrock by becoming frozen onto the ice.

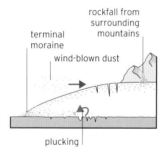

rockfall from surrounding mountains

terminal moraine

wind-blown dust

plucking

#### △ ICY HIGHWAY
The surface of the lower part of the glacier contains deep crevasses, meltwater streams, rock fragments, and large piles of rocks. Lines of high mountains with tributary glaciers run down both sides.

**CROSSING THE CHASM**
A member of an Everest expedition traverses a bridge constructed from aluminium ladders above a large crevasse in the Khumbu Icefall.

S. Asia

# The Khumbu Glacier

*A large, partly rock-covered glacier in Nepal, famous mainly because it lies on some popular routes for approaching and climbing Mount Everest*

The Khumbu Glacier has two main parts (see panel, below). Its upper part, on the western flank of Mount Everest (see pp.226–27), is a colossal icefall – a steeply descending, deeply crevassed mass of broken ice. With its source area at 7,600 m (24,900 ft), the ice moves downhill relatively fast, at about 1 m (3¼ ft) per day. As a result, large crevasses can open up with little or no warning. Smaller ones can become buried under the snow, creating treacherous snow bridges through which unwary climbers can fall. Crossing the icefall is a dangerous exercise, but much of the route up and down is along ladders and ropes that are fixed each year by experienced guides.

▽ **SPIKY SHARDS**
Seracs (towers of ice) commonly develop on the Khumbu Glacier and can fall at any moment, sending large blocks of ice tumbling down the steep slopes.

▷ **MORNING DESCENT**
A line of climbers can be spotted descending the icefall. Early morning, when the ice is at its most stable, is a favoured time for moving on it.

# Crevasses in the Khumbu Icefall can be as deep as **50 m (165 ft)** or more

### THE KHUMBU GLACIER IN PROFILE

The glacier has two parts: the steeply declining, chaotically crevassed Khumbu Icefall at its top end, which forms part of one of the main routes to Everest; and the gently inclined lower part, which is covered in rock debris and descends down a long, straight-sided approach valley.

Khumbu Icefall
5 km (3 miles) long

7,600 m
(24,900 ft)

rock-covered valley glacier
10 km (6 miles) long

4,900 m
(16,076 ft)

# The Dead Sea

*An extremely salty Middle Eastern lake, with rapidly receding shorelines marking Earth's lowest point on dry land*

W. Asia

△ **SHORELINES OF THE PAST**
Concentric terraces shows how the shorelines of the Dead Sea have receded over time. As the water levels continue to fall, and the lake shrinks, it becomes saltier.

A landlocked salt lake on the border between Israel and Jordan, the Dead Sea sits in a depression, called a graben, between two tectonic plates in the Jordan Rift Valley. At around 430 m (1,410 ft) below sea level, its shores are the lowest point on Earth's land surface. With a salinity of about 35 per cent – up to 10 times that of typical ocean water – the Dead Sea is one of the saltiest bodies of water on Earth and so can support very few life forms, hence its name.

▽ **SALT SUPPLY**
Crystallized salt is found all over the shores of the Dead Sea. Salt from the lake has long been valued for its culinary and medicinal uses.

## THE ELEVATION OF THE DEAD SEA

The low elevation of the Dead Sea is due to its position between two tectonic plates. A lake first formed when the land to the west rose up and cut it off from the Mediterranean Sea. As the plates moved apart, the floor of the lake gradually dropped to a level below that of the sea.

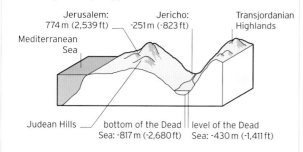

Jerusalem: 774 m (2,539 ft)
Jericho: -251 m (-823 ft)
Transjordanian Highlands
Mediterranean Sea
Judean Hills
bottom of the Dead Sea: -817 m (-2,680 ft)
level of the Dead Sea: -430 m (-1,411 ft)

### Shrinking sea

The Dead Sea is mainly fed by the River Jordan, and although it has no outflow, in recent years its surface level has fallen rapidly, largely due to water diversion for commercial use. Receding shorelines have caused the lake to become separated into two basins by the extension of what was previously the Lisan Peninsula. The southern basin has since been divided into a series of evaporation pools for the extraction of salt.

The **level** of the Dead Sea is **dropping** by **more than 1 m (3 ft)** a year

# The Ob-Irtysh

*A major river system of northwest Asia, crossing through some of the most remote and least populated regions on Earth*

NW. Asia

The Ob–Irtysh river system is a combination of two major Asian rivers that have their sources on different sides of the Altai Mountains (see p.224) and flow north, through the Siberian lowlands, to the Arctic Ocean.

## Freezing and floods

Measuring a total of 5,568 km (3,460 miles) from the source of the Irtysh to the mouth of the Ob, the river system is the longest in Siberia. Its catchment area is similar in size to the huge Mississippi–Missouri basin (see p.53). In its lower reaches, the river becomes braided across a vast, swampy plain that is subject to seasonal freezing and floods. The Ob–Irtysh empties through the Gulf of Ob, the world's longest estuary.

▷ **ICE FLOES**
The Ob-Irtysh starts to freeze in the autumn from its mouth, in the north, southwards. For more than half of the year, it is icebound and unnavigable.

N. Asia

# The Lena

*An epic trans-Siberian waterway, draining one-fifth of Russia's territory*

One of the three great Siberian rivers, along with the Ob-Irtysh and Yenisei, the Lena flows 4,400 km (2,700 miles) from its source in the Baikal Mountains, just west of Lake Baikal (see pp.238–39), across Siberia to its mouth in the Laptev Sea (an arm of the Arctic Ocean). The Lena has one of the largest basins in the world, draining one-fifth of Russia's territory. In its lower reaches, it is free of ice for only 4–5 months each year, following seasonal flooding.

At its mouth, the river forms the largest delta in the Arctic and the second largest on Earth. Spanning 280 km (170 miles) and pushing out 120 km (70 miles) into the ocean, it comprises lakes, channels, sand bars, and islands, with extensive peat bogs. In winter, it is a harsh wilderness, but in spring it supports millions of migratory birds.

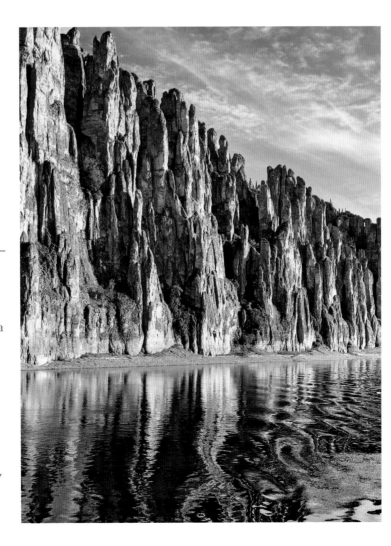

◁ **THE LENA PILLARS**
The limestone of these rock pillars lining the banks of the Lena in the Sakha Republic was formed in a sea basin around 500 million years ago.

NE. Asia

# Lake Baikal

*Earth's deepest freshwater lake, containing more water than any other and home to many endemic species*

Lake Baikal is a Russian lake situated to the south of the Central Siberian Plateau, near Mongolia. The largest freshwater lake on Earth by volume, it contains around 20 per cent of the world's total surface fresh water – similar to that of North America's Great Lakes combined (see pp.50–51). At 636km (395 miles) long and up to 79km (49 miles) wide, Baikal has the largest surface area of any freshwater lake in Asia. Its waters have a maximum depth of 1,637m (5,370ft), making it the deepest body of fresh water on Earth.

## Ancient and dynamic

Also the oldest lake in the world, Lake Baikal formed in a deep rift valley – where the Earth's crust is pulling apart – more than 25 million years ago. The southwest-to-northeast fault along which the lake runs is still active, and the lake is widening by more than 2cm (¾in) each year. Lake Baikal's water is renowned for its purity and clarity. Well-oxygenated and well-mixed throughout its depths, it supports a wide variety of plants and animals, and more than 80 per cent of these species are thought to be found only in Lake Baikal.

## HALF-GRABEN BASINS

The type of basin in which Lake Baikal sits is known as a half-graben. It was formed through the vertical movement of the Earth's crust along a fault, with the fault bordering just one side. In a full graben, a basin is formed by a depressed block of land bordered on both sides by faults.

sedimentary rock
upward movement
footwall
border fault

lake water
bottom of basin
downward movement
hanging wall

## ▷ LATE FREEZE

Lake Baikal is completely covered with ice for 4-5 months a year. The huge volume of water in its basin means the surface freezes later in the Siberian winter than the region's other lakes. Peschanaya Bay and Baikal's other bays are the first parts of the lake to freeze over.

## ▶ THREE BASINS

Lake Baikal is divided into three basins: North, Central, and South. The deepest parts of the lake are in the Central and South basins. Both the lake's main source and its single outlet are in the South Basin.

**Single outlet**
Most lakes have a single outflow into a river system. Lake Baikal drains into the Angara River, a tributary of the Yenisei.

*South Basin*

△ FRESHWATER SEAL
The endemic Baikal seal is the only exclusively freshwater seal species in the world. One of the smallest seals, it grows up to 1.4m (4½ft) in length.

**Protected area**
Several areas around the lake are protected. The Baikal Nature Reserve by the southeastern shore is part of the World Network of Biosphere Reserves.

**Western side**
The western side of the lake is flanked by high mountain ranges. The Baikal Mountains rise sharply along the northwestern shore and are the source of the Lena River (see p.237).

**Upper Angara River**

*North Basin*

**Eastern side**
The eastern shoreline slopes more gently than the west. The Barguzin Range is covered in larch forest and features one of the lake's nature reserves on its western slopes.

**Svyatov Nos Penisula**

**Lake Arangatuy**

*Central Basin*

**lake surface**

KM MILES

Depth below lake surface

0 — 0
2 — 1
4 — 2
6 — 4 — 3
8 — 4 — 5

bottom layer of sediment laid down 16-4 million years ago

sediment laid down 4-1.7 million years ago

top layer of sediment laid down from 1.7 million years ago to present

fault line    bedrock

△ **LARGEST ISLAND**
Of Lake Baikal's 27 islands, Olkhon is the largest by far. At 72 km (45 miles) long and 21 km (13 miles) wide, it is the fourth largest lake-bound island in the world.

△ **DEEP SEDIMENTS**
In parts of Lake Baikal's Central Basin, the sediments on the lake bed are up to 7 km (4 miles) thick. Laid down over millions of years, they contain a record of climatic variation much valued by scientists.

◁ **MAJOR INFLOW**
Most inflow into lakes is surface water that carries a lot of sediment. The Selenga River, Lake Baikal's main inflow, forms one of the world's largest inland deltas as it enters the lake.

Lake Baikal's water is **so clear** that it is possible to see objects **40m (130ft) below the surface**

# The Indus

*One of Asia's major river systems, crossing the Himalayas
and running the entire length of Pakistan*

W. Asia

One of the greatest Asian rivers, the Indus is about
3,180 km (1,980 miles) long, and its catchment basin
drains an area of more than 1.1 million square km
(425,000 square miles).

From its source near Lake Manasarovar on the
Tibetan Plateau, the Indus flows northwest through
the disputed territory of Jammu and Kashmir, turns
southwest to enter Pakistan, and passes through a series
of huge canyons near the Nanga Parbat Massif on
its journey across the Himalayas. Emerging from the
mountains as a fast-flowing river, the Indus descends
onto the Punjab Plain, where it is joined by some of its
major tributaries and begins to move slowly across a
vast floodplain. During the monsoon season, the Indus
can grow to several kilometres wide in places. As it
approaches the Arabian Sea, the river branches into the
many distributaries that form the Indus River Delta.

▷ RIVER CONFLUENCE
In a mountainous region in its upper reaches,
the Indus (bottom) meets the Zanskar, a major
tributary, near Nimmu in northern India.

# The Ganges

*One of the world's great waterways, a holy river and
a place of pilgrimage*

S. Asia

The Ganges is the spiritual and physical lifeblood of
India. Considered sacred and worshipped by Hindus,
it is thought to support up to one-tenth of the world's
population. After the Amazon and the Congo, it is the
third largest river on Earth, measured by discharge.

## Winding journey

The Ganges flows more than 2,500 km (1,600 miles)
across the Indian subcontinent. It forms in the north
Indian state of Uttarakhand after the confluence
of several ice-fed Himalayan rivers. After passing
through narrow mountain valleys, it emerges on the
vast Indo-Gangetic Plain to the east of Delhi. From
here, it slowly meanders across a wide alluvial plain,
passing many centres of holy pilgrimage including
Allahabad and Varanasi, towards Bangladesh, where
it is joined by the Brahmaputra and Meghna rivers.
The river system subsequently fans out into the largest
delta system on Earth, depositing the greatest
sediment load of any river into the Bay of Bengal.

△ BROAD FLOODPLAIN
The middle course of the Ganges winds
sluggishly across a vast floodplain. Over a
distance of more than 1,600 km (1,000 miles),
the river falls by only about 180 m (590 ft).

△ COASTAL NETWORK
Seen here in dark
green, the Sundarbans
ecoregion stretches
inland about 80 km
(50 miles). It is largely
bordered to the north
by farmland.

▷ EXPOSED ROOTS
These black mangrove
trees have developed
aerial roots called
pneumatophores,
which project into
the air during low tide.

# The Sundarbans

*A vast stretch of mangrove forest and swampland lying on Earth's largest delta*

S. Asia

Stretching more than 250 km (160 miles) across the Bangladesh–India border, where the Ganges Delta merges with the Bay of Bengal, the Sundarbans is a complex network of intertidal waterways, estuaries, mud flats, and islands.

## MANGROVE ZONES

There are three main types of mangrove trees, with their own adaptations to particular tidal zones. Red and black mangroves, which live in the tidal zone, have specialized roots to absorb oxygen from the air. White mangroves, common in higher and drier areas, lack these adaptations.

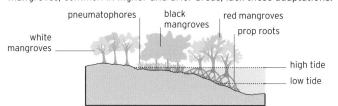

pneumatophores

black mangroves

red mangroves

white mangroves

prop roots

high tide

low tide

### Mangrove forest

The main habitat of the inland parts of the Sundarbans is seasonally flooded freshwater swamp forest consisting of broadleaf trees. The seaward side is dominated by the world's largest continuous tract of mangrove forest. One of the most common mangroves, known as *sundri* or *sundari*, is thought to have given the region its name.

An ecosystem rich in wildlife, the Sundarbans is home to many rare and threatened species, including the Bengal tiger and the Ganges river dolphin.

▽ **VARIED DIET**
As well as eating fish, the black-capped kingfisher varies its diet with prawns and small crabs.

# The Yellow River

*One of Asia's longest rivers, and the world's muddiest, the*
*"Mother River" of the Chinese people*

E. Asia

The most silt-laden river on Earth, the Yellow River takes its name from its coloration by the fine, windblown sediment, known as loess, that it carries along its lower course. It is the second longest river in China, after the Yangtze, and one of the longest in the world. After rising high on the Tibetan Plateau, the Yellow River follows an arching route eastwards across the northern plains of China to the Yellow Sea, a journey of 5,460 km (3,393 miles).

## Cradle to grave

Known as both China's Mother River and the Cradle of Chinese Civilization, the Yellow River historically sustained one of the country's most fertile and productive regions. However, the river basin is extremely prone to flooding, and a series of devastating floods has also earned the river the names the Ungovernable and China's Sorrow (see panel, right). As a result, much of the lower course of the Yellow River has been ramparted by levees and many of its tributaries have been dammed to control its flow.

△ **FIRST BEND**
The Yellow River's upper course in Sichuan Province forms a huge S-shape, which is known as the First Bend.

◁ **WIDE-RANGING TOAD**
The Asiatic toad is common in China, living in a range of high-humidity habitats that include valleys and floodplains of the Yellow River. It eats beetles, bees, ants, and molluscs.

# The Yangtze

*Asia's longest river, flowing across eastern China from the Tibetan Plateau*
*to the East China Sea; China's most commercially important river*

E. Asia

Known in China as the Chang Jiang – "Long River" – the Yangtze is the longest river in Asia and the third longest in the world, after the Nile and the Amazon. From its glacier-fed source in the Tanggula Mountains on the Tibetan Plateau, to its mouth in the East China Sea, it flows 6,300 km (3,900 miles). The catchment area of the Yangtze covers one-fifth of the area of China and is home to one-third of the country's population.

## Epic journey

On its journey across eastern China, the Yangtze flows through steep valleys, towering gorges, lake-strewn plains, and farmland, before forming a delta, where the megacity of Shanghai lies. Since 2006, the river's middle course has been regulated by the world's largest hydroelectric dam, the Three Gorges Dam, which has created a reservoir 600 km (370 miles) long. The Grand Canal – the longest human-made waterway on Earth at 1,700 km (1,100 miles) – links the lower section of the Yangtze to the Yellow River and Beijing to the north.

△ **GOLD DUST RIVER**
In its upper course, the river passes through the mountainous Yunnan Province. Here the Yangtze is known as the Jinsha Jiang, "Gold Dust River", due to the alluvial gold panned from its waters.

## CHANGING COURSES

The Yellow River has changed course many times over the centuries, sometimes as the result of human activity. It moved 26 times and flooded nearly 1,600 times between 602 BCE and 1938 CE, resulting in the loss of millions of lives.

2278 BCE
602 BCE
1853 CE
1324 CE
1938–1947 CE

Yellow River | Yangtze

**KEY**

〰 Current channel
〰 Former channel
▨ Former shoreline

---

SE. Asia

# The Mekong

*Southeast Asia's greatest river system, passing through six different countries*

△ **DELTA FLOODPLAINS**
This aerial view shows part of the huge Mekong Delta in southern Vietnam. The flat floodplains in the south contrast with the hilly areas in other parts of the delta.

Like the Yellow River and the Yangtze, the Mekong has its source on the Tibetan Plateau. It runs in a southeast direction through western China, Myanmar, Laos, Thailand, Cambodia, and Vietnam, crossing some national borders and forming others. At its mouth, in the South China Sea, it forms a delta just south of Ho Chi Minh City. With a length of about 4,350 km (2,700 miles), the Mekong is the longest river in Southeast Asia. The very long river basin includes many climatic zones and a diverse range of habitats including upland plateaus, forests, savannas, grasslands, wetlands, and mangroves. As a result, it is an area of extremely high biodiversity.

▷ **MEKONG GIANT**
The Mekong is home to many large fish species. The giant freshwater stingray can be over 5 m (16 ft) long and 2 m (6 ft) wide, and has the longest stinging spine of any stingray.

stinging spine used in defence

**INSIDE A DOLINE**
Explorers are dwarfed by the first doline in the cave system, Watch out for Dinosaurs - it is so big a skyscraper could fit inside it.

SE. Asia

# Hang Son Doong

*A recently discovered limestone cave system of immense scale, situated within jungle-covered mountains in a remote part of Vietnam*

One of the largest cave systems discovered on Earth, Hang Son Doong is up to 250 m (820 ft) high, 200 m (660 ft) wide, and stretches a total of 9 km (5½ miles). Located near the Laos border in central Vietnam, it is part of a network of around 150 caves within the Annamite Mountains. First discovered in 1991, Hang Son Doong was unexplored until 2009.

### Mountain river cave

Meaning "mountain river cave", Hang Son Doong has been carved out of some of the oldest limestone in Asia by the fast-flowing Rao Thuong River over the last 2–5 million years. Giant openings (dolines), created as massive sections of the ceiling collapsed, have allowed areas of vegetation to develop, including trees up to 30 m (100 ft) tall. These are dwarfed by some of the stalagmites found within the cave, which, at more than 75 m (245 ft) high, are the largest on Earth.

### CHAMBERS AND PASSAGES

The Hang Son Doong system is straight, relatively flat, and lacks side passages. This is due to the fact that it has formed along a fault in the limestone. It has many discrete sections and remarkable features, often with evocative names. These include the Hand of Dog, a huge, dog-paw-shaped stalagmite, and the two forested areas, Watch out for Dinosaurs and the Garden of Edam.

## Hang Son Doong is **so large** that it has its **own microclimate,** including **wind and clouds**

◁ **EARTHLY TREASURES**
Calcite deposits called cave pearls are found in parts of the cave. Usually less than 1 cm (½ in) wide, here they are the size of oranges.

△ **GREEN COVER**
Around the borders of the open sections, where dim light reaches the ground, colonies of ferns and algae cover terraces of limestone.

E. Asia

# South China Karst

*One of the world's largest and finest examples of a karst landscape, featuring a variety of dramatic rock formations*

Covering an area of about 500,000 square km (190,000 square miles), mostly within the provinces of Guizhou, Guangxi, Yunnan, and Chongqing, South China karst is the world's largest karst terrain – a landscape characterized by an array of weathered limestone formations. The diversity, number, and spectacular nature of the geological features make South China karst the greatest example of a humid tropical and subtropical karst landscape on Earth. Since 2007 it has been a UNESCO World Heritage Site.

### Caves, towers, and pinnacles

South China karst is perhaps best known for its isolated stone towers, often rising more than 100 m (330 ft) from a broad mosaic of farmland, and its clusters of cone-shaped hills covered in lush, cloud-draped

▷ **TRAINED TO FISH**
Cormorants are used in a traditional method of fishing on the rivers that cross South China karst. The great cormorant is a favoured species.

forests (see pp.248–49). In fact, the region is regarded by geologists as being home to the defining examples of formations such as tower karst (fenglin), cone karst (fengcong), and pinnacle karst (shilin), which creates dense stone forests (see below). It also features many giant sinkholes (or dolines), gorges, natural bridges, and table mountains. Beneath the ground there are extensive cave systems containing subterranean rivers and vast caverns rich in stalactites, stalagmites, and other cave deposits.

### Mature landscape

The limestone rock in which South China karst is formed was laid down on the ocean floor over a long period before being uplifted. This produced an exceptionally thick, strong, and horizontal bedrock from which the massive and relatively stable landforms have been carved. Although the region's warm and wet climate has enabled the chemical weathering of the limestone to occur at a relatively rapid rate (see pp.248–49), it has been estimated that the fengcong and fenglin karsts of Guizhou and Guangxi have been evolving over a period of 10–20 million years. The result is what is often described as the ultimate mature karst landscape.

# One of the region's **vast caves** houses a village of more than **100 people**

▷ **STONE FOREST**
These tightly bundled limestone pinnacles forming stone forests near the village of Shilin, in Yunnan Province, are considered classic examples of what has become known as shilin karst.

### △ COLLAPSED CAVE

Moon Hill, a huge stone arch near Yangshuo in Guangxi, is all that is left of what was once a limestone cave. It is now a popular climbing site.

## SOUTH CHINA KARST TYPES

Often viewed as a precursor to the development of fenglin karst's high towers, fengcong karst dominates areas like the Yunnan-Guizhou plateau in the north and west of the region, where the bedrock has been more recently uplifted and subjected to a new wave of erosion. Fenglin karst is mainly found in areas that have been stable over long periods, such as the lowlands of Guangxi in the south and east.

Yangtze

Chongqing •

fengcong karst dominant

Wu

• Guiyang

Hongshui

• Kunming

Guilin •

Yu

Yangshuo •

fenglin karst dominant

Nanning •

### △ BROAD PLAINS

Although the plains between the towering formations of South China karst have been farmed for hundreds of years, the relatively thin soils and lack of water retention in some areas present major challenges to farmers.

# The formation of a
# KARST LANDSCAPE

Karst landscapes are formed by the weathering of a terrain's underlying carbonate rock, such as limestone or dolomite. Through a process known as dissolution, rain, surface water, and groundwater made acidic by carbon dioxide in the atmosphere and organic matter in soil start to dissolve the rock down cracks and faults, and along its bedding planes. As the rock degrades over time and the cracks become larger, a variety of surface features are produced and an underground drainage system may start to form.

## Underground erosion

Percolating water may create subterranean rivers that eventually carve out extensive cave systems. Sinkholes, or dolines, a defining feature of karst landscapes, either develop gradually as surface openings enlarge or are formed suddenly by cave collapses. As they expand and merge, they create sunken regions, and any high ground left between them becomes linked cone-shaped hills, like the fengcong formations of South China karst. Over time, erosion by surface water can transform these hills into the isolated towers of fenglin karst. The formation of karst landscapes is helped by moist climates and lush vegetation – providing carbon dioxide as it decays – so the most developed karst terrains occur in tropical regions.

## The **bedrock** of South China karst **formed on the seabed** more than **250 million years ago**

▶ **SOUTH CHINA KARST**
This cross-section of a karst landscape combines many of the features present in South China karst. It represents a well-developed karst system that has been created under perfect geological and environmental conditions.

△ **CAVE DWELLER**
Caves provide shelter for some animals. Large colonies of great roundleaf bats roost in South China karst caves.

**Fenglin tower**
Fenglin towers are steep-sided and stand isolated on the flat karst plain.

land converted to agriculture

river meanders on surface

sinkhole

**Sediment**
A layer of sediment acts as a barrier between the river and the limestone, enabling the river to stay above ground.

**Foot cave**
Water streams from a cave, known as a foot cave, at the base of a conical hill.

**Losing stream**
A losing stream first runs along the surface before being captured by an underground drainage system.

**Gour**
A series of stone dams called gours, formed by mineral deposits, has created a tiered cascade pool.

## EVOLUTION OF KARST FEATURES

Karst features can be seen to evolve in a sequence from a flat karst plain to fengcong then fenglin karst. Once a drainage system is established, sinkholes get bigger and join together, producing the hills of fengcong karst. If surface water erodes their bases and the sunken areas between, they develop into the isolated towers of fenglin karst.

sinkhole captures surface water

drainage system established

**KARST PLAIN**

groups of hills are left

sinkholes merge to create sunken areas

**FENGCONG KARST DEVELOPS**

surface water creates plain

isolated tower created

**FENGLIN KARST FORMS**

◁ **TREE-COVERED PEAK**
Patches of monsoon forest cling to the steep and craggy limestone rock faces of even the tallest fengcong cones. The forests of the South China karst region are key habitats for several species of rare and endangered animals.

**Fengcong cone**
Fengcong karst is characterized by linked conical hills that share a common bedrock and form a continual terrain. The fengcong cones of South China karst reach more than 300 m (980 ft) high.

karst plain

remnant of old cave

△ **DEEP DEPRESSION**
Sinkholes, or dolines, are depressions in the ground often caused by the rock layer above a cave network suddenly collapsing. Sinkholes vary greatly in size, with some of those in South China karst reaching widths and depths of more than 600 m (2,000 ft).

**Limestone rock**
The limestone bedrock of South China karst is extremely thick and broken into blocks by bedding planes, joints, and faults.

**Fissure**
A fissure created by vertical faults in the limestone allows water from the surface to feed into the cave below.

submerged cave

dry cave, now without running water

◁ **CAVE DECORATIONS**
Over time, underground caves formed within karst terrains will often become decorated with calcite deposits, known as speleothems. Many of the ancient caves in South China karst are renowned for their elaborate speleothems, including stalactites and stone pillars.

**Underground river**
Fed by losing streams and water seeping through sinkholes and fissures, an underground river continues to carve out a passage in the limestone.

rock debris from collapsed roof

debris washed in by stream in past

# Ha Long Bay

*A large inlet into the Gulf of Tonkin, containing a scattering
of hundreds of spectacular karst islands*

SE. Asia

Covering an area of about 1,500 square km (600 square miles) off the coast of Vietnam, Ha Long Bay was created about 8,000 years ago when sea-level rise drowned an existing karst landscape. Karst consists of limestone blocks that have been partially dissolved by rainwater over tens of millions of years. In tropical areas, it typically creates conical hills and towers with internal caves and sinkholes (see panel, right). When the sea flooded into Ha Long Bay, these hills and towers were turned into some 1,600 islands, most of which are fringed by vertical cliffs. There are hundreds of caves within the islands, but none is long or deep because their extent is limited by the

sizes of the islands. Because of their precipitous nature, most of the islands are uninhabited by humans. However, several communities live on the water in floating houses, sustaining themselves by fishing and aquaculture. The bay's shallow waters are biologically highly productive and support hundreds of species of fish, molluscs, and crustaceans. Animals living on the islands include lizards, monkeys, bats, and many bird species.

◁ **PERILOUSLY LOW**
Cat Ba, a relatively large island on the southwestern edge of the bay, is home to the Cat Ba langur, a critically endangered monkey, with a population estimated at less than 100.

△ **INTRICATE SHORELINES**
Many of the islands in the bay have convoluted shapes. Some have been given distinctive names, such as Man's Head Island, Rocky Dog, and Fighting Cocks Islet.

Some caves have **tunnels** leading into **inland lakes** that have **no other access**

## CONES, TOWERS, AND CAVES

Much of the Ha Long Bay karst consists either of large, cone-shaped blocks (known as fengcong), or isolated, tower-like islands (called fenglin). Cave types include wave-eroded notch caves at sea level, horizontal foot caves just above sea level, and larger, much older caves called phreatic caves, which are well above sea level.

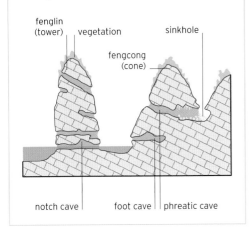

fenglin (tower) · vegetation · sinkhole · fengcong (cone) · notch cave · foot cave · phreatic cave

### ▽ TRANQUIL WATERS

A group of kayakers ply their way between karst islands that rise up to 200 m (660 ft) above the waters of the bay.

# Shiraho Reef

*A Japanese coral reef of world-class importance in terms of biodiversity, home to a unique type of coral*

SE. Asia

Stretching to 3 km (2 miles), Shiraho Reef first came to notice in the 1980s as an example of biodiversity, with some 120 species of coral and 300 fish species. It also contains the world's largest colony of an unusual species of coral known as blue coral. This belongs to a group known as octocorals, whose colonies mostly have flexible, branching skeletons. However, blue coral has a hard, rigid skeleton. Unfortunately, the reef's coral coverage has greatly decreased in recent years, but efforts are being made to conserve it.

### ◁ BRANCHING CORAL

The vertical plates of blue coral can form massive colonies. Despite its name, its colour varies from blue through turquoise and green to yellow-brown.

# Dragon Hole

*The world's deepest underwater sinkhole, located on a reef in the South China Sea*

SE. Asia

Dragon Hole is just over 300 m (980 ft) deep. Its great depth was realized only when it was measured for the first time in 2016 – previously, a sinkhole called Dean's Blue Hole in the Bahamas, which is 202 m (663 ft) deep, held the record for the world's deepest blue hole. Blue holes are sinkholes that form through erosional processes in blocks of limestone – while the limestone is on dry land and before the holes are flooded by sea-level rise. Local fishermen believe this is where the Monkey King found his golden cudgel in the 16th-century Chinese novel *Journey to the West.*

### ▷ BLUE EYE

Viewed from above, the hole looks dark blue due to its great depth compared to the light blue of the water on top of the reef.

SE. Asia

# The Krabi Coastline

*An area of southern Thailand notable for its fantastic karst formations*

The limestone rock of the Krabi coastline originally formed about 260 million years ago. At that time, a shallow sea covered what is now south Asia and slowly built up deposits of shells and coral fragments. These eventually formed layers of limestone, which were later pushed upwards and tipped over when India began to collide with Eurasia some 50 million years ago. Around Krabi and Phang Nga Bay to its north, chemical erosion of these limestone layers by rainwater (which combines with dissolved carbon dioxide to form carbonic acid), followed by a rising sea level, created a landscape of thousands of craggy karst islands and hills. These include a number of isolated towers that rise to heights of up to 210 m (690 ft).

## HOW THE COASTLINE FORMED

The Krabi coastline is a drowned karst landscape - one in which an existing landscape of limestone hills and towers became inundated by the sea (see panel, p.248).

group of karst hills   seawater

cave   isolated karst tower

△ **HIDDEN PARADISE**
At the island of Phi Phi Le, the drowning of the coastline has created a hidden lagoon and secret beach.

▷ **HANGING ROCK**
Near the entrance to a small cave, stalactites covered in tropical vegetation hang from an eroded mass of limestone.

# Andaman Sea Reefs

*Some of the largest continuous areas of coral reefs in south Asia, with an enormous variety of fish, corals, and other invertebrates*

SE. Asia

Situated in a northeastern corner of the Indian Ocean, the Andaman Sea bathes the coasts and coastal islands of Thailand and Myanmar (Burma) on its eastern side and the Andaman and Nicobar Islands on its western side. Most of the coral reefs here are fringing reefs, and altogether they cover an area of some 5,000 square km (2,000 square miles). The reefs contain a wide diversity of marine life, with over 500 fish species and 200 coral species recorded. The reefs and nearby islands are also important feeding and breeding grounds for endangered sea turtles.

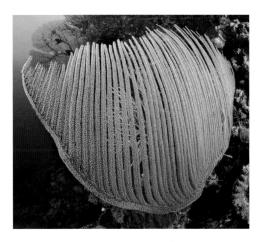

△ **COMB CORAL**
This soft coral is known for its distinctive comb-like shape. It has long, unbranched stems and can grow up to 1.5 m (5 ft) wide.

# Nusa Tenggara

*A chain of coral islands in southern Indonesia, fringed by reefs that contain a high diversity of marine life*

SE. Asia

Stretching eastwards from Bali, Nusa Tenggara includes several large islands such as Lombok, Sumbawa, and Flores, as well as 500 smaller islands. Many of the fringing reefs around these islands have barely been explored, but as a whole the area is known to include around 500 species of reef-building coral. Common animals seen on the reefs include eagle rays, humphead parrotfish, and various species of nudibranch (sea slug) and octopus.

△ **DENSE HABITAT**
Just beneath the surface, a mass of stony corals covers a fringing reef a few hundred metres off the coast of Flores in eastern Nusa Tenggara.

# Siberian Taiga

*A massive forest on the edge of the Arctic, ruled by conifers, where winter lasts most of the year*

N. Asia

The vast boreal forest known as Siberian taiga covers around 6.7 million square km (2.6 million square miles) from the Ural Mountains east to the Pacific Ocean. It stretches from northern Russia's Arctic fringes, where the taiga merges into the tundra, as far south as Mongolia. Like all boreal forests, Siberian taiga is primarily a conifer habitat, beginning as a mixture of spruce, pine, and larch in the west before giving way to unbroken swathes of larch in its eastern Russian and Mongolian reaches.

## Changing canopy

In more southerly areas, dense tree growth results in "closed canopy" forests, with little light penetrating to the moss-covered forest floor beneath. However, with increasing distance northwards, the forest cover becomes sparser, and lichen replaces moss on the ground. In more temperate parts, dominant conifers become mixed with broadleaf species such as birch, aspen, and willow.

▷ **PINPOINTING PREY**
The great grey owl mainly eats tiny rodents, which it locates with exceptional hearing - pinpointing them even under snow.

In the east, marsh and bogland support understorey shrubs such as cranberry and bilberry, while in the west, poor drainage and permafrost mean such areas are mainly given over to spongy, shallow bogs with no tree cover at all.

## Cold climate

Winter in the Siberian taiga lasts six to seven months, with the climate dominated by cold Arctic air. The average temperature range is broad – from -54°C (-65°F) in winter to 21°C (70°F) in summer – but extremes of -60°C (-76°F) and 40°C (104°F) are not uncommon, and the yearly average is below freezing.

# The taiga contains **as many trees as all the rainforests** combined

◁ **SNOW FOREST**
Snow lasts into May in the taiga, returning in September or earlier. However, snow-covered trees serve as an insulator, keeping the ground warmer than the air above the forest.

△ **CHANGING TERRAIN**
In many parts of the taiga, conifers cluster around rivers, where richer soils offer better growing conditions. Forested sections are divided by mountains at higher elevations.

**FROM TAIGA TO TUNDRA**

Taiga tree growth varies greatly according to latitude. In southern areas, trees are taller and more closely packed together, but density and height both diminish towards the north. Eventually, only a few small trees survive before the landscape changes entirely from taiga to tundra.

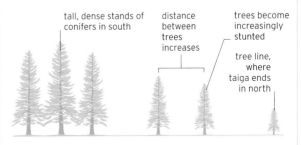

tall, dense stands of conifers in south

distance between trees increases

trees become increasingly stunted

tree line, where taiga ends in north

# Eastern Himalayan Forest

*A wealth of trees growing in the world's highest mountain range, supporting
a wide variety of plant and animal life*

S. Asia

The eastern Himalayas region is characterized chiefly
by temperate broadleaf evergreen and deciduous
forests. These cover an area of about 83,000 square km
(32,000 square miles) at altitudes of 2,000–3,000 m
(6,600–10,000 ft) and stretch from central Nepal
eastwards through Bhutan into northeast India.

## Plant treasury

This is an ecoregion rich in plant life, where tree type
varies according to altitude and geography. In the
temperate evergreen forests, oaks and other trees such
as magnolia and cinnamon grow alongside thickets
of rhododendrons – some areas, such as Bhutan, may
contain up to 60 species of rhododendrons alone.
Temperate deciduous forest areas are dominated
by maples, birches, and walnuts, which give way
to magnolias and maples in the wetter
eastern reaches of Nepal, together with
more tropical shrubs such as schefflera
and pockets of bamboo. At least 125
mammal species are found in the ecoregion,
including some, such as the Namdapha
flying squirrel, that live nowhere else in
the world. Threatened species include the
stump-tailed macaque and clouded leopard.

▷ **FLORAL DISPLAY**
Rhododendron blooms
dot south-facing
slopes, where they
flourish alongside oaks,
epiphytic orchids, ferns,
and mosses.

▽ **RAINBOW PLUMAGE**
The Himalayan
monal pheasant is
one of around 500
bird species that live
in the forest. The
iridescent male is the
national bird of Nepal.

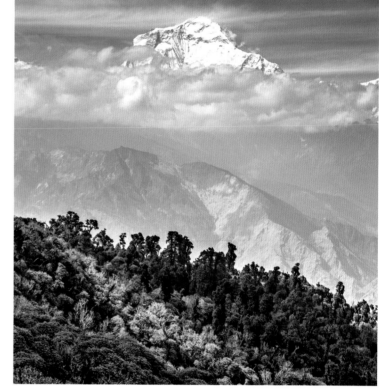

# Taiheiyo Montane Forest

*High and hilly, one of Japan's seven forest ecoregions, covering parts of three
islands with both broadleaf and fir trees*

E. Asia

Hardy deciduous trees mixed with firs and bamboo
characterize Taiheiyo montane forest. Running along
the Pacific side of Japan's main island of Honshu, as
well as Shikoku and Kyushu islands, this temperate
ecoregion covers an area of about 42,000 square km
(16,200 square miles).

In this forest, the climate is humid all year round,
but trees must withstand sharp seasonal temperature
changes. Winters are cold and snowy, with average
temperatures below freezing, but during summer,
they rise to 25°C (77°F) or more. Beech and fir are
dominant species, growing alongside maple and
oak, with an understorey of sasa, a type of dwarf
bamboo. The forest's seeds, nuts, and bark provide
food for mammals such as the Asiatic black bear and
sika deer, and birds such as the Oriental white-eye, all
of which help replenish the forest by dispersing seeds.

◁ **PINK TRAILS**
The Taiheiyo's beech
and maple trees are
joined in some areas
by other hardwoods
such as oaks and
Japanese cherries,
which stand out
against pockets of firs.

# Upper Yangtze Forests

*A landscape of evergreen and deciduous forests between the Yangtze and Yellow rivers, home to China's iconic giant panda*

E. Asia

The forests of the Upper Yangtze extend eastwards from the Hengduan Mountains and across the provinces of Sichuan and Shaanxi in south-central China. Covering an area of 390,000 square km (150,600 square miles), this ecoregion comprises three subregions: the evergreen broadleaf forests of the Sichuan Basin, the Daba Mountains evergreen forests, and the deciduous forests of Qinling.

The forests in the north of the region are made up of deciduous and mixed conifer species that thrive in the area's cooler, more temperate climate. The forests of the lower-lying Qinling Mountains support a dense bamboo understorey, providing food and shelter for species such as the rare giant panda. In the warmer Sichuan Basin, subtropical evergreen broadleaf trees flourish, and it is here that the dawn redwood, a deciduous conifer previously known only from fossil remains, was discovered growing in the 1940s.

▽ **AUTUMN LEAVES**

In autumn, the oak, walnut, and maple trees of the Qinling Mountain foothills create a colourful patchwork landscape. The area is also home to the primitive ginkgo, whose leaves turn golden in autumn.

### CLIMATE

Due to its higher altitude, the climate in the Upper Yangtze is generally temperate, with mild winters, but temperatures in the lower-lying Sichuan Basin can reach 29°C (84°F). Humidity in the entire region is generally high.

| Temperature | | Wolong, Sichuan Province | Rainfall | |
|---|---|---|---|---|
| °F | °C | | MM | IN |
| 104 | 40 | | 220 | 8½ |
| 86 | 30 | | 165 | 6½ |
| 68 | 20 | | 110 | 4¼ |
| 50 | 10 | | 55 | 2¼ |
| 32 | 0 | | 0 | 0 |

Jan Feb Mar Apr May Jun Jul Aug Sep Oct Nov Dec

KEY ▮ Temperature ▮ Rainfall

## Together, the Upper Yangtze forests contain **one-fifth** of all **China's mammal species**

▽ **BAMBOO DIET**

Once common in lowland areas, the giant panda has moved to the mountains due to human expansion. As bamboo makes up 99 per cent of its diet, it can only live where bamboo forests thrive.

SE. Asia

# Bornean Rainforest

*A threatened habitat governed by three nations, home to more than 15,000 plant species and one of the oldest and most biodiverse rainforests on Earth*

Shared by Malaysia, Brunei, and Indonesia, the island of Borneo is home to Asia's largest rainforest, which is also one of the world's most ancient at 130 million years old – around 70 million years older than the Amazon Rainforest (see pp.120–21). Borneo's rainforest is hugely biodiverse: while the island amounts to just 1 per cent of Earth's landmass, its rainforest contains around 6 per cent of the planet's plant and animal species.

## A wealth of hardwood

The plants include a family of tropical hardwood trees called dipterocarps, many of which reach heights of 60 m (200 ft). Most of the 600-plus species of dipterocarps grow in Southeast Asia at elevations up to 1,000 m (3,300 ft), and Borneo's lowland rainforests contain more of these trees than any other location. The island's 270 species of dipterocarp include the highly prized Borneo ironwood, whose wood is so dense it never needs treating. Borneo's rainforests, including those covering its more mountainous interior, are also rich in many other types of life. More than 360 new plant species have been discovered here

since 1995, and the island is home to more than 1,400 species of amphibians, mammals, birds, reptiles, and fish – many found nowhere else.

## Rainforest under threat

However, this species treasury is in peril. Until the early 1970s, more than three-quarters of Borneo's 743,330 square km (287,000 square miles) were covered in thick tropical rainforest that was densest in the lowlands. Since then, at least one-third has been destroyed. Fires and the planting of oil palm plantations are partly to blame, but it was demand for the highly valuable dipterocarps that resulted in logging on an industrial scale, particularly in the Malaysian states of Sabah and Sarawak in the north of the island, where an estimated 80 per cent of rainforest has been lost. To preserve this habitat, in 2007, 220,000 square km (85,000 square miles) of rainforest in the centre of the island were designated as a protected area, known as the Heart of Borneo.

◁ **CALLING CARD**
The nocturnal tree-dwelling tokay gecko is just one of many reptiles in Borneo and is named for its "to-kay!" call. It grows up to 40 cm (16 in) long.

# Up to **240 tree species** can be found in just **1 hectare (2½ acres)** of **Borneo's rainforest**

△ **PROTECTED TREES**
The crowns of the tallest trees, known as emergents, rise above the morning mist that covers the rest of the rainforest in the Danum Valley, a conservation area of lowland dipterocarp forest in Sabah.

## HOW A STRANGLER FIG GROWS

In most cases, a sticky strangler fig seed germinates high on a branch of a rainforest tree, where it was left by a monkey, bird, or bat that had eaten the fruit. The seedling sends long roots down the trunk of its host to enter the soil. Eventually, a scaffolding of roots encases the trunk, and the host's root system must compete with the strangler fig's. The fig's thick foliage also shades the host tree's crown. Ultimately the host dies, leaving the fig tree behind.

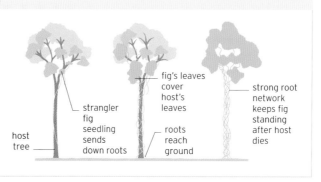

host tree

strangler fig seedling sends down roots

fig's leaves cover host's leaves

roots reach ground

strong root network keeps fig standing after host dies

▷ **LIFE SUPPORT**
Borneo's rainforests support many highly endangered animals, including the critically endangered Bornean orangutan, which depends on dipterocarp trees for food and shelter.

◁ **GIANT ORCHID**
The queen of orchids, believed to be the world's largest orchid, may be found growing in a fork of tall trees. This giant orchid is one of Borneo's more than 1,700 orchid species.

# The Terai-Duar Savannas

*A fertile band at the base of the Himalayas, where Earth's tallest grasses grow*

S. Asia

Forming a narrow belt of savanna and marshy grassland dotted with remnants of old forests, the Terai-Duar stretches from the southern Himalayan foothills in Nepal eastwards to Bhutan and India. Several rivers, including the Ganges, flow into the region, creating vast alluvial fans of sand, silt, and gravel – ideal for grass and reed growth.

## A forest of grass

The Terai's grasses are the tallest in the world, with some species, collectively known as elephant grass, reaching heights of 7 m (22 ft) or more. These grass forests, as they are known, provide perfect cover for hoofed mammals such as swamp deer, pygmy hogs, and buffalo, and the region's national parks are a sanctuary for the Indian rhinoceros and Bengal tiger. The ecoregion also supports mugger crocodiles and rare gharials, as well as many types of birds, including three species found nowhere else on Earth.

△ **FIRST FLOWERS**
Kans grass is the first grass to germinate after the floodwaters retreat in the Terai-Duar.

# The Eastern Steppe

*The largest intact temperate grassland on Earth, encompassing a vast, unbroken, wind-blown plain and prone to seasonal temperature extremes*

E. Asia

Part of the huge Eurasian Steppe, the Eastern Steppe is dominated by grasslands across most of its 887,330 square km (342,600 square miles). It extends from southern Siberia to the coastal hills of northeast China and from the Altai Mountains east to the Greater Khingan range.

## Harsh land of extremes

This is a much harsher region than its counterpart, the Western Steppe (see p.175), and so has not been subjected to as much agricultural activity. However, the grazing of sheep and goats remains a constant threat to what is a fragile and easily degraded habitat. The region receives just 250–500 mm (10–20 in) of precipitation a year, but hardy grass species manage to grow, despite a temperature range of -20°C (-4°F) in winter to 40°C (104°F) in summer. At higher elevations, rainfall increases, and snow accumulated on mountain peaks descends as streams into the dry lands below.

◁ **STEPPE COUNTRY**
The Eastern Steppe's vast grasslands in Mongolia look empty, yet are rich enough to feed a diverse range of wildlife, from marmots and grazing animals to many birds of prey.

## CLIMATE

Heat and humidity characterize the Terai-Duar all year round, but temperatures often rise to 40°C (104°F) in the late dry season. Annual monsoon rains result in floods throughout the region, renewing vital deposits of fertile silt.

| Temperature | | | Rainfall | |
|---|---|---|---|---|
| °F | °C | Dehradun, India | MM | IN |
| 140 | 60 | | 570 | 22½ |
| 122 | 50 | | 475 | 18¾ |
| 104 | 40 | | 380 | 15 |
| 86 | 30 | | 285 | 11¼ |
| 68 | 20 | | 190 | 7½ |
| 50 | 10 | | 95 | 3¾ |
| 32 | 0 | | 0 | 0 |

Jan Feb Mar Apr May Jun Jul Aug Sep Oct Nov Dec

KEY    ■ Temperature    ■ Rainfall

◁ **RIVER CROSSING**
A rare Indian rhinoceros cools off in a river in Nepal's Royal Chitwan National Park. Of all five rhino species, this one is most at home in water.

# Siberian Tundra

*A cold and often frozen landscape lying mostly above the Arctic Circle, where life persists even in the most inhospitable of conditions*

N. Asia

Northeastern Russia's Siberian tundra ranges from the edge of the taiga in the south (see pp.254–55) through the central region of the north coast and eastwards as far as the Chukchi Peninsula.

Long winters where temperatures plunge to -40°C (-40°F) are the norm in Siberian tundra, while summers are short and cool, warming to just 12°C (54°F). Winds blow at speeds of up to 100kph (60mph), and the frozen soil layer known as permafrost (see p.175) can be as thick as 600m (2,000ft). Yet clumps of grasses, fungi, and low-growing shrubs manage to survive, providing a vital source of nourishment for swarms of insects and migrating bird and mammal species.

▷ **FIELDS OF COTTON**
Arctic cotton grass grows in wet ground and blooms during the short summers, when it provides food for reindeer calves.

▷ **SNOW SURVIVOR**
Insulated by two coats of hair, reindeers survive on sedges, mosses, and lichens, which they find even beneath snow.

# The Arabian Desert

*Asia's largest desert, shared among nine nations, contained within a ring of mountains, and home to one of the largest continuous expanses of sand on Earth*

W. Asia

Covering 2.3 million square km (900,000 square miles), the Arabian Desert is so large that it spans almost the whole of the Arabian Peninsula. The bulk of it lies within Saudi Arabia, but it also extends southwest into Yemen, southeast into Oman, east along the Persian Gulf into the United Arab Emirates and Qatar, northeast into Kuwait and Iraq, and northwest into Jordan, just nudging into the tip of Egypt.

## Region of extremes

With vast stretches of sand in the south, gravel or salt-encrusted plains (*sabkhahs*) in the east, volcanic fields to the west, and temperatures ranging from 50°C (122°F) on summer days to below freezing on winter nights, this environment at first seems too harsh for survival. However, a surprising number of creatures live here, particularly insects such as locusts and dung beetles, as well as spiders and scorpions. These serve as food for many snakes and lizards, while mammals, including jerboas, goats, and gazelles, find enough vegetation in oases and fringe areas to sustain them.

Beneath the desert surface lies a huge groundwater reserve that was trapped here during the Pleistocene Epoch, 2.6 million–11,700 years ago. This, together with the area's vast oil reserves, has been tapped in recent years: the former used for irrigation, the latter for export of petroleum products.

both sexes have long, ringed horns

wide, shovel-shaped hooves aid walking on loose sand

▷ **DESERT SURVIVOR**
Once extinct in the wild, the rare Arabian oryx has been reintroduced and now roams the desert, foraging on grasses and roots. Its brilliant white coat reflects the worst of the desert Sun's heat.

◁ **SHIFTING SEA OF SAND**
In the south of the desert lies Ar Rub 'al Khali, or the Empty Quarter, an area of sand roughly the size of France. Strong winds called *shamals* move huge amounts of sand twice a year, reshaping the dunes.

## HOW A STAR DUNE FORMS

While most dunes are crescent-shaped, almost 10 per cent of the planet's sand dunes are star-shaped. Winds blowing from several directions push piles of sand together, forming three or more "arms" of sand radiating from a high central point. The star dunes of the Arabian Desert are found in the eastern part of the Empty Quarter.

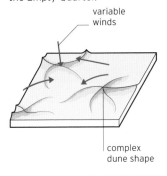

variable winds

complex dune shape

W. Asia

# The Dasht-e-Lut

*A salt desert with spectacular dunes and one of the hottest places on the planet*

Southeastern Iran's Dasht-e-Lut, or Lut Desert (*lut* means "completely barren"), is a torrid salt desert. In a satellite study of global land surface temperatures, the Lut's readings were the hottest for five out of seven years. In 2005, a temperature of 70.7°C (159.3°F) was recorded.

## Black sand, high heat

The extreme heat is due partly to the desert's geology. Much of the Lut's surface is made up of black sand – mainly magnetite, a type of iron ore. Magnetite absorbs more radiant energy than light-coloured sand, causing the surface temperatures to rise. Other distinguishing features of the Lut include yardangs. These huge, high rock ridges created by seasonal wind erosion give the desert a corrugated appearance, particularly on its western edge.

△ **LUNAR LANDSCAPE**
Southern Jordan's Wadi Rum, or the Valley of the Moon, reveals a volcanic past that created granite and sandstone mesas rising 800 m (2,620 ft) above the desert's surface.

## Average rainfall in the Arabian Desert is less than 100 mm (4 in) a year

△ **MOBILE DUNES**
In the eastern part of the Lut's estimated 51,000 square km (20,000 square miles), desert winds shape the more sandy areas into massive mobile dune fields.

# The Karakum Desert

*East of the Caspian Sea and south of the Aral Sea, a sparsely vegetated desert of three terrains – and a crater that has burned for decades*

C. Asia

The Central Asian Karakum Desert covers about 350,000 square km (135,000 square miles), occupying around 70 per cent of Turkmenistan.

## Three deserts in one

Southern salt marshes give way to a mineral-rich central plain, then to barren, higher northern areas, constantly eroded by winds. The centre contains sand ridges that are 75–90 m (245–295 ft) high as well as barchans (crescent-shaped dunes) and clay-lined depressions where the average annual rainfall of 70–150 mm (3–6 in) accumulates to form temporary lakes. Grasses, wormwood, and saxual trees provide food and shelter for wildlife.

▷ **DOOR TO HELL**
In 1971, a crater created by a Soviet drilling accident released toxic methane gas near Darvaza village. The fire that was lit to burn off the gas has never gone out.

yellow-brown coat provides camouflage

blunt claws aid digging

◁ **PROTECTED FEET**
The sand cat's foot pads are covered in thick fur, helping it to walk on the hot, rocky terrain of the Karakum.

# The Taklamakan Desert

*China's largest desert, composed of fields of shifting dunes that are sporadically convulsed by tornado-like sandstorms*

C. Asia

The second-largest shifting sand desert on Earth, surpassed only by the Sahara, the Taklamakan is known as the Sea of Death by the Chinese. Until the 1950s, when the huge Tarim Basin oilfield was discovered beneath it, few people braved the desert's extremely hostile conditions.

## Black hurricanes

Constantly shifting dunes (see panel, right) make the 320,000 square km (123,550 square miles) of the Taklamakan difficult to navigate, despite a series of isolated oases scattered at the base of the Tian Shan foothills on its northern border. Mobility aside, the dunes themselves are formidable; some dune chains are up to 500 m (1,640 ft) wide and 150 m (490 ft) high, with the distance between the chains varying from 1 to 5 km (½ to 3 miles). Fierce winds also cause *karaburans*, or "black hurricanes": sudden violent storms of sand and gravel so dense that they block out sunlight, resulting in dust clouds that rise as high as 4,000 m (13,000 ft).

Although several rivers feed into the desert, bringing snowmelt from the mountains that surround it on three sides, actual precipitation is minimal – only 10–38 mm (½–1½ in) a year. What groundwater there is lies just 3–5 m (10–16 ft) below the surface, and even that is increasingly being compromised by the transportation of oil and gas out of the Tarim Basin via a highway constructed in 1995.

Although classed as a **cold desert**, Taklamakan **summer temperatures** can soar to **38°C (100°F)**

# The Thar Desert

*The most densely populated desert in the world, named for its iconic sand ridges accumulated over the past 1.8 million years*

C. Asia

▷ **AT RISK**
The Thar Desert provides a haven for the Egyptian vulture, endangered elsewhere partly due to habitat destruction.

The Thar, or Great Indian Desert, belongs to two nations. Pakistan lays claim to around 15 per cent, but the greater part lies in India, mainly in Rajasthan.

Undulating dunes give the Thar Desert its name, derived from *thul*, the local word for "sand ridges". The region is made up of metamorphic gneiss, sedimentary rocks, and alluvial deposits that, over about 1.8 million years, have been covered by wind-borne sand. Parts of the Thar consist of sandy plains, others of continually shifting dunes. Barren hills and vast saline lakebeds are also scattered throughout the desert.

Seasonal dry southwesterly winds blow from March to July, when temperatures rise to 50°C (122°F). From July to September, monsoons bring an average 100 mm (4 in) of rainfall to the western Thar and as much as 500 mm (20 in) to the east. This is enough to support sparse scrub and grasses, as well as drought-tolerant tree species such as acacia. Grasses provide food for animals and birds, including the critically endangered great bustard.

△ **SEA OF SAND**
Huge crescent-shaped dunes resembling rolling ocean waves cover the landscape of the arid western Thar.

## HOW SEIF DUNES FORM

Most dunes are formed by multidirectional winds. When winds vary between two different, usually acutely angled, directions, longitudinal or "seif" dunes are created. These may extend for 100 km (60 miles) or more.

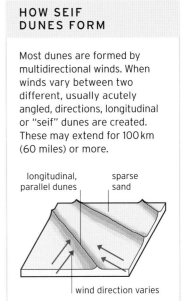

longitudinal, parallel dunes

sparse sand

wind direction varies

◁ **ANCIENT POPLARS**
The Taklamakan's relatively temperate climate does support some vegetation, including Euphrates poplars, which grow near the Tarim River and stabilize the sand dunes.

## DUNE MIGRATION

Taklamakan dunes move up to 9 m (30 ft) per year, due to prevailing northwesterly winds. Sand blown to the top of a dune falls to the other side, gradually shifting the position of the dune.

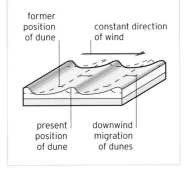

former position of dune

constant direction of wind

present position of dune

downwind migration of dunes

E. Asia

# The Gobi Desert

*High, harsh, and dry, a desert of climatic extremes,*
*rugged beauty – and very little sand*

The Gobi Desert stretches a staggering 1.3 million square km (500,000 square miles) across Asia's remote heartland, making it the continent's second largest desert. Once part of the Mongol Empire, today it straddles northern China and southern Mongolia.

## The waterless place
Bordered by Siberia to the north and the Tibetan Plateau to the south, the Gobi lies in the shadow of the Himalayas, which block most clouds laden with life-giving rain. Indeed, the Mongolian name for the Gobi is *gēbi* – the "waterless place". Seasonal monsoons occasionally moisten its southeast corner, but the desert is mainly arid – average annual rainfall is just 100–150 mm (4–6 in), although snow sweeps into the north from the Siberian Steppes.

Summers are brutally hot, with temperatures soaring to 50–66°C (122–151°F) during the day; in winter, night temperatures can plummet to -38°C (-36°F). Within just 24 hours, temperatures may shift by as much as 33°C (60°F).

Most of the Gobi's terrain is rocky, forming a pavement-like surface (see panel, below), which tempted merchants to cross it as part of the ancient Silk Road. Rugged valleys, chalk or gravel plains, and rocky massifs abound, but sand, for the most part, is thin on the ground. Exceptions include the massive dunes at the base of the Gobi Altai Mountains in southern Mongolia.

◁ **BORN SURVIVOR**
Wildlife adapts to the Gobi Desert in remarkable ways. The long-eared hedgehog, which shelters in burrows during the day, can go without food or water for up to 10 weeks.

## Only **five per cent** of the Gobi Desert is made up of **sand**

### HOW A DESERT PAVEMENT FORMS

Water and wind together form hard desert pavements. Over millennia, silt particles on extremely dry ground are either blown away by wind or filtered beneath the surface during the rare occasions when moisture reaches the desert floor. This leaves gravel behind, which fuses into a conglomerate whenever the minerals in it, such as gypsum, absorb enough soil moisture to liquefy, then set. The result is a "pavement".

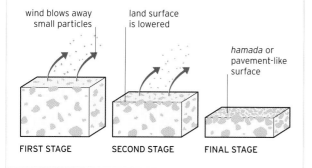

wind blows away
small particles

land surface
is lowered

*hamada* or
pavement-like
surface

FIRST STAGE    SECOND STAGE    FINAL STAGE

△ **RED BEDS**
The colour red dominates the Gobi due to the large amounts of iron oxide present in its rocks. "Red bed" formations are typically made up of sandstone, siltstone, and shale.

▷ **SINGING SAND**
At Khongoryn Els, Mongolia, "Singing Dunes" at the foot of the Altai Mountains rise as high as 300 m (980 ft). Their name refers to the sound made when wind sweeps the dunes.

**Ancient life**
South Australia's Kati Thanda-Lake Eyre shrinks and grows according to the continent's erratic rainfall. The pink colour comes from pigment in the cell-membranes of salt-loving bacteria, some of the oldest forms of life on Earth.

# Australia & New Zealand

# AN ANCIENT LAND
## Australia and New Zealand

The smallest continent, Australia is surrounded by water and sits in the middle of the huge Australian Plate. This plate is largely inundated by the Pacific and Indian oceans but is dotted with islands – some large like New Guinea, others tiny like the Fijian islands. New Zealand straddles the boundary between the Australian and adjacent Pacific plates.

Australia is dominated by the vast and flat outback, scorched by the tropical sun. In the east, it rises to the Great Dividing Range of mountains, and beyond lies the moist and mild coastal plain. Off the northeast coast stretches the world's biggest coral reef, the Great Barrier Reef. Temperate New Zealand's two islands lie far to the east – a mix of lofty, snowcapped mountains, rolling plains, and active volcanic features.

*Ne*

*Arafura Sea*

*Timor Sea*

*Torre*

Melville
Island

*Arnhem
Land*

*Barkly
Tableland*

*Kimberley
Plateau*

*Tanami
Desert*

*Great Sandy
Desert*

*Macdonnell Ranges*

*INDIAN*

*OCEAN*

*Hamersley
Range*

A    U    S    T    R    A    L    I    A

*Gibson
Desert*

△ *Uluru
(Ayers Rock)
863 m*

*Simpson
Desert*

▼ *Lake Eyre
shore
-16 m*

*Lake Eyre
Basin*

*Great Victoria
Desert*

*Lake
Torrens*

*Flinders
Ran*

*Nullarbor Plain*

*Darling Range*

### KEY

- ■ Precambrian (pre-541 million years ago)
- ■ Paleozoic (541-252 million years ago)
- ■ Mesozoic (252-66 million years ago)
- ■ Cenozoic (66 million years ago to present day)

### GEOLOGY

Australia is the lowest, oldest, and flattest continent. The ancient eastern mountains rise to only 2,228 m (7,310 ft). Far from any plate boundaries, the continent is geologically stable and has no volcanoes.

*Kangaroo
Island*

| ELEVATION | | | |
|---|---|---|---|
| M | FT | | |
| 1,000 | 3,300 | | |
| 0 | 0 | | |
| -5,000 | -16,000 | | |

Hamersley Range

Gibson Desert

Flinders Ranges

Great Dividing Range

Southern Alps

seafloor of the
Tasman Sea

A ◄ ............................................ CROSS-SECTION ............................................ ► B

_Guinea_

_Solomon Islands_

△ Mount Wilhelm
4,509m

_Strait_

_Cape York Peninsula_

_Great Barrier Reef_

_Great Dividing Range_

_Coral
Sea_

Fraser
Island

Bourke
52°C

_Darling_

Charlotte Pass
-23°C
△ Mount
Kosciuszko
2,228 m

_Murray_

_Bass   Strait_

Flinders
Island

_Tasman
Sea_

Tasmania

## KEY DATA

▲ **Highest point**  Mount Wilhelm,
Papua New Guinea: 4,509 m (14,794 ft)

▼ **Lowest point**  Lake Eyre shore,
Australia: -16 m (-53 ft)

● **Hottest record**  Bourke,
Australia: 52°C (125°F)

● **Coldest record**  Charlotte
Pass, Australia: -23°C (-9°F)

## CLIMATE

Most of Australia is very dry and often scorching. Only along
the east coast is there much rainfall, with a Mediterranean
climate in the south and a tropical monsoon climate in the
north. New Zealand's climate is cooler and more moderate.

**AVERAGE TEMPERATURE**

°C  30 20 10 0 -10 -20 -30 -40
°F  80 60 40 20 0 -20 -40

**AVERAGE RAINFALL**

MM  10,000  7,500  5,000  2,500  0
IN  400  300  200  100  0

## ECOSYSTEMS

Most of Australia's interior is desert, dry grassland, or scrub.
Tropical rainforests grow in the north; while temperate forests
grow in the south. New Zealand is mostly temperate forest.

**KEY**

■ Tropical broadleaf forest
■ Temperate broadleaf forest
■ Mediterranean woodland and scrub
■ Tropical and subtropical grassland
■ Temperate grassland
■ Desert and scrub
■ Montane grassland

North Island

_New
Zealand_

Aoraki (Mt Cook)
3,724m
△

_Southern Alps_

South Island

Km
0    150    300

0    150    300
Miles

▷ **OLD IRON**
The banded iron formations (alternate layers of iron-rich deposits), exposed here in the Hamersley Range on one of Australia's cratons, are 2.5 billion years old.

▷ **ANCIENT RIVER**
The Finke River first started to flow over 350 million years ago. At that time, it flowed freely over the plain above its present-day gorge.

◁ **SMOKING ISLAND**
Mount Tavurvur is one of the most active vents on the highly volatile Rabaul caldera on New Britain, Papua New Guinea.

# THE SHAPING OF AUSTRALIA AND NEW ZEALAND

Australia is a flat, time-worn continent. It holds some of Earth's oldest rocks, and its remarkable landscapes have been carved out over hundreds of millions of years.

## Ancient crust

The western half of the continent is the Australian Shield of old crustal rock, built around three ancient sections of stable crust called cratons. Zircon crystals in the Jack Hills, discovered in rocks on one of the cratons, are 4.4 billion years old – the oldest datable material on Earth. Another craton is, along with South Africa's Kaapvaal Craton, one of only two unaltered chunks of Archean crust on Earth. The Australian Alps formed over 360 million years ago – as a result, they have been worn lower than other ranges and have no jagged peaks. Instead, they are made of plateaus and gorges.

## Australia has the **most ancient landscape** and holds **Earth's oldest rocks**

## Volcanic islands

Along its northern edge, Australia has crunched into a complex melee of small tectonic plates to the north of Papua New Guinea. There are more than five small plates here, including the South and North Bismarck plates under New Britain. These small plates jostle to and fro as they are squeezed between the giant Australian and Pacific plates converging from either side. Some of the small plates will not last long, as they are subducting beneath other plates and melting into Earth's mantle. Molten magma from the mantle bursts through the overriding plates in a series of volcanoes to make the islands of Melanesia – the most volcanically active region in the world. There are 22 volcanoes here that have erupted to form islands, while others erupt under the sea. The Rabaul caldera on the northern tip of New Britain contains several violently explosive vents.

## Solo journey

Australia was attached to India and Antarctica as part of the great southern continent of Gondwana 150 million years ago, but about 130 million years ago a rift opened inside Gondwana and the three landmasses split away. About 100 million years ago, India broke

## KEY EVENTS

**300–280 million years ago**
The Great Dividing Range forms when Australia collides with landmasses that will become South America and Zealandia.

**85 million years ago** A landmass that will eventually fall below sea level and become Zealandia splits away from Australia while it is still attached to Gondwana.

**50 million years ago** Australia and Papua New Guinea break away from Antarctica and drift north, while Antarctica drifts southwards to the South Pole.

**500,000 years ago**
The Great Barrier Reef first forms, but it has receded and expanded since then.

**450 million years ago** An ancient ocean forms where Australia will eventually be. The sandstone seafloor will later be pushed upwards by tectonic movement – both Uluru and Mount Olga are remnants of this seafloor.

**180 million years ago**
Rainforests arise that will eventually survive to become the Gondwanan rainforests in southeastern Australia.

**80 million years ago**
A landmass that will become Australia and Papua New Guinea breaks away from the rest of Gondwana.

**25 million years ago**
What was once an ancient seafloor thrusts upwards due to tectonic activity and becomes the Nullarbor Plain.

**6,000 years ago**
New Guinea becomes an island as sea levels rise after the ice age.

▷ **UNLIKELY PEAKS**
There has been enough vertical movement along the usually horizontal fault that cuts through New Zealand to throw up the Southern Alps.

△ **UNIQUE WILDLIFE**
As a result of Australia's isolation from other continents, much of its wildlife has evolved independently and is unique to the continent.

away too and headed north – but Australia was hauled to the chilly South Pole in Antarctica's embrace. From 85 million years ago, Australia began to crack away from Antarctica until, about 45 million years ago, it finally broke free and started creeping northwards – and it is still heading in that direction.

Although the Australian Plate is now pressing up against the Eurasian Plate, unlike the Indian Plate it has never come into direct contact by land – Australia and the nearby islands remain separated by deep water. Australia's wildlife has evolved independently, and some species are completely unique to the continent.

## Hidden continent

The landmass that was to become New Zealand split away from Antarctica and Australia 80 million years ago. A great rip formed along Australia's western edge and along the edge of Marie Byrd Land in Antarctica. New Zealand drifted north, and the Tasman Sea opened to leave the two islands isolated.

In 2017, scientists confirmed what had long been suspected – that the split moved not just present-day New Zealand but an entire continent, over 90 per cent of which is now submerged.

This submerged continent, known as Zealandia, is now considered the world's eighth continent due to the type of crustal rock that it is made of. New Zealand is, in fact, half on the Australian Plate and half on the Pacific Plate, and for 25 million years has been torn apart along the Alpine Fault (which has thrown up the Southern Alps) as the two plates move in opposite directions.

**HIDDEN CONTINENT**
The islands of New Zealand and New Caledonia, although now separated by sea, are just the highest points of the now submerged continent of Zealandia.

extent of Zealandia

---

**BUILDING THE AUSTRALIAN AND ZEALANDIAN CONTINENTS**  KEY ▬ Convergent boundary ▬ Divergent boundary ▬ Transform boundary

Australia has started to split from Antarctica

Zealandia

land above sea level

**94 MILLION YEARS AGO** Australia and Zealandia are still attached to Antarctica as part of the same continental fragment of Gondwana, but the rift between them all will soon start to open.

Papua New Guinea rises above sea level

Australian Plate subducts beneath Pacific Plate

most of Zealandia submerged

**50–40 MILLION YEARS AGO** Australia and Zealandia are now separate from Antarctica. Australia is gliding quickly north towards the tropics, while Zealandia is moving northeast, largely underwater.

Papua New Guinea and Australia are linked by land due to low sea level

more of Zealandia emerges above sea level

**18,000 YEARS AGO** Australia is close to the south of Asia. The drop in sea levels during the ice age causes parts of Zealandia to emerge and Papua New Guinea to connect to Australia.

# The Great Dividing Range

*The main mountain range of Australia, stretching for more than 3,500 km (2,200 miles) down the country's eastern side*

E. Australia

The third longest land-based mountain range in the world, the Great Dividing Range has many parts, including, for example, the Australian Alps in the south, the Blue Mountains and Warrumbungles in New South Wales, and the Clarke Range in Queensland. Its origins are complex and partly connected to the break-up of the ancient supercontinent of Gondwana that started more than 100 million years ago.

south. The mountains are the source of many of the country's major rivers, including the Murray–Darling (see p.282). The highest peak, Mount Kosciuszko (2,228 m/7,310 ft), is in the Australian Alps. The region's wildlife is as varied as its landscapes and includes kangaroos and platypuses. The area also boasts rich agricultural, timber, and mining resources.

△ **RAZOR SHARP**
This rock pinnacle, known as the Breadknife, and similar outcrops in the Warrumbungles subrange are remnants of ancient volcanoes.

### Diverse vistas
Landscapes across the range vary from hills covered with lush tropical rainforest in Queensland to snow-covered peaks further

◁ **INDUSTRIAL TREASURE**
The valuable metalloid element antimony, which has a variety of industrial uses, is mined at a few different sites in the Great Dividing Range.

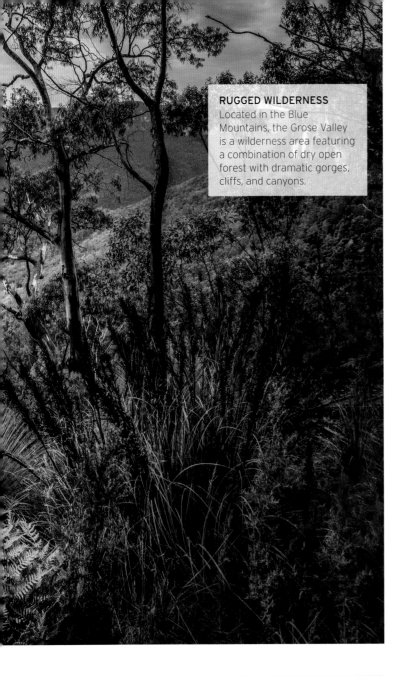

**RUGGED WILDERNESS**
Located in the Blue Mountains, the Grose Valley is a wilderness area featuring a combination of dry open forest with dramatic gorges, cliffs, and canyons.

N. New Zealand

# Tongariro Volcanic Park

*A volcanic wonderland in the centre of New Zealand's North Island*

Tongariro is a large national park in New Zealand centred on three active stratovolcanoes. These are Mount Tongariro, which has a complicated structure consisting of at least 12 overlapping volcanic cones and numerous craters; Ngauruhoe, a single, highly symmetrical volcanic cone; and Ruapehu, the largest active volcano in New Zealand. The summits of Ngauruhoe and Ruapehu – which at 2,797 m (9,177 ft) is the highest point on the North Island – are considered sacred by the local Maori people. The whole park, which is dominated by volcanic features such as ancient craters and solidified lava flows, has a stark beauty.

## TONGARIRO'S VOLCANOES

Much of the park is volcanic terrain (grey) surrounded by forested areas (green). The main concentration of small volcanic features lies at the northern end, around Mount Tongariro, while Ruapehu, which tends to produce a large eruption every 50 years, looms in the south.

Mount Tongariro

Red Crater

volcanic lakes

Mount Ruapehu with summit crater lake

Mount Ngauruhoe (stratovolcano)

▽ **CRATERS AND LAKES**
In this view of Tongariro volcano, several craters and lakes are visible. The lake colours come from minerals leached out of surrounding rocks.

## HOW THE AUSTRALIAN ALPS FORMED

The history of the Great Dividing Range is very complex, but one part of it, the Australian Alps, is theorized to have originated when a microcontinent called Zealandia (see pp.272-73) rifted away from southeastern Australia. Today, Zealandia is mostly submerged (the parts above land are New Zealand), while the rift margin on the Australian side has been eroded down to form the Australian Alps.

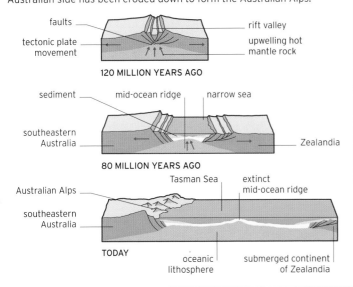

faults

tectonic plate movement

rift valley

upwelling hot mantle rock

**120 MILLION YEARS AGO**

sediment

mid-ocean ridge

narrow sea

southeastern Australia

Zealandia

**80 MILLION YEARS AGO**

Australian Alps

Tasman Sea

extinct mid-ocean ridge

southeastern Australia

**TODAY**

oceanic lithosphere

submerged continent of Zealandia

N. New
Zealand

# Rotorua

*A region of New Zealand's North Island that boasts an array of geothermal features, such as geysers, colourful hot springs, and bubbling mud pools*

A large part of New Zealand's North Island has a volcanic nature and history. Near its centre is Lake Taupo, the site of a supervolcano that last erupted about 1,800 years ago, and to the southwest of this is the Tongariro Volcanic Park (see p.275). To the northeast of Lake Taupo is a hotbed of geothermal activity often referred to as Rotorua, although only some of the geothermal features it contains are close to the city of Rotorua itself. In addition to geysers, hot springs, fumaroles, and mud pools, more unusual sights in the region include a hot water waterfall (the Kakahi Falls), acidic boiling lakes, sinter and marble terraces, a mud volcano, and a mud spa.

▽ BRIGHT DEPOSITS
The orange coloration at the edges of this hot spring, one of several in an area called Wai-o-tapu, are caused by deposits of arsenic- and antimony-containing minerals.

coloured hot spring

water percolates downwards through cracks in rock

spouting geyser

terraces
of limestone-like
material

▽ SULPHUR TERRACES
Some flowing hot springs create stepped terraces by depositing limestone-like minerals and, in this case, yellow sulphur.

**Mud pool**
A mud pool is created by a combination of hot water, steam, and acidic gases from a fumarole dissolving nearby rocks.

**Dormant fountain geyser**
This is currently inactive due to a fracture lower down in the geyser system.

**Fumarole**
A hole in the ground that emits steam and gas is called a fumarole.

**Rising hot water**
Heated water expands and moves upwards after contact with hot rocks.

**Feeder channel**
This channel is dry because the rock fracture has blocked the water supply.

rising steam and gases, rather than hot water, occupy some channels

**Rock fracture**
Caused by an earthquake, this fracture has broken the channel supplying hot water to a geyser above.

heated water saturating rock

**Reservoir of superheated water**
When high pressure allows the water temperature to rise above boiling point, the water is called superheated.

porous rock can hold large quantities of heated water

**Geyserite**
This watertight and airtight mineral lines all the plumbing of a geyser.

**Hot rock**
Underlying magma provides the energy that heats the overlying rock and powers the geothermal features.

**△ SPOUTING GEYSER**
On the outskirts of Rotorua City is a geyser field containing seven active geysers. The Pohutu geyser, shown here, produces the biggest and loudest eruptions.

The Pohutu geyser erupts up to 20 times a day, and eruptions can reach heights of up to 30 m (100 ft)

**▷ MUD SWIRLS**
Bubbling mud pools occur at several sites around Rotorua. As the bubbles burst, they create intricate swirling patterns on the mud surface.

**▲ GEOTHERMAL FEATURES**
This illustration shows a typical geothermal landscape that includes many of the features seen at Rotorua. The geothermal areas lie over volcanic calderas. These still have magma under them. The magma heats overlying rock, which in turn heats water that has percolated downwards. This heated water supplies the geothermal features.

## HOW A GEYSER ERUPTS

A geyser can form where hot water occupies a system of underground channels and chambers lined by a mineral called geyserite. There also needs to be a constriction in the channel to the surface. Superheated water flows up from below, increasing pressure in the system until the high pressure blasts water out from the constricted outlet. Pressure then drops underground, allowing some superheated water to flash into steam. As this expands, it sustains the eruption, which continues until pressure drops to zero. The cycle then starts again.

geyser plume

geyser outlet

steam expands as water pressure drops

**PRESSURE RELEASE**

steam circulating at high pressure

constriction

feeder channel to surface

rising water

geyserite lining

super-heated water

bedrock

**PRESSURE BUILD-UP**

W. Australia

# Hyden Rock

*An ancient granite mound in Western Australia,*
*with a wave-like cliff on one side of its base*

Hyden Rock is famous for an unusual rock formation, which looks like a tall, breaking ocean wave, along a section of its base. Known as Wave Rock, it is thought to have been created by weathering and erosion of the base of the mound. The initial stage of weathering took place underground through slightly acidic groundwater breaking down the granite. This produced a concave pocket of fragmented granite within the formerly solid base. When the land surface around Hyden Rock was later lowered by erosion, the pocket of fragmented granite was also removed, leaving behind Wave Rock. Crystals from it have been dated at 2.7 billion years old, making it one of the oldest formations in Australia.

### ◁ WAVE ROCK
The formation is around 110 m (360 ft) long and 14 m (46 ft) high. The grey and amber stripes are the result of rainwater dissolving out minerals.

### ▽ SMOOTH FORMATION
The wave illusion is largely the result of the rounded overhang. Above the overhang, the rock's top surface is smooth and domed.

### ◁ ICY HEIGHTS
Aoraki, or Mount Cook, is exceptionally challenging to climb. The surrounding area is an extensive, high crevassed ice-field from which several major glaciers descend.

## THE ALPINE FAULT

Along the Alpine Fault, the Pacific Plate is sliding past and also slightly towards the Australian Plate, moving 3–4 m (10–12 ft) per century. This has caused the crust of the Pacific Plate to be pushed up along the fault, leading to the creation of the Southern Alps.

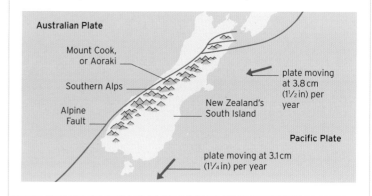

Australian Plate

Mount Cook, or Aoraki

Southern Alps

Alpine Fault

New Zealand's South Island

plate moving at 3.8 cm (1½ in) per year

Pacific Plate

plate moving at 3.1 cm (1¼ in) per year

# The Southern Alps

*A high, steep mountain range running almost the whole length of New Zealand's South Island, with a pristine environment of glaciers, lakes, and forests*

S. New Zealand

The Southern Alps are the most significant mountain range in New Zealand, extending for 500 km (310 miles) across the country's South Island, from southwest to northeast. Aoraki, or Mount Cook, the highest point in New Zealand, rises from the centre of the range, peaking at 3,724 m (12,218 ft). Another 16 peaks exceed 3,048 m (10,000 ft) in height. Most of the rivers of South Island, including the longest, the Clutha River, flow from the Southern Alps.

### Inching skywards

A boundary between two tectonic plates, called the Alpine Fault, runs to the west of the main line of the Southern Alps. It is of great significance to the mountains' origins and continuing uplift (see panel, left). The Southern Alps are continuing to rise, at a rate of up to several millimetres annually, but are also being eroded at an equally fast rate, aided by the sheer steepness of the mountains, high rainfall on the western slopes, and the presence of several glaciers. Due to high rainfall – which can be up to 10,000 mm (390 in) a year – the western slopes of the Alps are covered in temperate rainforest up to a height of about 1,000 m (3,300 ft). This leads to the unusual appearance of glaciers terminating a short distance away from lush rainforest. The eastern slopes are much drier and contain several large glacial lakes, such as Lake Tekapo and Lake Pukaki, which are both a brilliant powder blue colour. Endemic birds in the region include the kea, the rare and flightless southern brown kiwi, and the New Zealand rock wren.

▽ **CLEVER BIRD**
The kea, the world's only alpine parrot, is well known for its curiosity and intelligence, including an ability to use tools and solve logical puzzles. It is found only in the Southern Alps.

# The **Southern Alps** were **so-named** because of their **resemblance** to the **European Alps**

W. New
Zealand

# The Franz Josef Glacier

*One of two glaciers that descend steeply from the Southern Alps towards the west coast of New Zealand's South Island*

Named in 1865 after the Emperor of Austria, the Franz Josef Glacier is about 12 km (7 miles) long. Its lower parts used to flow through a temperate rainforest near the coast. However, since 2008 its terminus has retreated rapidly and now hardly reaches the tree line.

The glacier is known to advance and retreat cyclically according to changes in regional climate. It is currently shrinking, but between 1983 and 2008 it advanced 1.5 km (1 mile). The driving force behind the glacier's whole existence is a high rate of precipitation in the upper area of the Southern Alps – up to 30 m (100 ft) of snow falls onto the glacier's ice accumulation area each year – and annual changes in this precipitation rate can affect the glacier's behaviour for years to come.

△ **AQUA TUNNEL**
One part of the glacier contains tunnels carved out by meltwater. This one, dubbed the Aqua Tunnel, is a few metres across, large enough to walk through.

△ **BROKEN FLOES**
The surface of the glacier's meltwater lake sometimes freezes over in winter. The floes (flat expanses of ice) later break up into angular slabs.

▷ **MELTWATER LAKE**
At the foot of the glacier stands a 7-km- (4-mile-) long meltwater lake. A century ago, the valley was still filled with the glacier itself.

W. New
Zealand

# The Tasman Glacier

*New Zealand's largest glacier, with a skiing
arena at its top and a silty lake at its foot*

The Tasman Glacier flows down a long valley to
the east of New Zealand's highest mountain, Aoraki,
or Mount Cook, originating in a large ice accumulation
area at an altitude of 2,800 m (9,200 ft). In total, it is
24 km (15 miles) long.

### Ski runs, crevasses, and lakes

The upper part of the glacier is smooth and supplies
one of the world's longest ski runs, about 11 km
(7 miles) in length. Further down, it breaks up into
a maze of crevasses and ice tunnels. The lower third
is covered in a thick layer of rock debris – the result of
ice melting. At the foot of the glacier is a large meltwater
lake that has enlarged considerably in recent decades. It
is up to 250 m (820 ft) deep and is grey-coloured due to
suspended silt. The lake has a steady temperature close
to freezing because icebergs are continually discharged
into it from the glacier's terminus.

## In 2011, millions of tonnes of ice fell from the glacier into its lake after an earthquake

△ **BRAIDED FLOW**
An intricate braided river (a network of
branching channels with intervening
sand and gravel bars) carries water
away from the melting glacier.

## FEATURES OF A RETREATING GLACIER

A retreating glacier such
as the Tasman leaves behind
some characteristic landscape
features below its terminus
(or snout). These may include
drumlins (elongated hills lined
up in the direction of retreat),
kettle lakes (water-filled
depressions left behind
when partially buried ice
blocks melt), and eskers
(long, winding ridges of
sand and gravel).

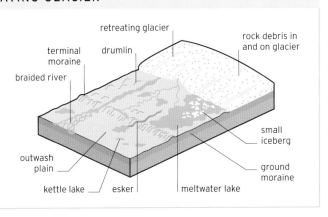

retreating glacier

rock debris in
and on glacier

terminal
moraine

drumlin

braided river

small
iceberg

outwash
plain

ground
moraine

kettle lake

esker

meltwater lake

# The Murray-Darling

*The largest river system in Australasia with one of the most extensive catchment areas on Earth, responsible for irrigating the "food basket" of Australia*

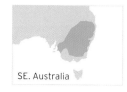

SE. Australia

Consisting of the Murray, Australia's longest river, and its main tributary, the Darling, the Murray–Darling river system is the largest in Australasia. Draining the southeast of Australia, its basin covers an area of more than 1 million square km (386,000 square miles) – around 14 per cent of the country's land area. The Murray–Darling basin is the most important agricultural region of a largely arid country, growing three-quarters of Australia's irrigated crops and producing one-third of its total food supply.

### From mountains to mouth
Rising in the Australian Alps, at the southern end of the Great Dividing Range, the Murray River flows across flat inland plains, forming the border between New South Wales and Victoria for much of its route. The river receives most of its main tributaries from the north along this section, including the Darling, which flows southwest

## The Murray-Darling is thought to be the **driest major river system** in the world

across almost the whole length of New South Wales from its source near the Queensland border. Both the Murray and Darling have very low gradients over most of their lengths, resulting in slow, meandering waterways.

After the Murray–Darling confluence, the river enters South Australia and flows through ancient gorges with towering sandstone cliffs separated by a series of shallow lakes. Having flowed a distance of 2,530 km (1,570 miles) from the source of the Murray, the river discharges into the Southern Ocean through a dynamic, sand-dune-flanked point known as Murray Mouth, at the eastern end of the Great Australian Bight.

△ **FRESHWATER TURTLE**
The eastern long-necked turtle has webbed feet and is able to stay underwater for long periods. Unlike other turtles and tortoises, it pulls its neck sideways into its shell.

# Jenolan Caves

SE. Australia

*A complex Australian cave system containing elaborate formations, the oldest caves ever discovered*

Situated on the western flank of the Blue Mountains in New South Wales, the Jenolan Caves are the best-known limestone cave system in Australasia. They sit within a narrow, uplifted ridge of limestone formed 500 million years ago, when the area was submerged beneath the ocean. The Jenolan Caves comprise a series of water-carved passages and caverns that stretch for more than 40 km (25 miles), over different levels, with more than 300 entrances. One of the largest caves, the Lucas Cave, has several vast sections including the 54-m- (177-ft-) high Cathedral Chamber, which is famous for its excellent acoustics. Scientists have estimated that the Jenolan Caves date back more than 340 million years, making them by far the oldest caves ever discovered.

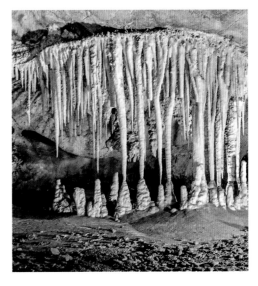

△ RICH DECORATIONS
The caves are renowned for their rich and varied deposits, including stalactites, stalagmites, columns, arches, and other intricate calcite formations.

▷ LARGE CRYSTALS
Deposits of a form of calcite called dogtooth spar, consisting of large crystals formed over long periods, are found in the pools of the Jenolan Caves.

△ STILL WATERS
An Australian pelican glides along the Murray River in the Riverland region of South Australia. The Murray-Darling basin is home to almost 100 species of waterbirds.

▽ RIVER'S END
Just before emptying into the ocean, the Murray-Darling connects with the Coorong, a narrow channel of lagoons lying behind the Younghusband Peninsula.

# Waitomo Caves

N. New Zealand

*An underground network of New Zealand caves, many decorated by the luminescence of glowworms*

In the area of Waitomo, in the west of New Zealand's North Island, runs a 45-km- (28-mile-) long network of caves, passages, and grottoes beneath fields of open farmland. The Waitomo Caves formed within a section of 30-million-year-old limestone rock up to 100 m (330 ft) thick. Fractures and fault lines in the rock allowed water to percolate through, creating an underground drainage system that gradually increased in size and complexity. Many of the caves and passages have streams with waterfalls running through them, and most feature a combination of calcite formations including stalactites, stalagmites, and columns in shades of brown, pink, and white. The best known of the 300 caves in the region are the Glowworm Caves, Ruakuri Cave, Aranui Cave, and Gardner's Gut.

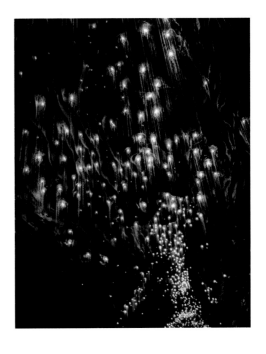

▷ GLOWWORM GROTTO
The ceilings of many of the Waitomo Caves are lit up by thousands of tiny glowworms, the larvae of a species of gnat endemic to New Zealand.

# The Great Barrier Reef

*The world's most extensive coral reef system, often described as the largest structure ever built by living organisms*

NE. Australia

The Great Barrier Reef stretches for 2,300km (1,400 miles) over a part of the Coral Sea, which lies off the coast of Queensland, Australia. Rather than a single unbroken reef, it is composed of more than 2,900 individual small reefs of various types and 900 coral islands, covering a total area of some 344,400 square km (133,000 square miles). The whole structure has been built through the activity of animals called coral polyps (see panel, below). The reef gets its name from the fact that it acts as a barrier between the coast of Australia and the large waves of the Pacific Ocean.

### Biological diversity

Teeming with marine life, the Great Barrier Reef boasts an astonishingly high biological diversity, including about 350 species of stony corals and several hundred species of soft corals, as well as 1,500 species of fish, 17 species of sea snake, 30 species of whales, dolphins, and porpoises, and 6 species of sea turtle. Over 200 bird species visit the reef, or nest or roost on the islands.

Unfortunately, in recent years the Great Barrier Reef has become severely threatened, the biggest cause of concern being rising ocean temperature as a result of global warming. This leads to coral bleaching (see p.286), a process that can ultimately kill corals permanently.

◁ **LIVING FOSSIL**
The chambered nautilus is one of over 4,000 species of mollusc that inhabit the reef. It has evolved little over many millions of years.

## The Great Barrier Reef is the **size of** about **48 million** international **football pitches**

△ **BUDDING LIFE**
A coral forms when a polyp attaches itself to a rock underwater. It then divides to form a colony – this one contains hundreds of polyps.

◁ **STRUCTURAL COMPLEXITY**
A variety of forms exist in even a small reef section, with patches and platforms of limestone as well as ribbon-like structures.

### STONY AND SOFT CORALS

Corals are made up of many individual animals called polyps. There are two main types of coral, called hexacorals (stony corals) and octocorals (soft corals). A significant difference is that it is mainly hexacoral polyps that secrete calcium carbonate, making them the principal reef builders.

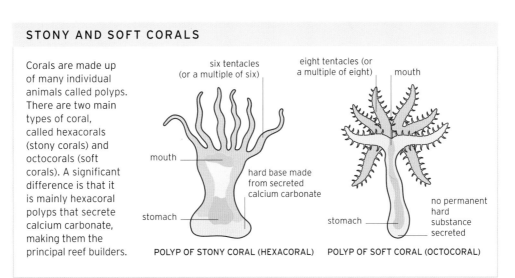

six tentacles (or a multiple of six)

mouth

mouth

stomach

hard base made from secreted calcium carbonate

**POLYP OF STONY CORAL (HEXACORAL)**

eight tentacles (or a multiple of eight)

mouth

stomach

no permanent hard substance secreted

**POLYP OF SOFT CORAL (OCTOCORAL)**

# The structure of a
# BARRIER REEF

A barrier reef is a type of coral reef widely separated from a continental mainland or island by a deep channel or lagoon. One circumstance in which such a reef can develop is when a volcanic island in the tropics begins to sink (as typically occurs when its volcanic activity ceases). As the island subsides, any fringing reef hugging its shores continues to grow upwards. A gradually widening lagoon then develops between the reef and the island. A second way a barrier reef can develop is on and around the edge of a continental shelf when sea level rises. This is how the world's largest barrier reef – the Great Barrier Reef off Queensland, Australia – formed.

## Development of the Great Barrier Reef

The Great Barrier Reef has taken some 20 million years overall to develop, although much of the structure visible today has grown since the last ice age on top of an older platform of continental shelf. Around 18,000 years ago, the sea was about 120 m (390 ft) lower and the outer edge of today's barrier reef was the edge of continental Australia. As sea level rose, ocean water and then corals encroached on what had been hills of the Australian coastal plain. By 13,000 years ago, many of these hills had become islands on the continental shelf. Eventually, they were submerged and corals grew over them. Sea level on the Great Barrier Reef has not changed significantly in the last 6,000 years.

## CORAL BLEACHING

Bleaching occurs when zooxanthellae (tiny algal organisms), which give corals their colours, leave the coral polyps due to a change in seawater temperature. This is reversible if the temperature change is not too prolonged or severe. But bleached coral becomes more vulnerable to disease. If it becomes colonized by a mat of other types of algae, called "turfing" algae, this signifies irreversible damage.

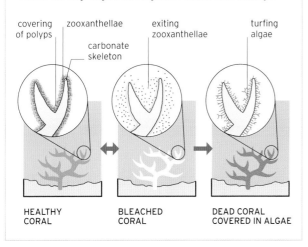

covering of polyps | zooxanthellae | exiting zooxanthellae | turfing algae

carbonate skeleton

HEALTHY CORAL | BLEACHED CORAL | DEAD CORAL COVERED IN ALGAE

▽ **REEF BUILDERS**
Some polyps, called stony corals or hexacorals, secrete small amounts of calcium carbonate (limestone) as part of their body structures, building on the substrate underneath. When polyps die, their limestone skeletons remain, building up the reef.

▶ **THE GREAT BARRIER REEF**
A 250-km- (155-mile-) long section of the reef, off northern Queensland, shows the overall structure. There are many individual reefs in a variety of forms ranging from ribbon-like structures making up the outer reef edge to rounded and crescent-shaped patch and bank reefs.

Mainland of Queensland, Australia

spur of the Great Dividing Range

Normanby River

Cooktown

**Bank reef**
A reef of this sort is larger than a patch reef and often linear or semicircular in outline. It may support a cay, or small island.

Williamson Reef

◁ **BLEACHED CORAL**
A bleached coral appears white due to loss of algal organisms that supply the colour. The algae provide most of the coral's energy: after they leave, the coral begins to starve and may eventually die.

**Fringing reefs**
These reefs develop near coasts, around the edges of what were once hills on the mainland. Polluted or sediment-laden water flowing off the mainland is a threat to them.

Flinders Group National Park

Combe Island National Park

**Patch reef**
Reefs of this type are rounded patches that grow up from the continental shelf. Some are topped by forested or sandy islands, and others develop lagoons.

◁ **CORAL SPAWNING**
Once a year, male and female coral polyps across the reef release a blizzard of tiny sperm and eggs, called gametes. This synchronized release is crucial to chances of fertilization, as the gametes can survive in the ocean for only a few hours.

seamounts

**Ribbon reefs**
Reefs of this type follow the edge of the continental shelf, tracking the coastline of the last ice age. Some are up to 25 km (15 miles) long.

**Cay**
This is a small, low-elevation, sandy island formed on the surface of a reef from eroded material that piles up.

Lizard Island

◁ **DROP-OFF**
Ribbon reefs form the outer edge of the barrier reef, bearing the brunt of Pacific storms. They are populated by robust corals and have steep drop-offs on their oceanic sides. These merge with the continental slope below, which descends to 2,000 m (6,600 ft).

continental shelf

continental slope

continental rise

Three Islands National Park

About **one-third** of the reefs making up the **Great Barrier Reef** are **fringing reefs**

# Shark Bay

*A coastline with exceptional natural features, including*
*seagrass beds, living stromatolites, and rich animal life*

W. Australia

Shark Bay lies 800 km (500 miles) north of Perth, on the west coast of Australia, and has a coastline that is more than 1,500 km (900 miles) long, including about 300 km (190 miles) of spectacular limestone cliffs. Another feature of Shark Bay is a dazzling white beach consisting almost entirely of the shells of one species of cockle. The bay contains one of the world's largest seagrass beds, which provides food for about

▷ **BLUE STINGER**
A sting from this local *Mastigias* jellyfish can be painful but it is not dangerous unlike that from other smaller species called irukandji.

10,000 dugongs – large marine mammals related to manatees (or sea cows). In turn, these are preyed on by various species of shark, including the great white and tiger sharks, which give the bay its name. The area is also home to more than 100 species of reptiles and amphibians, 240 species of birds, 820 species of fish, and more than 80 types of corals.

But Shark Bay is probably most famous for its living stromatolites – modern equivalents of some of the earliest types of life on Earth. These are colonies of microbes that form hard dome-like accretions (see panel, below).

◁ **LAYERED LIFE**
In an area called Hamelin Pool lies the world's most abundant collection of living stromatolites. They are thought to have started growing here about 1,000 years ago.

▷ **SHORELINE SPECTRUM**
Mangrove-lined creeks, emerald waters, dunes, and mineral-stained pools come together in this aerial view of a part of the bay.

## HOW STROMATOLITES GROW

The surface of a living stromatolite is covered in a mat of microbes called cyanobacteria, which in some cases develop into long, sticky filaments. These trap and bind together sediment particles into a hard layer. The stromatolite grows through the build-up of alternating layers of this hard substance and dead filaments.

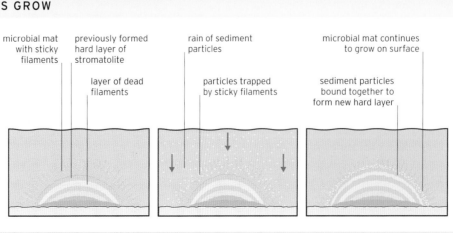

microbial mat with sticky filaments

previously formed hard layer of stromatolite

layer of dead filaments

rain of sediment particles

particles trapped by sticky filaments

microbial mat continues to grow on surface

sediment particles bound together to form new hard layer

# The Twelve Apostles

*An iconic group of large sea stacks, formed through the erosion of 20-million-year old limestone cliffs*

SE. Australia

The Twelve Apostles are located in Port Campbell National Park, a coastal area of Victoria. The sea stacks are strung out along about 5 km (3 miles) of coastline, with some reaching up to 50 m (165 ft) in height. The Twelve Apostles are one of Australia's best-known geological landmarks, but their name is something of a misnomer, as when they were originally named only nine large stacks existed and one has since collapsed, leaving only eight. However, several prominent headlands and islands nearby are continuing to erode through wave action (see panel, right), so new sea stacks are expected to form in the future. Like all coasts, the shoreline around the stacks is constantly changing.

## Until the **1920s**, the stacks were known as the **Sow and Piglets**

▷ **NEW STACKS IN THE MAKING**
This aerial view shows one existing isolated stack, at left, and other headlands and islands that, as they erode, are expected to form new stacks.

## HOW SEA STACKS FORM

Wave action creates caves in the sides of a headland, eventually tunnelling through it. The arch over the tunnel collapses to leave an island, which is then eroded to form stacks.

arch
cave in headland
tunnel
wave action

**TUNNEL FORMS THROUGH HEADLAND**

remains of collapsed arch
island

**ARCH COLLAPSES**

straightened coast
sea stack

**SEA STACKS FORM**

---

# Moeraki Beach

*A beach in southeastern New Zealand strewn with some unusually large, almost spherical boulders*

E. New Zealand

Some 70 km (45 miles) northeast of Dunedin, Moeraki Beach appears at first sight as though it might have been the location for a giants' game of bowls or petanque. Almost perfectly spherical grey boulders lie scattered across a 50-m (165-ft) stretch of the beach. The scientific view is that the huge boulders actually formed within clay and silt sediments on the seafloor around 60 million years ago by a process called concretion, in which a cementing mineral binds together other mineral particles into a hard, erosion-resistant mass. This process can take millions of years. Later, the concretions, now embedded within mudstone rock, were lifted up, and today most of them are buried within a cliff at the back of Moeraki Beach. There the boulders are gradually exposed and released onto the beach by erosion of the mudstone (see panel, right).

△ **SUPERSIZED BOULDERS**
The boulders come in two main sizes – around 0.9 m (3 ft) and 1.8 m (6 ft) in diameter – and weigh several tonnes. Many are half-buried in the sand.

## HOW THE BOULDERS FORMED

The boulders originally grew within marine sediments, which hardened into rock. This was later lifted above sea level and now forms a coastal cliff. From time to time, a boulder is freed up and rolls onto the beach.

spherical mineral concretions forming around seed nuclei

**FORMATION UNDER THE SEA**

boulders on Moeraki Beach

fully formed concretions embedded in mudstone

**TODAY**

# New Zealand Fiords

*A group of fiords in the southwestern corner of New Zealand, featuring abundant marine life and spectacular cliffs and waterfalls*

SW. New Zealand

New Zealand's fiords formed some 15,000 years ago when glaciers retreated at the end of the last ice age and the sea flooded in to fill the deep, U-shaped coastal valleys carved out by the glaciers.

## Dangerous discovery

Of some 14 major fiords, two of the best known are Doubtful Sound – one of the largest at 30 km (20 miles) in length – and Milford Sound. The former was named in 1770 by Captain James Cook, who refrained from sailing into it as he doubted that he would ever be able to sail out again. The smaller but more famous Milford Sound, which runs 15 km (9 miles) inland, is notable for the spectacular waterfalls that rain down into it, and an enormous mountain – the 1,692-m- (5,551-ft-) high Mitre Peak – that soars above it. Both these fiords, and others nearby, are home to bottlenose and dusky dolphins, New Zealand fur seals, little penguins, and the rare Fiordland crested penguin. Prevailing westerly winds blowing moist air in from the Tasman Sea produce exceedingly high rainfall across the region, and this supports lush temperate rainforests close by.

◁ **CURIOUS MINERAL**
Small deposits of arsenopyrite are found in the region. If heated, the mineral gives off toxic fumes and becomes magnetic. If struck with a hammer, it emits a garlic-like odour.

▽ **MISTY CASCADE**
The 151-m- (495-ft-) high Stirling Falls is one of two permanent waterfalls in Milford Sound. Its Maori name is Wai Manu, meaning "cloud on the water".

# The Daintree Rainforest

*Australia's largest tropical rainforest, one of the oldest and most biodiverse in the world, with a striking array of plants and animals*

NE. Australia

Covering 1,200 square km (460 square miles) in northeastern Queensland, the Daintree is the largest tropical rainforest in Australia and one of the oldest on Earth. Estimates of its age vary from 135 to 180 million years. Part of the Wet Tropics World Heritage Site, it contains the closest living relatives of prehistoric rainforest plant species that covered the continent millions of years ago. These include the ribbonwood tree, also known as the idiot fruit or green dinosaur. A primitive evergreen dating back 120–170 million years, it was believed to be extinct until 1971. Half of Australia's bird species, a third of its frogs and mammals, a quarter of its reptiles, and more than 12,000 insect species have been catalogued in the Daintree, many of them living nowhere else.

◁ **RAINFOREST WARDEN**
The endangered flightless southern cassowary is among the Daintree's residents and is an important disperser of tree seeds.

## The Daintree contains the **greatest concentration of birds** in all of **Australia**

# Eastern Australian Temperate Forest

*Home to more than one-fifth of Australia's eucalypt trees, the oils of which give nearby mountains their blue haze*

E. Australia

▽ **TOWERING TREES**
Near the Great Dividing Range, which forms the ecoregion's inland boundary, eucalypts rise above smaller trees such as acacia.

Stretching from the New South Wales central coast into southeast Queensland, the Eastern Australian Temperate Forest ecoregion covers 222,100 square km (85,750 square miles). Varying altitudes and microclimates support a wide range of vegetation, but the eucalypt is the region's dominant plant – more than 100 species of eucalypts, also known as gum trees, grow here. Airborne oil droplets from eucalypt leaves are believed to refract blue light, lending the area's Blue Mountains their colourful haze. More importantly, these gum tree forests are home to many endangered plant and animal species, including the iconic koala.

▷ **HOLDING TIGHT**
The koala initially carries its young, or joey, in a pouch. Eventually, the joey clings to its mother's back as she feeds on eucalyptus leaves.

**△ STANDING GUARD**

The granite summit of Mount Pieter Botte, or Ngalba-bulal, keeps watch over the Daintree National Park's rare black kauri pines.

**◁ PREHISTORIC LANDSCAPE**

Because of its isolated location and climate, the Daintree has remained largely unchanged since its origin. Today, it is home to 12 of the world's 19 families of primitive flowering plants.

# Waipoua Forest

*An ancient native New Zealand forest, home to the world's largest stand of kauri trees*

N. New Zealand

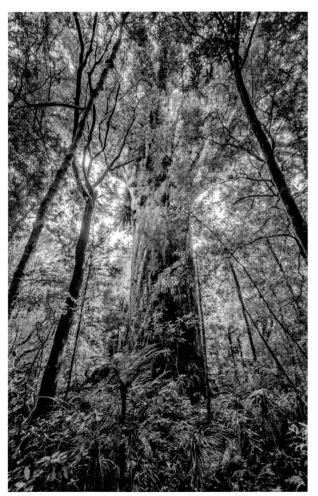

The subtropical Waipoua Forest lies on the west coast of the Northland Region of New Zealand's North Island. Declared a sanctuary in 1952, Waipoua contains Earth's largest collection of kauri trees – giant, cone-bearing, hardwood conifers that are among the oldest trees in the world and are also some of its most endangered. While ancestors of the kauri were growing during the Jurassic Period, Waipoua's most ancient kauri, called Te Matua Ngahere, or Father of the Forest, is merely 2,000 years old. However, with a girth of 16 m (52 ft) and a height of 37 m (121 ft), it has the largest diameter of any living kauri – and it is still growing.

**◁ IMPOSING HEIGHT**

Kauris can reach heights of up to 50 m (165 ft), making them dominant emergent trees that rise above other conifers, such as toatoas and monoaos, in the surrounding forest.

**▽ SEED-BEARERS**

A single kauri produces both cylindrical male and round female cones. Not until a tree is 25-30 years old will its female cones release fertile winged seeds.

# The North Australian Savannas

*A diverse tropical grassland of two seasonal extremes, home to hundreds of native plant and animal species*

N. Australia

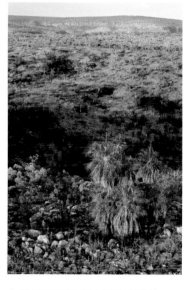

North Australia's grassland is one of the world's largest remaining tropical savannas and shares much in common with those of Africa. However, where Africa's savannas have large grazing animals, insects such as termites eat much of the plant material in Australia's savannas, alongside marsupials ranging from the small short-beaked echidna to large red kangaroos. Waterfowl, wading birds, and reptiles live in the wetter parts of the region, and the Cape York Peninsula alone contains around 60 per cent of Australia's butterfly species.

Due to its mid-latitude location, the grassland is subject to the El Niño Southern Oscillation climate cycle. This produces a pattern of high rainfall and drought that gives the area its two contrasting seasons.

◁ **DESIGNED TO RUN**
At up to 2.1 m (7 ft) high, the emu is the savanna's – and Australia's – largest bird. It cannot fly, but long, powerful legs enable it to run at speeds up to 48 kph (30 mph).

## CLIMATE

Australia's savannas have two distinct seasons. During the hot, humid wet season (December–March), northwesterly winds blow in heavy rainstorms. From May to October, the dry season sets in, with low temperatures and humidity.

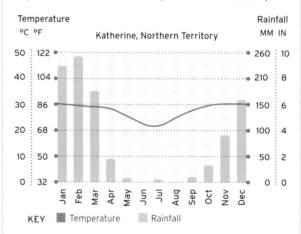

Katherine, Northern Territory

Temperature °C °F — left axis (50/122, 40/104, 30/86, 20/68, 10/50, 0/32)
Rainfall MM IN — right axis (260/10, 210/8, 150/6, 100/4, 50/2, 0/0)

Months: Jan Feb Mar Apr May Jun Jul Aug Sep Oct Nov Dec

KEY ▮ Temperature   ▯ Rainfall

△ **PERFECTLY ADAPTED PALM**
Highly drought-resistant Livistona palms can cope with seasonal extremes as they are able to survive frequent fires as well as heavy downpours.

▽ **MAGNETIC MOUNDS**
The magnetic or compass termite builds impressive mounds 3–4 m (10–12 ft) high. The wedge-shaped structures are constructed so that their longest sides are aligned north–south, keeping them cool.

**Giant grazing animals** roamed the savannas about **40,000 years ago**

# The Great Sandy Desert

*An active sandy desert, wetter than the continent's central region,
where heat and evaporation steal rain almost as fast as it falls*

NW. Australia

△ **LIFE SUPPORT**
An escarpment just above the Oakover River marks the Great Sandy Desert's western edge. The river provides a lifeline for a range of animals.

▽ **EVER-CHANGING DUNES**
The dunes of this desert are thought to have formed about 10,000 years ago. However, constant winds continue to reshape their crests.

Northwest Australia's Great Sandy Desert is bordered to the east by the rocky hills of the Tanami Desert and to the south by the gravelly Gibson. Besides its sand, what makes the Great Sandy different from its neighbours is its climate: the region gets a surprisingly high amount of rain. The average annual rainfall is 250 mm (10 in), even higher in northern sections, and this would normally result in a reasonable amount of plant growth. In the Great Sandy, however, summer daytime temperatures of around 40°C (104°F) mean that evaporation rates are so high that most moisture never has a chance to penetrate the soil. The prevailing winds blow mainly east to west, creating a landscape of constantly shifting red linear sand dunes.

## AUSTRALIA'S DESERTS

Deserts make up around one-fifth of Australia's landmass, making it the driest inhabited continent on Earth, but due to their challenging conditions, they are home to less than 3 per cent of Australians. The five largest are shown here.

Great Victoria Desert
348,750 square km
(134,650 square miles)

Great Sandy Desert
267,250 square km
(103,200 square miles)

Tanami Desert
184,500 square km
(71,250 square miles)

Simpson Desert
176,500 square km
(68,150 square miles)

Gibson Desert 156,000 square km
(60,250 square miles)

◁ **SMALL WONDER**
The tiny spinifex hopping mouse has the most efficient kidneys of any mammal, so it can survive without water for long periods.

C. Australia

# The Simpson Desert

*Australia's most centrally located desert, where sand ranges from pink to deep*
*red and forms the largest parallel dune system in the world*

The Simpson Desert forms part of Australia's red centre: the colloquial name for the Northern Territory's southern desert region that reflects the dominant colour of its sands. Stretching south from the Northern Territory's famous outback town of Alice Springs to cross the border of South Australia, the Simpson covers an area of around 176,500 square km (68,150 square miles). It contains the largest parallel sand dune system on Earth. These dunes vary in height from 3 m (10 ft) in the western desert to 30 m (100 ft) on its eastern side, and can extend for 200 km (120 miles) or more.

### A desert in bloom

Rainfall in the Simpson is low and irregular, averaging just 125 mm (5 in) annually. However, spinifex grass, shrubs, and acacia trees grow year round, joined by colourful wildflowers that bloom suddenly after a rare rain shower. Rainfall in nearby areas can also cause some parts of the Simpson to flood as inflowing rivers burst their banks. The resulting vegetation provides a haven for various animals, including the rare fat-tailed marsupial mouse. National parks and conservation areas have been set up in parts of the desert to ensure both flora and fauna are protected.

△ **PLAINS DWELLER**
The gibberbird's name reflects its main habitat – the gibber plains of Australian deserts, a type of desert pavement. It rarely flies, preferring to feed, nest, and roost on the ground.

# Uluru is **higher than** the **Statue of Liberty,** the **Eiffel Tower,** and the **Great Pyramid**

△ **PURPLE BLOSSOM**
Part of the nightshade family, bush tomato shrubs thrive in the Simpson Desert's arid conditions.

▷ **PARALLEL DUNES**
The true size and extent of the Simpson Desert's sandy expanses, as well as the vegetation that grows between the dunes, can be seen from the air.

## BENEATH ULURU

The rocks that make up Uluru began in the Amadeus Basin, a depression that formed around 900 million years ago. Collected sediment was compressed into a sandstone layer that folded, fractured, and rotated as tectonic activity pushed it upwards. Uluru is just the tip of a formation thought to extend up to 6 km (4 miles) underground.

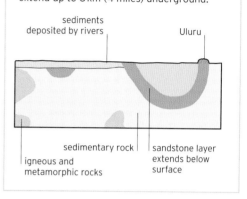

sediments deposited by rivers — Uluru

sedimentary rock | sandstone layer extends below surface
igneous and metamorphic rocks

▽ **ULURU**
Uluru rises 348 m (1,142 ft) above the surrounding desert surface. Around 3.6 km (2¼ miles) long and 1.9 km (1¼ miles) wide, it takes about 3½ hours to walk around.

# The Gibson Desert

*A hostile, largely untouched sandy desert in the centre of Western Australia*

W. Australia

Bordered by three other deserts – the Great Sandy (to the north), the Little Sandy (west), and the Great Victoria (south) – the Gibson is one of Australia's most pristine deserts, protected from human impact by its hot, harsh conditions. Few people live here – not surprisingly, given that it is named for an explorer, Alfred Gibson, who perished trying to cross it in 1874. The desert is 156,000 square km (60,250 square miles) of mostly low, undulating dune fields and sandy plains. There are also gravel ridges, formed of pebbles coated in iron oxide, and small saline lakes. Despite summer temperatures of 40°C (104°F) and little water, animals such as kangaroos and emus manage to survive.

△ **BIG RED**
Australia's largest living marsupial, the red kangaroo, is common in the Gibson Desert. Adaptations such as hopping conserve energy and minimize contact with hot sand.

▷ **YELLOW BLOOM**
When rain does fall, the Gibson Desert's sand supports thousands of tufts of yellow spinifex grass, one of the few plants able to germinate here.

# The Pinnacles

*A jagged landscape, with thousands of stone pillars rising out of yellow desert sand, continually covered and uncovered, and all born of rainwater and plant roots*

W. Australia

In the centre of Western Australia's Nambung National Park lies a coastal desert with a difference. The Pinnacles is the name given to thousands of limestone pillars, rising out of a landscape otherwise covered by sand dunes. How the pillars formed, or why they formed here and not in any other coastal dune environment, is still a subject of debate. However, what is known is that about 30,000–25,000 years ago, they began as crushed seashells, which eventually turned into limestone beneath a dune system.

## Root cause

As vegetation developed in the area, rainwater leached through enriched acidic soil, causing a hardened layer to form on top of the softer limestone. Plant roots eventually forced their way down through cracks in this hardpan, setting off a vertical erosion process in the limestone. As winds shifted the sand dunes in a drying climate, the first pinnacles were revealed.

◁ **ROCK SPIRES**
Most of the pinnacles are 1–2 m (3–6 ft) high, but some reach a height of 5 m (16 ft). Winds continually shift the dunes, covering some pinnacles and revealing others.

# The Great Victoria Desert

*Australia's largest sand-dune desert, shared between two states, dotted with rock pavement, and transected by shrubland*

S. Australia

Stretching more than 700 km (440 miles) from west to east across parts of Western Australia and South Australia, the Great Victoria is the continent's largest, yet possibly least desert-like, desert. Although its average annual rainfall is just 162 mm (6⅜ in) and daytime temperatures are as high as 30–45°C (86–113°F), a surprising amount of plant life still manages to thrive here.

## Under protection

Eucalyptus trees grow in parts of the region, while acacias and shrubland form a narrow, continuous strip known as the Giles Corridor that stretches across the desert's entire width. Clumps of spinifex, aristida, and other dry-adapted grasses also break up its dune fields and gibber plains (tightly packed layers of pebbles often covered with a layer of iron oxide). The desert has enough vegetation to attract many animals. It is famous for its reptiles, which number over 100 species, with geckos and skinks being particularly diverse. The region is also home to threatened mammals such as the marsupial mole and sandhill dunnart, and predators such as the dingo and Gould's goanna, a large monitor lizard that averages 1 m (3 ft) in length. The Great Victoria Desert's biodiversity has led to large sections being protected from development.

▷ **POWERFUL PREDATOR**
The perentie is Australia's largest lizard and the largest goanna species. Growing up to 2 m (6 ft) long, this formidable predator charges at its prey at a speed of up to 32 kph (20 mph).

## LUNETTE DUNES

Lunettes are fixed, crescent-shaped dunes found along the edges of temporary saline lakes, as in the Great Victoria Desert. Some are thought to occur due to wind-driven waves, which carry sediment into vegetation, where it becomes trapped.

wind pushes waves ashore in crescent shape

lunette horns face into the wind | lake bed dries out

▽ **MORE THAN SAND**
Near Kati Thanda-Lake Eyre, Australia's largest salt lake, dunes and spinifex give way to mesas topped by silcrete, a hard layer of gravel and sand cemented together by silica.

**Crossing the gap**
As the Matusevich Glacier flows towards the coast of East Antarctica, it passes through a narrow gap between hills, causing cracks, or crevasses, to form in the ice. After traversing the gap, the glacier ice floats out onto the sea.

# Antarctica

## ECOSYSTEMS

The extreme cold means most forms of life are confined to the surrounding waters, but animals such as midges and penguins do survive on the ice.

**KEY**
- Tundra
- Ice

**KEY**
- Precambrian (pre-541 million years ago)
- Paleozoic (541–252 million years ago)
- Mesozoic (252–66 million years ago)
- Cenozoic (66 million years ago to present day)
- Unknown age

## GEOLOGY

Most of East Antarctica comprises ancient igneous and metamorphic rocks such as gneiss. However, the west is formed from various younger plate fragments. The age of some of West Antarctica's rock is unknown due to the thick, impenetrable ice-sheet that conceals it.

### Map labels

Scotia Sea
South Orkney Islands
Fimbulisen
Dronning Maud Land
Maudheimvidda
Bransfield Strait
Esperanza Base 17.5°C
Graham Land
Antarctic Peninsula
Palmer Land
Weddell Sea
Filchner Ice Shelf
Berkner Island
Ronne Ice Shelf
Pensacola Mountains
Alexander Island
Bellingshausen Sea
Mount Vinson 4,897 m
South Pole
Peter I Øy
Ellsworth Land
West Antarctica
Whitmore Mountains
Transantarctic
Thurston Island
Bentley Subglacial Trench, -2,538 m
Queen Maud Mountains
SOUTHERN OCEAN
Amundsen Sea
Marie Byrd Land
Rockefeller Plateau
Ross Ice Shelf
Roosevelt Island
Mount Erebus 3,794 m
Ross Sea

### Cross-section

**ELEVATION**
| M | FT |
|---|---|
| 3,500 | 11,500 |
| 2,000 | 6,600 |
| 0 | 0 |

Marie Byrd Land
West Antarctic Ice-sheet
Ross Ice Shelf
Transantarctic Mountains
East Antarctic Ice-sheet

A — CROSS-SECTION — B

### Scale bar

Km
0   250   500
0   250   500
Miles

S O U T H E R N   O C E A N

*Thorshavnheiane*

*Enderby Land*

*Kemp Land*

**Mac. Robertson Land**

*Prince Charles Mountains*

*Amery Ice Shelf*

*Mackenzie Bay*

*Princess Elizabeth Land*

*West Ice Shelf*

A N T A R C T I C A

*E a s t Antarctica*

● East Antarctica -95°C

*Wilhelm II Land*

*Davis Sea*

*Shackleton Ice Shelf*

*M o u n t a i n s*

*V i c t o r i a   L a n d*

*W i l k e s L a n d*

*Vincennes Bay*

*Terre Adélie*

*Porpoise Bay*

*George V Land*

*Dumont d'Urville Sea*

# THE FROZEN CONTINENT
## Antarctica

Entirely surrounded by the Southern Ocean, Antarctica is the most remote continent. South America and Australia are 1,000 km (620 miles) and 2,500 km (1,600 miles) away respectively – but it wasn't always so isolated. Antarctica was part of the Gondwana supercontinent 200 million years ago, before it broke away to drift and settle at the South Pole some 35 million years ago.

The Antarctic ice-sheet is the world's largest, averaging over 1.6 km (1 mile) thick – and weighing so heavily on the rock beneath that some parts have been pushed 2.5 km (1½ miles) below sea level. Antarctica is divided into two parts, East and West, separated by the Transantarctic Mountains. To the east, an enormous, flat plateau of ice overlies the ancient continental shield. To the west, the icescape and geology are more varied and have much in common with the South American Andes.

### KEY DATA

▲ **Highest point** Mount Vinson: 4,897 m (16,066 ft)

▼ **Lowest point** Bentley Subglacial Trench: -2,538 m (-8,327 ft)

● **Hottest record** Esperanza Base: 17.5°C (63.5°F)

● **Coldest record** East Antarctica: -95°C (-139°F)

### CLIMATE

Antarctica is the coldest continent; 99 per cent of its land is covered by ice, and freezing temperatures persist all year. High winds can cause violent snowstorms that may last for days.

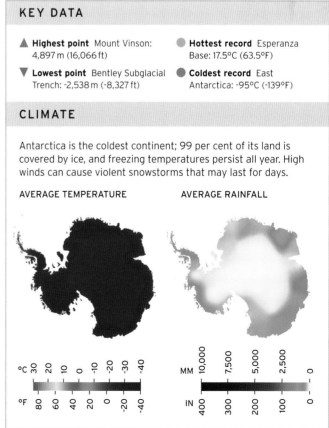

AVERAGE TEMPERATURE

AVERAGE RAINFALL

°C 30 20 10 0 -10 -20 -30 -40

°F 80 60 40 20 0 -20 -40

MM 10,000 7,500 5,000 2,500 0

IN 400 300 200 100 0

C. Antarctica

# The Transantarctic Mountains

*Antarctica's longest mountain range, which was first sighted by people only in 1841*

The Transantarctic Mountains are a curved belt of mountains, about 3,500 km (2,200 miles) long, that cross Antarctica. One section divides the continent into its two main parts: East and West Antarctica. Much of the rest of the belt runs along a part of the coast of East Antarctica, including a long stretch where it borders the Ross Ice Shelf. Some large glaciers flow through gaps in the range in this area.

The mountain belt has many subranges, including the Queen Maud Mountains and the Royal Society Range. Eight out of its 10 highest peaks are in just one subrange, the Queen Alexandra Range. Parts of the Transantarctic Mountains were first sighted by the British naval explorer Captain James Ross in 1841, but the Queen Alexandra Range was not discovered until 1908 (by a British Antarctic expedition). Life in the range's interior is limited to microorganisms – such as bacteria and algae. However, in the ancient past a wide variety of animals and plants thrived there, as evidenced by fossils of primitive amphibians and reptiles in rocks laid down 400–180 million years ago.

### HIGH PEAKS OF THE TRANSANTARCTIC MOUNTAINS

1 Mt Kirkpatrick 4,528 m (14,856 ft)
2 Mt Elizabeth 4,480 m (14,698 ft)
3 Mt Markham 4,351 m (14,275 ft)
4 Mt Bell 4,303 m (14,117 ft)
5 Mt Mackellar 4,297 m (14,098 ft)

▽ **BREAKING THROUGH THE ICE**
Among the few features not buried under ice, the tops of the mountains consist of sedimentary rock layers lying on top of granites and gneisses.

△ **WHITE DOME**
This view of Mount Erebus, which is spouting steam at its summit, is from the ice-covered Ross Sea. Erebus has a moderately steep, irregular dome shape.

C. Antarctica

# Mount Erebus

*Earth's most southerly volcano known to have erupted, with a lava lake and ice caves*

Antarctica's most active volcano, Mount Erebus is a 3,794-m- (12,448-ft-) high stratovolcano, located on Ross Island, just off the coast of East Antarctica. Ross Island also contains three other, apparently inactive, volcanoes.

## Fire and ice

Erebus was erupting when it was first sighted by British explorer Captain James Ross in 1841, and it is still erupting frequently, if not particularly explosively, today. The volcano is one of just a handful on Earth to contain a long-lasting lava lake. This occupies a vent in the main summit crater and produces regular small explosions of lava. The temperature of the swirling pool of fiery magma is 900°C (1,700°F), and it may be several hundred metres deep. Riddling the sides of the volcano are several ice caves, carved out of its thick blanket of ice and snow by steam and other warm gases escaping from fumaroles (gas vents) on its rocky surface. Above some of the caves are ice chimneys, formed by escaping steam turning to trickles of liquid water, and then ice, on encountering the cold air outside.

△ **HOUSE OF ICE**
This ice cave on Mount Erebus is about 12 m (39 ft) high. It is continuously being reshaped as warm gases melt the ice in some parts, while condensing steam deposits new ice in other parts.

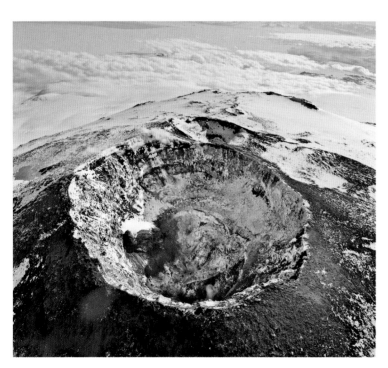

◁ **GAPING HOLE**
Erebus' summit crater is 400 m (1,300 ft) wide and 120 m (400 ft) deep. At its base are several pit-like openings, one of which contains the volcano's lava lake.

## UNDERNEATH MOUNT EREBUS

Mount Erebus, as well as Mount Terror and two other currently inactive volcanoes on Ross Island, are thought to sit over plumes of hot material rising up and forming magma under the Antarctic Plate. These plumes account for the past and present volcanism.

Mount Erebus — magma chamber
Ross Island — McMurdo Sound — Transantarctic Mountains
Mount Terror
Ross Sea — East Antarctic Ice-sheet
— crust of Antarctic Plate
— sediments
asthenosphere — lithospheric mantle
plume of rising, partially melted rock — slow movement of mantle rock

Antarctica

# The Antarctic Ice-sheet

*The world's largest glacier, having a volume of some 30 million cubic km (7.2 million cubic miles) and holding over 60 per cent of Earth's fresh water*

The Antarctic Ice-sheet is by far the most extensive continuous mass of ice on Earth, covering an area of 14 million square km (5.4 million square miles). It consists of two main, neighbouring, parts. The larger of these, the East Antarctic Ice-sheet, is over 4.5 km (2¾ miles) thick in places and lies on top of a large landmass. It has an area of about 12 million square km (4.6 million square miles). In contrast, the smaller West Antarctic Ice-sheet has an area of about 2 million square km (0.8 million square miles) and a maximum thickness of 3.5 km (2¼ miles), and rests on bedrock that is mostly below sea level. Both parts are slightly dome-shaped, with ice flowing from their

highest parts towards the coasts. The rate of ice movement varies from less than a metre per year in some of the higher regions to several hundred metres per year in outlet glaciers and ice streams that drain ice down to the coasts. In some places, the outlet glaciers merge to form vast platforms of floating ice called ice shelves, or extend over the sea as smaller ice tongues.

**Transantarctic Mountains**
Along a section of their length, the mountains form a boundary between the two main parts of the ice-sheet.

the Antarctic Peninsula is mountainous and about 80 percent ice-covered

Southern Ocean surrounds Antarctica on all sides

### ▶ THE ANTARCTIC ICE-SHEET
The ice-sheet, shown here with a cross-section through the Ross Ice Shelf, covers about 99 per cent of the Antarctic landmass. The only ice-free areas are the peaks of the highest mountains and a few coastal areas of limited extent.

### ▷ SPEEDY GLACIER
The massive Byrd Glacier, which drains part of the East Antarctic Ice-sheet into the Ross Ice Shelf, is about 136 km (84 miles) long and 24 km (15 miles) wide. Its flow rate is high for a glacier - about 750 m (½ mile) per year.

*West Antarctic*

**Lake Ellsworth**
This is one of many subglacial lakes. It is buried under about 3.4 km (2⅛ miles) of ice.

the Ronne Ice Shelf is the world's second largest ice shelf

### ▷ ICE FLOES
Sea ice occurs all around Antarctica, although its extent varies seasonally. Unlike ice shelves and icebergs, which originate on land, sea ice is frozen seawater and contains some salt. Flat, freely moving sheets of sea ice are called floes.

# Several hundred **lakes** lie **under the ice-sheet** and **on top** of the **continental crust** beneath it

## ICE FLOW LINES

Snow falling on the Antarctic Ice-sheet compacts to form ice. Under the force of its own weight, this ice slowly deforms and moves towards the coasts along the flow lines. Imaginary lines called ice divides separate regions of ice moving towards different coasts.

ice divide

ice flow line

◁ **ABOVE THE ICE**
In the upper Taylor Valley, which lies in the Transantarctic Mountains, the spectacular Taylor Glacier – an outlet glacier from the East Antarctic Ice-sheet – is channelled by the mountains down towards the Ross Sea.

**High point**
The highest area on the ice-sheet, called Dome Argus, is more than 4,000 m (13,000 ft) above sea level.

*East Antarctica*

△ **CLIFFS OF ICE**
The Mertz Glacier ends in an ice tongue that protrudes about 20–25 km (12–15 miles) over the Southern Ocean. At its terminus are cliffs that tower as much as 50 m (165 ft) above the sea surface.

**Bedrock**
Continental crust underlies the ice-sheet. In some areas, its surface is below sea level.

the Ross Ice Shelf floats with around 90 per cent of its bulk underwater

**Grounding line**
This marks the boundary between the floating part of an ice shelf and grounded ice feeding into it.

Icebergs calving from the Ross Ice Shelf

**Continental shelf**
This occupies a strip about 80–320 km (50–200 miles) wide around the whole of Antarctica.

△ **ROSS ICE SHELF**
The world's largest ice shelf, with an ice cliff along its edge that is more than 600 km (370 miles) long, the Ross Ice Shelf (see p.308) has a total area of about 487,000 square km (188,000 square miles).

# The Ross Ice Shelf

*The world's largest floating body of ice, approximately the same size as mainland France*

S. Antarctica

△ **THE BARRIER**
The near-vertical ice cliff at the front end of the Ross Ice Shelf is up to 50 m (165 ft) high. Underwater, it plunges down another 300 m (980 ft) or so.

The Ross Ice Shelf is a roughly triangular area of floating ice that extends up to 450 km (280 miles) from the shore of Antarctica, over the southern part of the Ross Sea. It ranges in thickness from about 350 m (1,150 ft) at its front end to around 750 m (2,460 ft) further back, close to where its base is in contact with land. Named after Captain James Ross, who discovered it in 1841, the ice shelf was originally called the Barrier. The ice comprising the shelf flows seaward at a rate of about 900 m (2,950 ft) per year and is fed by several outlet glaciers and ice streams that flow down into it from the East Antarctic Ice-sheet (see p.306–307).

These include the Byrd, Nimrod, Beardmore, Scott, Shackleton, and Amundsen glaciers. The upper surface of the Ross Ice Shelf is an inhospitable place, subject to strong cold winds that shape it into a series of ridges and troughs called sastrugi.

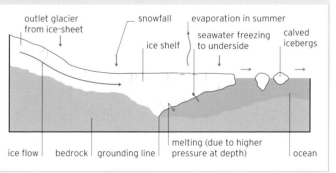

◁ **SHELF SHELTERERS**
Some emperor penguin colonies occupy the shores of the Ross Sea a short distance from the base of the ice shelf, where the steep cliffs provide some protection against wind.

## ICE SHELF GAINS AND LOSSES

An ice shelf gains ice from glaciers flowing into its landward end, from water freezing to its underside, and from new snowfall. It loses ice mainly from icebergs calving (breaking off) from its front edge, and also as a result of some melting and evaporation.

outlet glacier from ice-sheet · snowfall · evaporation in summer · seawater freezing to underside · calved icebergs · ice shelf · ice flow · bedrock · grounding line · melting (due to higher pressure at depth) · ocean

# Antarctic Tundra

*The world's most inhospitable continent for vegetation, where only 1 per cent is covered in plant life*

W. Antarctica

Given that around 99 per cent of Antarctica is covered in ice, it is surprising that any plants survive on Earth's southernmost, most elevated continent. Ice-sheets, glaciers, and mountains make up its interior, yet small patches of tundra exist, mainly on the Antarctic Peninsula and several sub-Antarctic islands, as well as on a few exposed inland rocks known as nunataks. Only two flowering plant species grow here – Antarctic hair grass and Antarctic pearlwort – joined by around 100 moss species, 25 liverwort species, and 300–400 lichen species. Due to the region's extreme climate, where icy winds blow year round and average annual temperatures range from -10°C to -60°C (14°F to -76°F), some algae and lichen species survive inside rocks, growing within tiny pores.

▽ **FLORAL CUSHION**
Antarctic pearlwort grows in cushionlike clumps, reaching a height of around 5 cm (2 in).

# The McMurdo Dry Valleys

*One of Earth's most extreme desert areas, where organisms survive severe cold, arid conditions*

S. Antarctica

The McMurdo Dry Valleys lie west of the McMurdo Sound, bordering the Ross Sea. Covering just 0.03 per cent of the continental landmass, they contain the largest ice-free region in Antarctica and constitute an extreme cold desert ecosystem, where temperatures can plunge to -68°C (-90°F) and winds rip through at up to 322 kph (200 mph). The surrounding mountains block the valleys from the East Antarctic Ice-sheet, and humidity levels are extremely low, yet microbial life survives here in all but the driest areas.

△ **STAINED LANDSCAPE**
The outflow of a saline lake, the Blood Falls gets its red colour from the water's high iron content.

◁ **SUPER-SALTY**
The shallow Don Juan Pond is so salty it does not freeze even in winter, unlike other water bodies in the dry valleys.

△ **BIGGEST ICEBERG**
In 2000, the largest iceberg ever recorded, called B-15 (to left of centre), broke off the Ross Ice Shelf (at bottom in this image).

**Ocean in bloom**
Under the perpetual sunshine of an Arctic summer, a
huge bloom of cyanobacteria - an ancient type of marine
bacteria - spreads out across the waters of the Baltic Sea
between Sweden and Latvia.

# The Oceans

Atlantic
Ocean

# The Mid-Atlantic Ridge

*Part of a global oceanic ridge system, a huge underwater mountain chain that is pushing continents apart*

Running down the middle of the Atlantic Ocean, the Mid-Atlantic Ridge is a slow-spreading submarine ridge system that extends around 16,000 km (10,000 miles) from the Arctic to beyond the southern tip of Africa. The ridge marks a boundary where the North and South American tectonic plates are diverging from the Eurasian and African plates.

## Mountains and valleys

The path of the ridge mirrors the contours of the continents it not only divides but is also pushing apart at a rate of 2–5 cm (¾–2 in) per year – much slower than the 16 cm (6¼ in) per year of the East Pacific Rise (see pp.322–23). The Mid-Atlantic Ridge rises 2–3 km (1¼ –2 miles) above the seafloor in a long chain of underwater mountains, and it has a deep central rift valley along its crest. Where the ridge has pushed above sea level, it has created landmasses such as the volcanic islands of Iceland and the Azores.

## MAGNETIC BANDS AT SPREADING RIDGES

Earth's magnetic field periodically reverses, changing the north magnetic pole to the south magnetic pole, and vice versa. Why this occurs is unknown, but it is recorded in bands of basaltic rock on either side of mid-oceanic ridges. Metallic elements within new crust created at a spreading ridge are aligned with Earth's polarity at the time the basaltic lava emerges. The study of these bands has been a crucial line of evidence in support of the theory of plate tectonics because it helped prove that Earth's plates move.

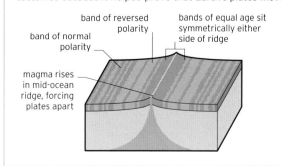

band of reversed polarity

bands of equal age sit symmetrically either side of ridge

band of normal polarity

magma rises in mid-ocean ridge, forcing plates apart

▶ **DIVERGENT BOUNDARY**
Mid-oceanic ridges like the Mid-Atlantic Ridge mark areas where tectonic plates diverge as new crust forms. Deep below the ocean surface, the upwelling of magma from the mantle causes volcanic eruptions, creating mountain ranges, rift valleys, and hydrothermal vents along the ridge's length.

▽ **CARBONATE TOWERS**
During 2000, an undersea research team filmed a cluster of ghostly spires 27–61 m (90–200 ft) high. This is the Lost City, the only known hydrothermal field made solely of white carbonate chimneys.

North America

**Transform fault**
Instead of forming a straight line, the ridge follows a stepped path, with sections offset along faults called transform faults. Movement along these faults often sets off shallow earthquakes.

South America

continental crust

westward movement of South American Plate

oceanic crust

**Asthenosphere**
Below the lithosphere lies the asthenosphere, part of Earth's mantle. Because it remains hot and fluid, it acts like a lubricant on the underside of tectonic plates, allowing them to move.

△ **SEABED PILLOWS**
Lava erupted along the ridge cools rapidly in the cold seawater to form mounds 0.5-1 m (20 in-3 ft) across called pillow lavas.

**Mid-oceanic ridge**
The Mid-Atlantic Ridge is part of a vast global network of mid-oceanic ridge systems – the longest mountain chain on Earth.

Greenland

Iceland

◁ **FRESHWATER FISSURE**
Iceland was formed by hotspot activity (see p.318) merging with the Mid-Atlantic Ridge. Further ridge spreading created a deep fissure, known as Silfra, at the bottom of Thingvellir Valley. Silfra is spring-fed with pure glacial meltwater, providing clear visibility for divers as they swim between the North American and Eurasian tectonic plates.

Africa

△ **ISLAND CHAIN**
The Azores are an example of an island chain formed by a combination of hotspot (see p.318) and ridge activity. When a hotspot occurs near a mid-oceanic ridge, the ridge segments nearest the hotspot are exposed to larger volumes of magma, leading to greater volcanic activity in the area.

continental crust

eastward movement of Eurasian plate

continental shelf

sediment on ocean floor

**Surface eruption**
Once magma erupts through the seabed, it emerges as lava and cools, creating new oceanic crust and causing seafloor spreading.

**Mantle upwelling**
Originating in the upper mantle and lower crust, magma is a mixture of molten and semi-molten rock. It rises at points of weakness in the oceanic crust.

lithospheric mantle

Most of the Mid-Atlantic Ridge's **central rift valley** is **deeper and wider** than the **Grand Canyon**

# The Seychelles

*More than 100 tropical islands of remarkable beauty, home to the world's rarest palm and one of the largest raised coral atolls on Earth*

W. Indian Ocean

△ **TIDAL CHANNELS**
On Aldabra Atoll, channelling tides have created many habitats, including a saline lagoon, mangrove swamps, and seagrass beds that support numerous aquatic species.

The Seychelles' 115 tropical islands lie off the east coast of Africa in the western Indian Ocean and are divided into two groups. Most of the 41 mountainous granite Inner Islands are located 4° south of the equator and are projections of the large undersea Mascarene Plateau that extends south to Réunion. Among the Inner Islands is Praslin, whose

Vallée de Mai World Heritage Site contains Earth's largest forest of endangered coco-de-mer palms. In contrast, the low-lying, coralline Outer Islands are situated beyond the plateau, 10° south of the equator, and include the Aldabra group, a raised coral atoll that has been a designated nature reserve since 1976.

# More than **152,000** Aldabra giant tortoises live on the atoll

▷ **DOUBLE COCONUT**
The coco-de-mer is endemic to the islands of Praslin and Curieuse. Its double coconut is the heaviest seed on Earth and takes seven years to ripen.

# The Great Chagos Bank

*The largest coral atoll on the planet, surrounded by some of the cleanest waters in the world*

C. Indian Ocean

Located a few hundred kilometres south of the Maldives (see below), the Great Chagos Bank is Earth's largest coral atoll in terms of area, although most of its 12,640 square km (4,880 square miles) lie underwater. Just 5.6 square km (2.2 square miles) of the bank's eight islands form low-lying sandy beaches, some covered with coconut palms and bushy vegetation. The bank is part of the vast 640,000 square km (247,100 square mile) Chagos Marine Reserve, which is a "no-take" marine protected area, meaning all fishing, netting, and removing of species is prohibited. With its extremely clean waters and some of the world's healthiest reef systems, and comprising seamounts and deep knolls as well as shallow plains, Chagos is one of the world's richest marine environments.

◁ **MARINE REFUGE**
More than 220 species of coral make up the Great Chagos Bank, creating a reef habitat for around 800 fish species, including anemonefish.

▽ **RACE TO THE SEA**
Bird Island is a centre of wildlife conservation. Here, endangered hawksbill and green sea turtles lay their eggs, which are protected until they hatch. The hatchlings then head out to sea.

# The Maldives

*Asia's smallest country, with more than 1,000 low-lying islands, all at the mercy of the sea*

C. Indian Ocean

Around 1,190 live coral reefs and sandbar islands make up the double chain of atolls that comprise the tropical Maldives, which lie off the southwestern coast of India. The island count fluctuates, for the Maldives has the lowest sea level of any country. With an average elevation of just 1.8 m (6 ft) above sea level, islands come and go as sea level and climate change. All, however, are projections of the underwater Chagos-Laccadive volcanic mountain range. Within each atoll are up to 10 inhabited islands and up to 60 uninhabited ones, most with white sandy beaches that are important nesting grounds for green sea turtles.

△ **GLOWING TIDE**
Bioluminescent plankton light up the surf as waves wash ashore in the Maldives. Marine life in the area ranges from microscopic organisms to whale sharks.

E. Pacific Ocean

# The Hawaiian Islands

*The longest island chain on Earth, where continual volcanic activity has created a unique tropical habitat*

Situated in the Pacific Ocean about 3,800 km (2,360 miles) southwest of the continental US and around 6,200 km (3,850 miles) southeast of Japan, the Hawaiian Islands are some of the most remote in the world. At 2,450 km (1,520 miles) long, they also form the world's longest archipelago.

## Expanding landscape

The eight main islands and more than 100 islets are part of the Hawaiian Island–Emperor Seamount chain of undersea volcanoes that began to erupt around 70 million years ago. Each main island has one or more volcanoes, and three are currently active on Hawaii itself (see pp.318–19). The constant outpouring of lava is still adding to Hawaii's landmass: it has gained about 2 square km (¾ square mile) of land since 1983, created by the continual eruption of Kilauea. Hawaiian animals and plants are as unique as their island homes – more than 90 per cent are endemic, but this makes them vulnerable, and the islands contain at least one-third of all listed endangered species of the US.

▷ **LAVA RIVERS**
Hawaii's Kilauea volcano releases 200,000-500,000 cubic m (7,000,000-17,650,000 cubic ft) of lava each day - enough to resurface 32 km (20 miles) of a two-lane road. Fortunately, most of it flows into the sea.

◁ **STATE FLOWER**
The endangered Hawaiian hibiscus is found on all main Hawaiian islands except Niihau and Kahoolawe, and blooms mainly from spring through early summer.

## THE WORLD'S TALLEST VOLCANOES

A mountain's height is usually measured from sea level to summit, but submerged volcanoes are measured from their bases on the sea floor. Hawaii's Mauna Loa is technically the world's tallest volcano. Its base is a depression it created in the seabed, which gives it a total height of 17,170 m (56,332 ft) - more than 8,230 m (27,000 ft) taller than Mount Everest.

Mauna Kea, Hawaii: 10,204 m (33,478 ft)

Mauna Loa, Hawaii: 9,170 m (30,085 ft)

Piton des Neiges, Réunion: 7,071 m (23,199 ft)

Teide, Tenerife: 7,500 m (24,606 ft)

Ojos del Salado, Andes: 6,893 m (22,615 ft)

sea floor

sea floor depressed by 8,000 m (26,000 ft)

# The formation of a
# VOLCANIC ISLAND CHAIN

**Kilauea** has released enough **lava** to **encircle** Earth three times

Like beads in a necklace, volcanic islands often form in arcs, or chains, where two oceanic tectonic plates converge or where an oceanic plate meets a continental plate. The volcanic island chains that encircle the Pacific Ocean, which are known as the Ring of Fire, are formed by both processes, as well as, in the case of the Hawaiian Islands, hotspot volcanism.

## Hot marks the spot

Hotspots are areas in Earth's mantle that allow magma to rise through the lithosphere until it erupts, often underwater on the seabed. The hotspot that formed the Hawaiian island chain is located in the middle of the Pacific Plate. While the hotspot is fixed, the plate moves continuously. As the plate migrated over the hotspot, the latter created the shield volcanoes responsible for all 132 islands, atolls, reefs, banks, and seamounts of the Hawaiian archipelago. Currently, the hotspot lies beneath the island of Hawaii, where three volcanoes continue to erupt. It also generates activity at the youngest member of the chain, the submarine volcano Loihi, about 30 km (19 miles) south of Hawaii. Loihi last erupted in 1996, following a series of minor earthquakes.

### HOTSPOTS AND FLOOD BASALTS

Hotspots are thought to be stationary. As tectonic plates move over them, chains of progressively older volcanoes form in what is known as a hotspot track. The track follows the direction of plate movement, with the youngest, most active volcanoes and seamounts found nearest the hotspot and the less active and extinct ones farthest away. Hotspot tracks are often marked by large areas of basaltic lava, which flooded the area during peak eruption periods.

| KEY | ● Hotspot | – – – Hotspot track |
| --- | --- | --- |
| | ■ Flood basalt | —— Plate boundary |

**Niihau**
Formed around 4.9 million years ago.

**Kauai**
Formed around 5.1 million years ago.

**Oahu**
Formed 3.7–2.6 million years ago.

oceanic crust

asthenosphere

**Lithospheric mantle**
The crust and uppermost layer of the mantle make up the lithosphere, which is broken up into mobile tectonic plates.

**Movement of plate**
The Pacific Plate moves over the asthenosphere from southeast to northwest at a rate of 5–10 cm (2–4 in) a year.

◁ **SCULPTED CLIFFS**
The fluted Na Pali cliffs dominate the northwest coast of the island of Kauai, towering up to 1,200 m (4,000 ft) above the Pacific Ocean. The flanks of shield volcanoes often collapse, resulting in jagged, water-permeable surfaces that are further sculpted by rain and wave action.

▲ **THE HAWAIIAN ISLANDS**
This cross-section shows how conditions in Earth's interior have given rise to the Hawaiian Islands. The Pacific Plate moves over a mantle hotspot, setting off volcanic activity at the surface. With increasing distance from the hotspot, the islands cool and activity diminishes.

## ▷ GREEN SAND

One of only a handful of green sand beaches in the world, Papakolea borders Mahana Bay, on the southern tip of Hawaii. The sand gets its colour from olivine, a mineral released by eruptions from a now-dormant volcano, whose cinder cone forms three sides of the bay.

## ▽ LAND BUILDER

One of the world's most active volcanoes, Kilauea's eruptions appear violent, but are tame compared to more explosive continental volcanoes. Its basaltic lava flows build layer upon layer of what will one day be fertile soil.

**Molokai**
Formed 1.9–1.8 million years ago.

**Maui**
Formed 1.3–0.8 millon years ago.

**Hawaii**
Also known as Big Island, Hawaii began to form less than 0.5 million years ago and is still forming today due to the activity of three of its five volcanoes.

movement of plate drags head of mantle plume

**Magma chamber**
A magma chamber forms as heat from the mantle plume melts part of the lithosphere. As the plate moves across a magma chamber, the molten rock pushes upwards in weaker areas, creating a volcano that punches through the plate.

**Magma intrusion**
The magma rises, or intrudes, into lower-density rock, creating dikes. Eventually enough dikes are created to allow magma to collect in a chamber.

**Mantle plume**
Mantle plumes begin as large columns of molten rock (magma), which form deep within Earth's interior and rise through the mantle. Upon reaching the base of the lithosphere, the magma spreads, forming a cap or plume.

## ◁ IRON SIDES

Bacteria that feed on iron deposits in volcanic material form yellowish-orange mats, or flocs, as they rust (oxidize) the metal. Many such flocs have been found on the submerged flanks of the Loihi seamount, Hawaii's youngest submarine volcano.

E. Pacific Ocean

# The Galápagos Islands

*A group of Pacific islands, whose unique ecosystem and wildlife provide*
*extraordinary insights into evolution*

Clustered around the equator about 1,000km (620 miles) off the coast of Ecuador, the rocky Galápagos Islands consist of 13 major islands, six smaller ones, and more than 100 small outcrops. All are situated on the Nazca tectonic plate, whose slow but steady movement eastwards results in frequent volcanic eruptions. Three major ocean currents converge here, bringing in a variety of marine creatures as well as creating a much cooler climate than expected at the equator.

## A window on evolution

Due to the islands' remote location and volcanic conditions, life in the Galápagos evolved in relative isolation, and many species, such as native iguanas and giant tortoises, have not changed radically since prehistoric times. The adaptations of animals on different islands that led to their divergence as separate species inspired the British naturalist Charles Darwin to develop his groundbreaking theory of evolution in the 19th century. Today, the Galápagos is one of the world's most scientifically and biologically important regions, where unique native species make up a very high proportion of the islands' wildlife.

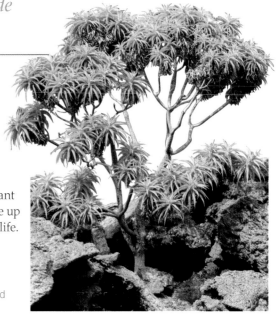

▷ **DAISY TREE**
The shrubby Stewart's scalesia grows only on Santiago and Bartolomé islands. Its daisy-shaped flower heads release dandelion-like seeds.

# Geologically, the Galápagos are **young islands**, the **oldest** having **emerged** just 4.2 million years ago

### GALÁPAGOS VEGETATION ZONES

On the Galápagos, rainfall generally increases with altitude, creating three main vegetation zones. The arid zone harbours cacti and other drought-resistant plants. On larger islands, small trees and shrubs dominate the higher transition zone, before giving way to lush forests in the humid zone.

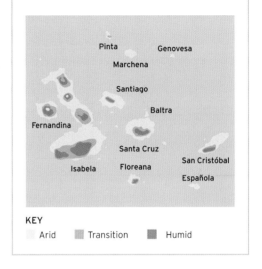

**KEY**

◻ Arid    ◼ Transition    ◼ Humid

△ **SPECIALIZED BILL**
The common cactus finch uses its specially adapted bill to feed on the pollen, nectar, fruit, and seeds of *Opuntia* cacti.

◁ **GIGANTIC CALDERA**
The Sierra Negra volcano is one of six volcanoes that formed Isabela Island. Five are still active, including Sierra Negra, which has an elliptical caldera.

▽ **VOLCANIC ISLET**
One of the archipelago's youngest islands, Bartolomé Island off the coast of Santiago is home to a breeding colony of Galápagos penguin, the world's second smallest penguin species.

E. Pacific Ocean

# The East Pacific Rise

*A volcanic ridge deep on the ocean floor, where life thrives in the harshest of conditions*

The East Pacific Rise is part of a volcanic mid-oceanic ridge system that extends along the floor of the southeastern Pacific Ocean. It lies at a depth of about 2.4 km (1½ miles) and rises about 1,800–2,700 m (5,900–8,900 ft) above the sea floor.

Oceanic ridges occur where tectonic plates move apart (see pp.312–13). The process, called rifting, creates fissures along the seabed, releasing lava that forms a ridge as it cools. New basaltic lava is added to either side of its crest as the plates spread apart. Compared to the Mid-Atlantic Ridge, the East Pacific Rise spreads relatively quickly and lacks a pronounced central rift valley. The rifting process produces geyser-like hydrothermal vents that discharge heated, mineral-laden seawater. Despite toxic sulphides, high temperatures, and enormous underwater pressures, the vents support an amazing community of creatures, from single-celled organisms to giant tubeworms.

## THE WORLD'S VENT FIELDS

The first hydrothermal vents were discovered in 1977 by an expedition exploring a spreading ridge near the Galápagos Islands. Vent fields have since been recorded in all of the oceans. They are formed in volcanically active areas, which is why so many are found along diverging plate boundaries that coincide with mid-oceanic ridges. Others are created by hotspots (see p.318) or where tectonic plates converge and one pushes beneath (subducts) the other.

KEY   ● Hydrothermal vent field   —— Plate boundary

◁ **STRANGE FLOWER**
The ocean dandelion is a type of deep-sea siphonophore, a group of animals that includes jellyfish and the Portuguese man o' war. It is not one, but a collection of individuals that share tissues.

**Vent clams**
These clams anchor themselves into basalt crevices on the seabed in clumps known as clambakes.

**Vent mussels**
Deep-sea mussels harbour symbiotic bacteria that convert hydrogen released by hydrothermal vents into energy.

**Bacterial mats**
Bacteria that harvest vent minerals for food form thick mats, which are eaten by other creatures.

▶ **HOW BLACK SMOKERS ARE FORMED**
Hydrothermal vents known as black smokers form when seawater penetrates fissures in new crust created by volcanic activity. Magma superheats the water to 400°C (750°F) or more, dissolving material from surrounding rocks. This superhot vent fluid rises back through openings in the seabed. As they cool, the dissolved minerals precipitate into solid form, creating mineral- and metal-rich chimney-like vents.

**Cold seawater descends**
Seawater at about 2°C (36°F) seeps down through fissures in the seabed created by volcanic activity, where it is heated by magma.

## More than **500 species** have been found living at **hydrothermal vents**

**△ BLACK FLUID**
Hydrothermal vents emit black or white "smoke" – heated fluids filled with dissolved minerals. Black emissions get their colour from iron monosulphide.

**Shaft**
As minerals accumulate, building the chimney walls, a central shaft allows the vent fluid to escape.

**Chimney**
As hot vent fluid meets cold seawater, mineral deposits create chimney-like channels up to 15 m (50 ft) high.

**△ COLOURFUL VENT**
Microbial mats also form on the sides of vents, where the action of bacteria on iron compounds released from vent fluid turns the walls a rusty orange.

**Vent crab**
This white crab feeds on bacteria, clams, mussels, and tube worms.

**Tube worms**
Giant tube worms, which reach lengths of up to 2 m (6 ft), anchor themselves around a vent. Bacteria inside each worm's body converts vent chemicals into food.

**△ TOP PREDATOR**
Vent eelpouts are believed to feed on shrimp and crustaceans in vent communities. Due to the water's toxic levels of sulphides, eelpouts have few competitors.

**Mound**
As dissolved minerals in the vent fluid meet cooler seawater, they fall out of solution, building up a mound of debris as well as the vent chimney.

**Heat source**
Far below the seabed, upwelling magma superheats the incoming seawater, dissolving minerals such as metals and sulphur from surrounding rocks.

**Heated water rises**
The superheated vent fluid rises back to the surface of the seabed through fissures in the ocean crust.

W. Pacific

# The Mariana Trench

*The deepest known point on the seabed of planet Earth, created by the collision of two oceanic plates*

On the floor of the western Pacific Ocean, around 322 km (200 miles) southwest of the island of Guam, the Mariana Trench plunges to a staggering 11,034 m (36,201 ft) at a section called the Challenger Deep – the deepest point yet discovered in any of Earth's oceans. If the world's highest mountain, Mount Everest, were placed inside the trench, it would still be covered by 2,183 m (7,162 ft) of water.

## When plates collide

The Mariana Trench marks the location of a subduction zone, the point at which one oceanic tectonic plate sinks under another. Here, the huge Pacific Plate has plunged so dramatically beneath the much smaller Philippine Plate and Mariana Microplate that it has created the steep-sided trench, which, at over 2,540 km (1,580 miles) long, is more than five times longer than the Grand Canyon yet averages just 69 km (43 miles) wide.

### PARTS OF THE OCEAN FLOOR

Just like landmasses, seabeds contain plains, ridges, mountains, and canyons. From the shoreline, the continental shelf inclines gradually until it drops away sharply into the deep, open waters of the continental slope. Where there is no mid-ocean ridge system, this levels out into the abyssal plain before plunging into deep-sea trenches such as the Challenger Deep.

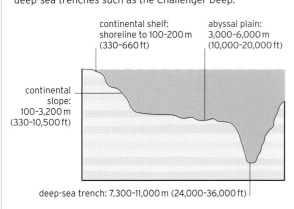

continental shelf: shoreline to 100–200 m (330–660 ft)

abyssal plain: 3,000–6,000 m (10,000–20,000 ft)

continental slope: 100–3,200 m (330–10,500 ft)

deep-sea trench: 7,300–11,000 m (24,000–36,000 ft)

▷ **WHITE SMOKE**
Undersea volcanic activity creates hydrothermal vents along the Mariana volcanic arc, such as this white smoker – the so-called Champagne Vent – which releases bubbles of liquid carbon dioxide.

▶ **SUBDUCTING PLATE**
These cross-sections of the Mariana Trench subduction zone show many of the geological processes involved. As the two oceanic plates converge, one pushes beneath the other, creating the trench itself but also setting in motion the processes that produce chains of seamounts and islands, and new ocean floor.

**Remnant arc**
The remnant arc once contained active volcanoes, but it has since been pushed away from eruptive areas and is now inactive.

**Philippine Plate**
Although the Philippine Plate is overriding the Pacific Plate, the upwelling mantle results in spreading, and the plate moves away from the subduction zone.

**Active back-arc basin**
Back-arc basins are created behind island arcs as a result of rifting and spreading of the sea floor due to the huge forces created by the presence of a deep ocean trench.

**Remnant arc**
When volcanic activity at an undersea ridge ceases, the ridge is known as a remnant arc.

seamount, an inactive submarine volcano

oceanic crust

lithospheric mantle

asthenosphere

**Mariana microplate movement**
The Mariana island arc rests on a tectonic microplate called the Mariana Plate. Located between sections of the Philippine and Pacific plates, it is diverging from the former, while the latter is subducting beneath it at the Pacific Plate's southeastern end.

**The Challenger Deep**
Despite total darkness, extremely cold water, and crushing pressure, the Challenger Deep is a habitat that supports a host of living creatures.

**Subducting plate**
As the Pacific Plate slides beneath the overlying plate at a rate of about 3 cm (1¼ in) per year, the plates rub against each other, generating tremors that can be felt across a wide area.

The **pressure** at the bottom of the **Challenger Deep** is **equal** to that underneath **220 Empire State Buildings** stacked on top of each other

**Seafloor spreading**
Behind the volcanic arc, there is an active spreading ridge where the Philippine and Mariana plates are moving away from each other. Here, intense volcanic activity creates new oceanic crust as magma rises from the underlying mantle.

Mariana Microplate

**Active arc**
Most volcanoes of the active volcanic arc are seamounts, but some have risen enough to form islands.

**Volcanic activity**
As the Pacific Plate sinks deep into the mantle, it releases water that triggers melting in the overlying rocks.

**Forearc**
The region between the volcanic arc and trench is called the forearc. Here, the forces of subduction are evident in the numerous faults that slice up the forearc rock sequences.

**Pacific Plate**
Faults in the Pacific Plate enable it to bend and plunge almost vertically beneath the Philippine Plate.

△ ARC ISLAND
The Mariana island arc parallels the curve of the Mariana Trench. Volcanic island arcs like this are created by the eruption of magma along a line roughly equidistant from the trench, which over time builds up the seafloor to create volcanic structures that rise above the waves. The island of Guam was formed by the union of two volcanoes.

**Guyot**
A guyot is a seamount that was once above sea level, where wave erosion planed off its top, creating a flat surface. As the seabed subsided, the guyot sank beneath the waves.

**Mariana Trench**
The sinking Pacific Plate pulls the overlying seafloor down with it, creating a deep trench, or groove, at the bottom of the ocean.

**Pacific Plate movement**
Both the Pacific and Philippine plates are moving in a northwesterly direction relative to the underlying asthenosphere, but the Pacific Plate is moving faster.

◁ COLOURFUL FISH
Small, brightly coloured fish such as this groppo are common at depths of 350–500 m (1,150–1,640 ft) around the seamounts closest to the trench.

**Dust on the move**
A storm sends dust into the air across hundreds of kilometres
of the Sahara. The clouds in the image suggest that the dust is
being driven by a desert wind called a haboob. More than half
of all the dust in the world's oceans blows off the Sahara.

# Extreme Weather

# A large **tropical cyclone** releases as much **energy** as **half of Earth's electrical generating capacity**

# Cyclones

*The intensity and scale of cyclones makes them the most destructive phenomena in Earth's atmosphere*

A cyclone is a rotating storm system centred around an area of low pressure. It forms as warm, moist air rises, causing a fall in pressure, which draws in more moist air from all directions. An effect known as the Coriolis force causes the air to spiral – anticlockwise in the northern hemisphere and clockwise in the southern hemisphere.

## Tropical cyclones

In temperate regions, cyclones are often called depressions and are only rarely destructive. They are at their most intense in the tropics. They are called hurricanes in the North Atlantic, cyclones in the Indian Ocean, and typhoons in the western Pacific. A tropical cyclone forms over the ocean in the presence of slack winds and surface temperatures of more than 26°C (79°F) – usually in late summer or early autumn. It becomes a full cyclone when winds exceed 118 kph (73 mph), although sustained speeds can exceed 320 kph (200 mph). Combined with torrential rainfall, this can cause flooding and mudslides and bring down buildings. During their most vigorous phase, cyclones track west. They begin to die only once they make landfall and their supply of warm ocean water is cut off.

## TROPICAL CYCLONE REGIONS

Most tropical cyclones develop in two bands between 10 degrees and 30 degrees north and south of the equator. At higher latitudes, ocean surface temperatures are unlikely to be warm enough to fuel the storm's development. Closer to the equator, the Coriolis force will not be great enough to set circulation in motion.

▼ **TROPICAL CYCLONE**
A section through a tropical cyclone reveals its central eye, where there is an area of sinking air, surrounded by a series of cloudbands.

▷ **MATURE STAGE**
Seen from overhead, a full-blown tropical cyclone, such as Hurricane Rita, will have a roughly circular cloud pattern as well as a distinct eye and spiral rainbands.

**Flat top**
The top of the storm may be as high as 15 km (9 miles) above sea level.

**Downpours**
The area beneath a cloudband is subjected to heavy rainfall, strong winds, and lightning.

◁ **FORCE OF NATURE**
Palm trees bend in intense winds as Hurricane Dennis hits the coast of Florida in July 2005, although the storm surge created more damage than the winds.

**Cloudbands**
Spiralling around the eye are cloudbands that can extend for hundreds of kilometres from the cyclone centre, gradually diminishing in height.

**▷ THE EYE OF THE STORM**
Air in the relatively calm centre of a tropical cyclone warms as it descends, leaving it clear of cloud. It is possible to see blue sky from the ocean surface as the eye passes.

**Cold exhaust**
Cooled air at high altitudes spirals away from the eye or descends into it.

**Cirrus shield**
Cooled air fans out into high-altitude cirrus cloud, spreading 1,000 km (620 miles) or more.

**STORM SURGES**

Hurricane-force winds blowing onshore bunch ocean water up to abnormal levels in shallow coastal waters. This creates a storm surge – an abnormal rise of water that is higher than the predicted astronomical tides. When combined with normal high tide, this creates a storm tide and can cause extreme flooding in coastal areas.

height of storm surge

height of storm tide

low tide      mean sea level      normal high tide

**Eye**
In the eye, typically 30-65 km (20-40 miles) across, air warms as it descends. The winds are lighter than anywhere else in the storm, but the ocean beneath the eye is rough.

**Eyewall**
The fastest winds are in the eyewall.

**Storm surge**
As the storm reaches the coast, a storm surge occurs where winds are blowing onshore.

**Moat**
Clear rings between concentric cloudbands are where cooled air is slowly sinking.

**△ STORM STRUCK**
Around 200,000 homes were damaged, mainly by flooding caused by the storm surge, when Hurricane Sandy made landfall on the eastern seaboard of the United States in 2012.

# Thunderstorms

*Thunderstorms are rain showers that produce powerful electrical discharges called lightning*

Thunderstorms come in different shapes and sizes. Single-cell storms typically last 20–30 minutes and may produce heavy rainfall and hail. Larger multicell storms may generate weak tornadoes. The most severe thunderstorms are supercells, which are bigger and may produce violent tornadoes. Severe thunderstorms are classed as those with hailstones 2.5 cm (1 in) in diameter, winds of more than 90 kph (56 mph), or tornadoes.

## Lightning strikes

As tiny water droplets and ice crystals in the cold tops of the clouds bump into each other to form bigger drops of rain and hailstones, these particles develop positive and negative electrical charges. It is this difference in charge (potential difference) between the particles that creates lightning, as a flow of electrons moves from the negatively charged particles to the positive. Lightning is extremely hot, and as it passes through air it causes the air to expand instantaneously. This expansion creates a shockwave – the familiar rumble of thunder. Lightning zigzags between one part of the cloud and another, between the cloud and air, or between the cloud and the ground.

**◁ BLUE JETS**
This type of lightning is found 40-50 km (25-30 miles) above the ground, projecting from the top of a large storm cell to the upper atmosphere.

**Cumulonimbus clouds**
The lower part is composed of water droplets but the upper regions are composed of supercooled droplets and ice crystals.

back sheared anvil

rear downdraft

interior cloud wall

**Cyclone core**
This is the area of the storm cell where warm air spirals upwards through the cloud.

flanking line, a mass of developing cumulus clouds extending from the main storm

circulating banks of wall cloud

**▶ SUPERCELL**
No two supercell storms are the same but they do all share most of the features shown here: a huge cumulonimbus cloud, wind blowing around an area of low atmospheric pressure, a large updraft of warm air, heavy rainfall and hail, and downdrafts of cooled air.

**Rain-free side**
Rain does not fall from every part of the cloud.

**Warm air inflow**
Rising air in the middle of the cell produces low pressure near the ground. This sucks in more warm air and keeps the system going.

rotating updraft

the rapid vortex of a tornado whips up soil and debris from the ground

**Tornado**
Destructive, rapidly spinning vortices may develop beneath about 30 per cent of large supercells.

**Cloud-to-ground lightning**
These powerful discharges reach a temperature of 30,000 °C (54,000 °F).

# At **any one time** about **2,000 thunderstorms** are **taking place**, mostly in wet tropical regions

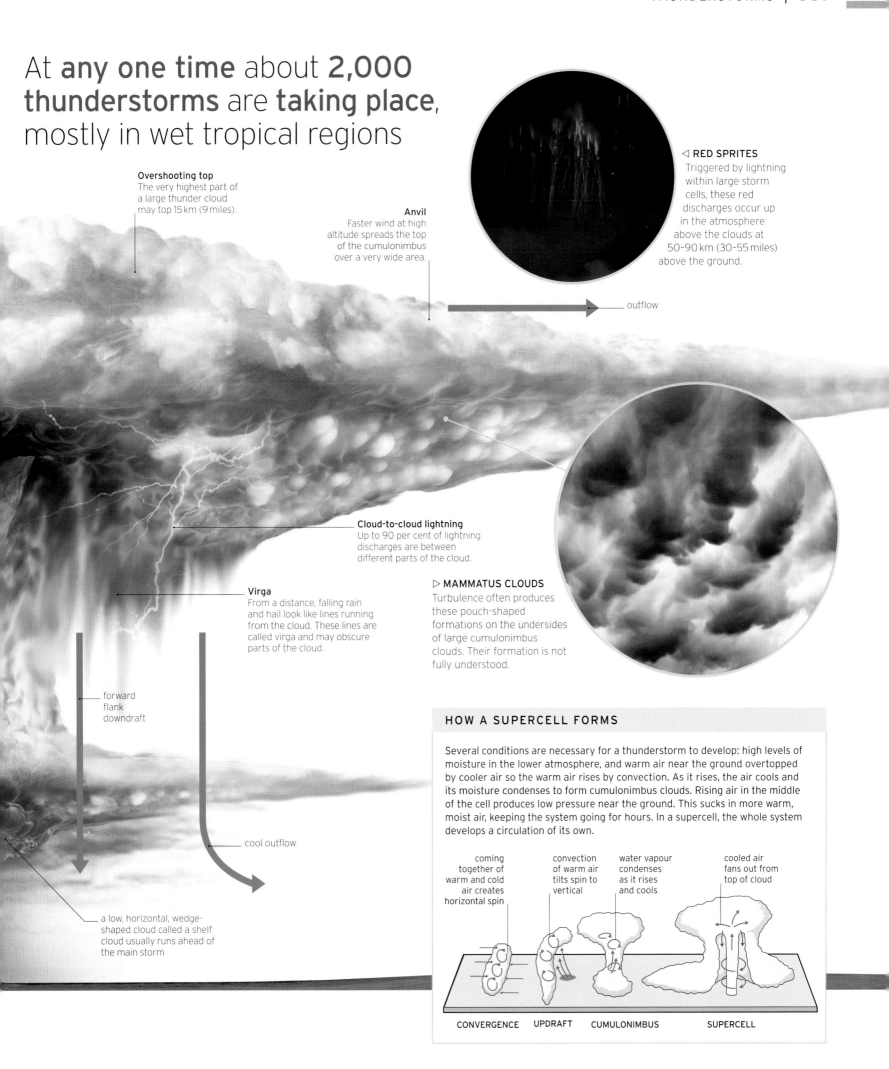

**Overshooting top**
The very highest part of a large thunder cloud may top 15 km (9 miles).

**Anvil**
Faster wind at high altitude spreads the top of the cumulonimbus over a very wide area.

⊲ **RED SPRITES**
Triggered by lightning within large storm cells, these red discharges occur up in the atmosphere above the clouds at 50–90 km (30–55 miles) above the ground.

outflow

**Cloud-to-cloud lightning**
Up to 90 per cent of lightning discharges are between different parts of the cloud.

**Virga**
From a distance, falling rain and hail look like lines running from the cloud. These lines are called virga and may obscure parts of the cloud.

▷ **MAMMATUS CLOUDS**
Turbulence often produces these pouch-shaped formations on the undersides of large cumulonimbus clouds. Their formation is not fully understood.

forward flank downdraft

cool outflow

a low, horizontal, wedge-shaped cloud called a shelf cloud usually runs ahead of the main storm

## HOW A SUPERCELL FORMS

Several conditions are necessary for a thunderstorm to develop: high levels of moisture in the lower atmosphere, and warm air near the ground overtopped by cooler air so the warm air rises by convection. As it rises, the air cools and its moisture condenses to form cumulonimbus clouds. Rising air in the middle of the cell produces low pressure near the ground. This sucks in more warm, moist air, keeping the system going for hours. In a supercell, the whole system develops a circulation of its own.

coming together of warm and cold air creates horizontal spin

convection of warm air tilts spin to vertical

water vapour condenses as it rises and cools

cooled air fans out from top of cloud

CONVERGENCE    UPDRAFT    CUMULONIMBUS    SUPERCELL

# Tornadoes

*Virtually impossible to predict accurately, tornadoes are the most capricious weather phenomena – and some of the most violent*

With a tight vortex of air spinning at up to 480 kph (300 mph), a grade F "twister" is the most terrifying sight in nature, capable of destroying almost anything in its path. To meet the definition of a tornado, the vortex must extend from the base of a cloud and be in contact with the ground.

## Types of tornadoes

Tornadoes can happen anywhere but most occur on the Great Plains of the US. They occur in different shapes and sizes. Rope tornadoes have a narrow funnel, while wedge tornadoes have a wide diameter at the ground and are even broader at the cloud base. Tornadoes change colour depending on the nature of the debris and dust that they pull into their violent vortex and may be near-white, brown, reddish, or nearly black. Their severity is measured on the Fujita scale. An F-0 twister – the most common type – is capable of breaking branches from trees, while F-5s are able to carry cars 100 m (330 ft) through the air.

## Tornadogenesis

The ideal situation for a tornado to form, a process known as tornadogenesis, is when a cold air mass converges with a warm, moist one, creating instability and towering cumulonimbus clouds. Where there is also wind shear (a change in wind speed or direction with height), a slowly rotating supercell storm may develop (see pp.328–29). If an input of warm, rising air is in contact with the rapidly descending air of the storm's rear flank downdraft, a narrow column of rapidly spinning air develops. The rotating air continues downwards, forming a funnel below the cloud. Air drawn into the funnel enters an area of much lower pressure, so it expands and cools, and moisture condenses as a result. The funnel becomes a tornado on touching the ground.

The biggest twister on record is the Tri-State Tornado of March 1925, which was in contact with the ground for at least 352 km (219 miles). It wreaked havoc across Missouri, Illinois, and Indiana, leaving 695 dead. Bangladesh has the highest annual death toll caused by tornadoes, averaging almost 200 each year.

## Bangladesh's **Daulatpur– Saturia tornado** of 26 April 1989 **killed 1,300 people**

### TORNADO ALLEY

Hundreds of tornadoes are recorded in this region of the central US every year, mostly in spring and early summer. The classic situation for tornadogenesis on the southern Great Plains is a combination of cold, dry air moving southeast from the Rockies and very warm, moist air advancing north from the Gulf of Mexico. This produces atmospheric instability and the growth of large storm cells, some of which produce tornadoes.

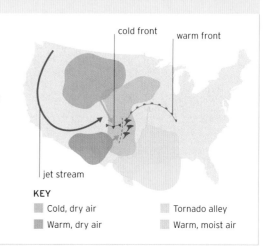

**KEY**

| | |
|---|---|
| ▨ Cold, dry air | ▨ Tornado alley |
| ▨ Warm, dry air | ▨ Warm, moist air |

### △ FORCE OF NATURE
Professional storm chasers monitor an approaching tornado in Kansas. Tornadoes grow from the base of large storm clouds and pick up dust and debris when they make contact with the ground.

### △ ROPE TORNADO
A sinuous rope tornado tears across the Kansas plains and ravages the countryside. Kansas is one of the states most frequently hit by tornadoes.

### ◁ OVER THE SEA
Waterspouts form when tornadoes pass over water. Most do not suck up water but kick up spray, like this pair over the Mediterranean Sea.

# Sandstorms and Duststorms

*In arid and semiarid areas, high winds can whip up sand and dust to form powerful storms, blocking out the Sun and stripping away soil*

A severe sand- or duststorm can create darkness in the middle of the day and reduce visibility to virtually nothing. These storms are formed when strong winds and dry conditions cause surface materials – dust, soil, or sand – to be lifted in a process called saltation. The wind carries particles a short distance before they fall back to the ground, dislodging other particles and accelerating the process. They remain suspended in the lower atmosphere.

The height and extent of the storm depends on the strength and persistence of the wind and the size of the particles. Wind gusts often exceed 80 kph (50 mph). Dust particles are smaller and are carried higher than sand – exceptionally to 6,100 m (20,000 ft). Some storms are driven by cold fronts (the leading edge of a mass of cold air pushing under warmer air) while others, called haboobs (see panel, below), are created by strong downdrafts from storm cells.

## Cold fronts

The most notorious duststorms struck the High Plains of the US on 14 April 1935, turning a warm, sunny afternoon into a choking, zero-visibility blackout in parts of Oklahoma, Texas, Kansas, Nebraska, and Missouri. A combination of an extended drought, poor soil management, and high winds produced by a rapidly moving cold front created several huge duststorms. The great Australian duststorm of 2009, also created by a cold front, carried 2.5 million tonnes (2.75 million tons) of the Australian continent into the ocean.

## HOW A HABOOB FORMS

Derived from the Arabic word *habb*, meaning "wind", a haboob is formed when strong winds blow out from a storm cell. The cell forms when warm, moist air rises, but air within that updraft is cooled by evaporating rain. This air may form a powerful downdraft and outflow, led by a gust front. As the front hits dry ground in advance of the storm, it kicks up surface dust or sand. The gust front produces a "wall" of dust, which may spread over a larger area than the storm itself.

Height
KM MILES
8 : 5
6 : 4
4 : 2½
2 : 1¼
0 : 0

cumulonimbus clouds

dust

warm air rising

cool downdraft

wake

strong wind

gust front

▽ **RED DUST**
Sydney Harbour Bridge was shrouded in red dust during the great duststorm of September 2009. This is recorded to have been the worst to hit eastern Australia in the last 70 years.

▷ **A CITY ENGULFED**
A billowing wall of dust can be seen as it is about to engulf downtown Phoenix, Arizona, in July 2011. The dust cloud was generated by a haboob, one of several to affect the city that summer.

At its greatest extent, the **front** of the **September 2009 Australian duststorm** stretched for **3,450 km (2,140 miles)** from north to south

◁ **PARCHED LAND**
Duststorms increased in frequency in southern China after a severe drought in 2010–11 dried out farmland and reservoirs, producing huge quantities of surface dust.

△ **ICE SCULPTURE**
Ice storms can produce some
beautiful structures. This is
St Joseph's lighthouse, on
the shore of Lake Michigan,
after it was glazed by an
ice storm in January 2015.

▷ **DEEP FREEZE**
This satellite photograph shows
the Great Lakes frozen during
an ice storm in February 2014.
The event left more than a million
people without power and
grounded thousands of flights.

# Ice Storms

*Unique meteorological phenomena that produce spectacular beauty and destruction in equal measure*

Unlike in other storms, conditions are often calm and quiet during an ice storm. But in terms of loss of life, disruption to transport and power networks, and general mayhem, a major ice storm can produce as much chaos as a tornado or hurricane. The vital elements that cause this phenomenon are a combination of heavy rain falling from a wedge of warm air, an underlying area of supercold air, and freezing ground temperatures. When the rain hits the ground, or any subzero structure close to it, it freezes and forms a layer of ice known as glaze ice. As the rain persists, this glaze ice grows thicker.

## Collapse under pressure

Glaze ice can be over 5 cm (2 in) thick. In 1961, a layer 20 cm (8 in) thick was measured in parts of Idaho. Since ice is ten times as heavy as the equivalent thickness of wet snow, the effects are often dramatic. Apart from turning everything white, power lines, trees, and unstable buildings are brought down. Roads become impassable, and aircraft are grounded. In January 1998, an ice storm hit a vast area of New England and southeast Canada – a region particularly susceptible to these events. At least 44 people died, with hundreds more injured. Some 129,000 km (80,000 miles) of power cables collapsed, leaving four million people without power.

◁ **FROZEN FRUIT**
Thousands of apple trees in New York State, which were damaged in the January 1998 ice storm, took years to recover.

## SLEET, SNOW, AND FREEZING RAIN

If snow passes first through a wedge of warm air then a shallow layer of cold air, it turns into rain, then freezes as it hits the ground. If it passes through a thicker layer of cold air beneath the warm, it partially refreezes as sleet. If it does not pass through the warm air, it lands as snow.

high-altitude cloud

wedge of warm air

warm air moving over cold air | rain freezes on impact | glaze ice on surface | sleet | snow

# Ice can **increase** the **weight** of **branches** by **30 times**

# Aurorae

*Spectacular displays of colourful, shimmering lights in the night sky*

The aurora borealis, or northern lights (in the northern hemisphere), and the aurora australis, or southern lights (in the southern hemisphere), appear at latitudes greater than 60° North and South, respectively. They are centred on Earth's magnetic poles and occur when electrically charged particles from the Sun are funnelled into Earth's magnetic field above the poles and collide with atoms in the upper atmosphere to trigger reactions that produce streams of light (see panel, right). Some displays are almost static patches of diffuse brightness, but others feature constantly moving lights suggestive of curtains waving in a breeze. Every fold of the curtain consists of many parallel rays, each lined up with the local direction of Earth's magnetic field. The colour of an aurora depends on its height and which gas in the atmosphere is being excited. Green is the most common colour, but displays of red, blue, violet, or even pink occur occasionally.

Cloudless conditions away from light pollution during the long nights of winter are ideal for watching aurorae. Alaska, Canada, northern Scandinavia, and northern Russia are the best places to see the northern lights. The southern lights can be seen from southern South America, Tasmania, and New Zealand's South Island, although near-inaccessible Antarctica has the best shows.

The **lights** of an aurora are **produced** mostly **90–150 km (56–90 miles) above Earth**, although very high (red) displays may be **up to 1,000 km (620 miles)** away

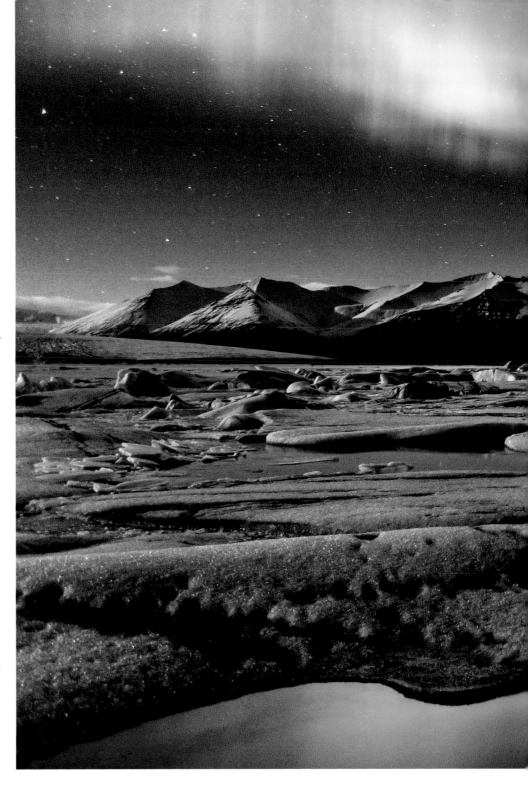

△ **CURTAIN OF COLOUR**
The dazzling display of an aurora can be seen reflected in the waters of the Jökulsárlón glacial lagoon in Iceland.

▷ **A VIEW FROM SPACE**
Astronauts orbiting in the International Space Station have a unique view of the aurora borealis.

## △ AURORA CORONA

The aurora borealis may sometimes appear overhead as a "corona" of rays. This spectacular display was photographed over Iceland.

## WHAT CAUSES AURORAE

Aurorae are caused by charged particles (electrons and protons) that speed out from the Sun. Earth's magnetic field provides a shield against these, but many electrons follow the magnetic field lines towards the poles, where they collide with oxygen and nitrogen atoms and molecules in the upper atmosphere. The electrons excite the atoms, and when the atoms return to their former state, they release energy in the form of photons, which create the dancing patterns of aurorae.

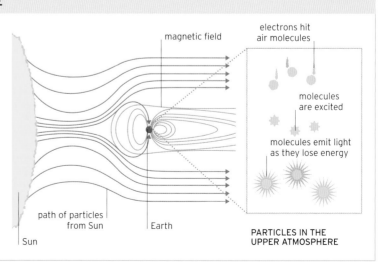

magnetic field

electrons hit air molecules

molecules are excited

molecules emit light as they lose energy

path of particles from Sun

Earth

Sun

**PARTICLES IN THE UPPER ATMOSPHERE**

**Salty shallows**
Water enters Tanzania's Lake Natron through a river delta.
Hot springs provide the lake's other inflow. High rates of
evaporation mean that the lake is shallow – just 3 m (10 ft)
deep – and extremely salty.

# Directory & Glossary

# MOUNTAINS AND VOLCANOES

Most mountains are part of chains that formed where two or more plates of Earth's crust collided. A nearly continuous range of mountains (the Circum-Pacific System) runs for 40,000 km (25,000 miles) around most of the Pacific Basin and includes three-quarters of the world's active and dormant volcanoes.

Its mountains include the Andes of South America, the Western Cordillera of North America, and the Japanese Alps. A second, almost unbroken mountain belt (the Alpine–Himalayan System) runs from Morocco through Europe and southern Asia to Southeast Asia, including Earth's greatest mountain range, the Himalayas.

# North America

## Mount Katmai

**Location** Alaska, northwest USA

This snow-covered volcano is about 10 km (6 miles) in diameter, with a massive, lake-filled caldera at its centre. It was once thought to be responsible for the 20th century's biggest eruption, in June 1912, but scientists later showed that the nearby Novarupta lava dome was the actual source. When Novarupta erupted, a magma chamber beneath Mount Katmai emptied, collapsing its summit and creating the caldera. Thousands of steam vents peppered the area for years after the eruption, inspiring the name Land of Ten Thousand Smokes.

## The Mackenzie Dikes

**Location** From the Great Lakes to the Arctic, western Canada

The Mackenzie Dikes are Earth's largest dike swarm – a large group of near-vertical, sheet-like formations of igneous rock. The dikes reach the surface across a vast area that is about 3,000 km (1,900 miles) by 500 km (310 miles). Mostly composed of basalt, they were intruded into the ancient rocks of the Canadian Shield about 1.3 billion years ago.

## The Palisades

**Location** Along part of the western shore of the Hudson River, eastern USA

The Palisades are dramatic, almost sheer cliffs that stretch 32 km (20 miles) along the west side of the Hudson River (see p.363) from near Jersey City to Nyack. The cliffs range in height from 90 to 165 m (295 to 540 ft) and are the eroded edge of the Palisades Sill, a large mass of an igneous rock called diabase. The sill is about 200 million years old.

## The Sierra Nevada

**Location** Mainly in California, western USA

The magnificent skyline and dramatic, glacier-eroded valleys of the Sierra Nevada range make it one of the most beautiful regions of the USA. Fast-flowing rivers, valleys cloaked in pine forest, mountain meadows, and granite peaks are all characteristic of the area. One of the Sierra Nevada's peaks, Mount Whitney, is, at 4,421 m (14,505 ft), the highest in the lower 48 states. The range extends about 640 km (400 miles) from south to north and 110 km (70 miles) west to east.

## The Devil's Postpile

**Location** The Sierra Nevada range, California, western USA

The Devil's Postpile National Monument is situated on the western slope of the Sierra

▶ AUTUMN FOLIAGE AT THE PALISADES SILL

Nevada range (see left) at an altitude of 2,300 m (7,500 ft). One of the world's finest examples of columnar basalt, the vertical, hexagonal columns of the Postpile tower 18 m (60 ft) above the valley side; some are more than 1 m (3 ft) in diameter. They were created 100,000–80,000 years ago when a huge outpouring of lava flooded the area to a depth of 125 m (410 ft), later cooling and solidifying. Glacial action subsequently eroded the basalt, exposing the columns in all their glory.

## The Coast Ranges

**Location** Between western British Columbia and southwest California, western North America

Stretching at least 2,700 km (1,680 miles), this series of mountain ranges runs roughly parallel to

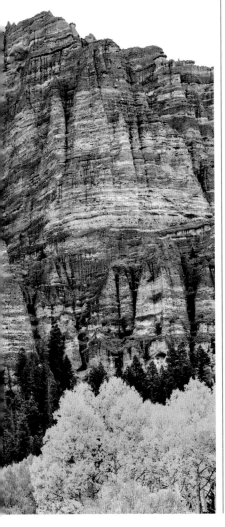

the Pacific coast from the Coast Mountains of British Columbia to the Transverse Ranges in southern California. The Olympic Mountains in Washington State are the most spectacular of the ranges. The folded and faulted igneous, metamorphic, and sedimentary mountains are the product of complex tectonic plate movements. Subduction produces volcanism in some regions, while major faults, such as the San Andreas, are active elsewhere.

## Parícutin

**Location** The state of Michoacan, west central Mexico

After several days of small earthquakes and underground rumbling, a crack opened up in a farmer's field in February 1943 and began spewing ash, lava, and pyroclastic bombs. Within 24 hours, its cinder cone was 50 m (165 ft) high, and by the time the eruptions stopped in 1952 the top of the volcano was 424 m (1,391 ft) above the surrounding landscape. Two nearby villages were evacuated before being engulfed in lava. Although Parícutin is now extinct, the crater is still hot and emits steam when rain falls.

## Volcán de Colima

**Location** Mostly in the state of Jalisco, west central Mexico

This massive stratovolcano is one of the most active in North America, having erupted more than 40 times since 1576, most recently in 2017. Eruptions produce flows of viscous lava, explosions of pyroclastic material, and enormous ash clouds. With 300,000 people living nearby, it is one of the most dangerous volcanoes on the continent. Volcán de Colima's summit rises 3,850 m (12,631 ft) above sea level.

▶ STEAM RISING FROM VOLCÁN DE COLIMA

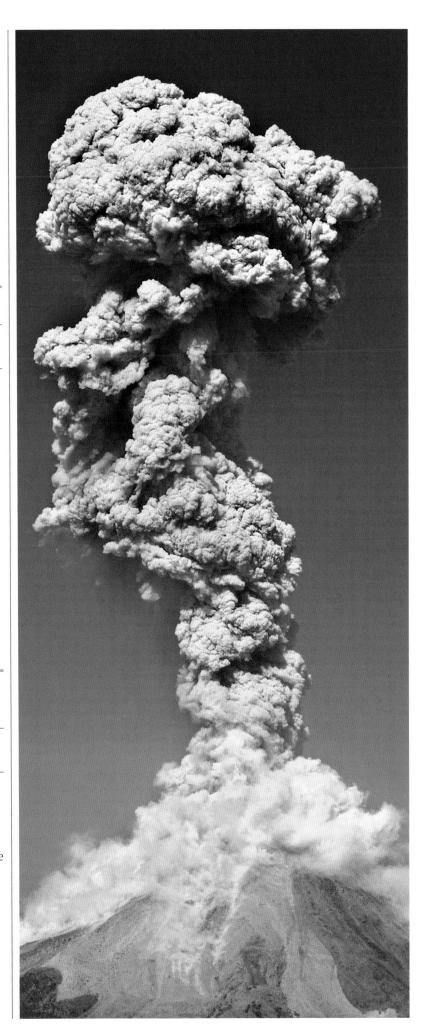

## El Chichón

**Location** The state of Chiapas, southern Mexico

Until late March 1982, El Chichón was thought to be an extinct volcano, with no eruption known for over 600 years. Then, in the space of a few days, three huge eruptions pumped millions of tonnes of gas and ash into the atmosphere, destroying nine villages and killing 2,000 people. The magma was very rich in sulphur, and sulphuric acid droplets forming in the stratosphere produced brilliant sunsets around the globe. A new crater was formed, 1 km (0.6 miles) across, which is now filled with an acidic lake. The peak of the volcano, which has been quiet ever since, is 1,150 m (3,772 ft) above sea level.

### Other mountains and volcanoes in North America

- The Alaska Range » p.24
- The Appalachians » p.35
- El Capitan and Half Dome » p.26
- The Cascade Range » p.27
- The Cave of Crystals » p.36
- Crater Lake » pp.28-29
- Devil's Tower » p.35
- Popocatépetl » p.37
- The Rocky Mountains » pp.24-25
- The San Andreas Fault » p.30
- Ship Rock » p.34
- The Sierra Madre » p.37
- Yellowstone » pp.30-33

# Central and South America

## Mount Pelée

**Location** Martinique in the Lesser Antilles, Caribbean Sea

Pelée is a stratovolcano (composite volcano) in the Antilles island arc. After many years of inactivity, in 1902 it produced the most devastating eruption of the 20th century in terms of casualties. A series of pyroclastic clouds swept down its slopes to destroy the city of Saint-Pierre, killing about 30,000 people. The Peléan type of eruption, characterized by explosive emissions of ash, gas, and pyroclastic flows, is named after this volcano.

## Arenal

**Location** Northwest Costa Rica, Central America

Arenal is Costa Rica's youngest and most active volcano. Situated over an active subduction zone where the Cocos Plate pushes beneath the Caribbean Plate, this stratovolcano was dormant for hundreds of years before July 1968, when an unexpected and violent eruption destroyed three small towns. Rocks weighing several tonnes were blasted more than 1 km (0.6 mile) away. Seven

▼ A LAVA FOUNTAIN AT ARENAL

more eruptions occurred over the next 30 years, but the volcano was declared dormant again in 2010.

## Galeras

**Location** Southwest Colombia, northern Andes

Colombia's most active volcano is also potentially its most dangerous, being situated near the city of Pasto's 450,000 inhabitants. It is a large stratovolcano, with a peak that is 4,276 m (14,029 ft) above sea level. The area's volcanism dates back at least 1 million years, with a particularly massive eruption 560,000 years ago. This activity results from the oceanic Nazca Plate plunging beneath the continental crust of the North Andes block.

## Ojos del Salado

**Location** Central Chile-Argentina border, southern Andes

This snow-covered stratovolcano has several claims to fame. At 6,893 m (22,615 ft), it is the highest active volcano in the world and the second highest peak of any description in the Americas. A small caldera not far from the summit is probably the highest lake on Earth. Although there are persistent fumaroles on the volcano, and a gas emission was reported in 1993, the last major eruption took place 1,000–1,500 years ago.

## Cerro Azul

**Location** Central Chile, southern Andes

Cerro Azul is located at the southern end of a volcano chain called the Descabezado Grande–Cerro Azul eruptive system. One of the largest explosive eruptions in South America in recorded history took place in April 1932, when the Quizapa crater of the Cerro Azul stratovolcano released 9.5 cubic km (2.3 cubic miles) of ash and lava in a classic Plinian eruption. At 3,292 m (10,801 ft), Quizapa is one of the world's highest Plinian craters. There has been no eruption since 1932, but small ash clouds are occasionally emitted.

### Other mountains and volcanoes in Central and South America

- The Altiplano » pp.96-97
- The Andes » pp.92-95
- El Tatio Geysers » p.97
- The Guiana Highlands » pp.90-91
- The Hornocal Mountains » p.98
- Masaya Volcano » p.89
- Santa María » p.88
- Soufrière Hills » p.89
- The Sugarloaf Mountain » p.98
- The Torres del Paine » p.99

▲ STEAM RISES FROM SURTSEY

# Europe

## Grímsvötn

**Location** Southeast highlands, Iceland

Iceland's most active volcano is, unusually, mostly subglacial, covered by the Vatnajökull ice-cap. When an eruption starts, large quantities of glacier ice melt and fill Grímsvötn's caldera. Pressure can then build up sufficiently to lift and puncture the ice-cap, releasing huge quantities of water in an episode called a jökulhlaup. During a 2011 eruption, Grímsvötn produced an ash cloud 12 km (7 miles) high, and air traffic was disrupted for several days.

## Surtsey

**Location** Off the south coast of Iceland

In November 1963, a series of Earth tremors, the odour of hydrogen sulphide, and finally a column of dark smoke rising from the sea signalled an underwater eruption off the Icelandic coast. Over the course of several weeks, a new volcano built up from the seafloor, and the brand new island of Surtsey was created, largely of a low-density volcanic rock called scoria. Sporadic eruptions continued until the summer of 1967, when the island reached its maximum size. Since then, wave action has gradually eroded it.

## The Rhine Rift

**Location** Southwest Germany and eastern France

The upper River Rhine (see p.393) flows through Europe's most impressive rift, a broad, flat-bottomed valley, for 350 km (220 miles) between Basel and Frankfurt. The rift valley, or graben, is bounded by faults on either side, beyond which are the uplands of the Vosges Mountains to the west and the Black Forest (see p.393) to the east. The rift formed about 30 million years ago, at the start of the Alpine mountain-building period, when the crust was being stretched and faulted on a massive scale.

## The Jura

**Location** Eastern France and western Switzerland

The Jura is a beautiful, part-forested massif of limestone deposits dating from 200–145 million years ago. The range, which runs for about 360 km (225 miles) along the French–Swiss border between the rivers Rhône (see p.372) and Rhine (see p.371), was folded and uplifted during the Alpine orogeny (mountain-building period), which also produced large-scale faulting. The highest peak in the Jura is Cret de la Neige, at 1,718 m (5,636 ft). The elevations of the crests are lower along the outer ridges of the arc.

## The Chaîne des Puys

**Location** The Massif Central, central France

This remarkable concentration of over 70 volcanoes in an area of the Massif Central about 40 km (25 miles) long and 5 km (3 miles) wide includes at least 48 cinder cones, eight lava domes, and 15 maars (broad craters that formed when lava came in contact with groundwater). Volcanic activity in the area spanned a period from about 95,000 to 6,000 years ago and was instigated by thinning of the crust and faulting related to the building of the Alps.

▼ PUY DE PARIOU CONE IN THE CHAÎNE DES PUYS

## The Cordillera Cantábrica

**Location** Northern Spain

The Cantabrian Mountains comprise two ranges extending for 300 km (190 miles) from the western end of the Pyrenees to the Galician Massif. The much smaller coastal range sometimes rises steeply from the sea, and features short, fast-flowing rivers. Inland, the southern range is much higher and more dramatic. It includes the Picos de Europa, a huge massif of mountain limestone, which has been subjected to heavy glaciation, although no glaciers survive. Its highest peak is Torre de Cerredo at 2,648 m (8,688 ft).

▲ HOT SPRINGS AND TERRACES AT TERME DI SATURNIA

## The Sierra Nevada

**Location** Andalucía, southern Spain

This domed massif of metamorphic rocks towers high above the Mediterranean and is dominated by Mulhacén, the highest peak in continental Spain, at 3,481 m (11,421 ft). Mulhacén is just one of 23 tops rising above 3,000 m (10,000 ft) in this 42-km- (26-mile-) long range. Generally, the Sierra's landforms are gentle apart from the heavily glaciated western peaks, which feature steep-sided, ice-sculpted valleys and cirques. No glaciers remain, but winter snow makes the area a popular skiing centre. With increased elevation, the vegetation on the massif's slopes changes from subtropical forest to alpine.

## The Apennines

**Location** From the Cadibona Pass to the Egadi Islands, Italy

Among the younger chains created during the Alpine mountain-building period, the Apennines are a series of parallel ridges extending for about 1,400 km (870 miles). They are mostly made of uplifted shales, limestones, and sandstones, with some igneous rocks, and there is plenty of evidence of volcanic activity, including the still-active Vesuvius and Etna. The only glacier here is Calderone, on the slopes of the highest peak, Corno Grande. Earthquake activity is common along the entire range.

## Terme di Saturnia

**Location** Tuscany, Italy

These are the largest and most beautiful of southern Tuscany's hot springs. Geothermally heated groundwater emerges at 37.5°C (99.5°F) and cascades over the waterfalls of Cascate del Mulino and Cascate del Gorello at more than 500 litres (132 gallons) per second into pools formed by the precipitation of travertine (a form of calcium carbonate) from the mineral-rich water. Despite the water's sulphurous odour, people have bathed in the pools since Roman times to take advantage of their curative properties. The groundwater is heated by hot rocks deep underground that are related to the volcanic activity at the nearby lava dome of Mount Amiata.

## Solfatara

**Location** Near Naples, Italy

The mythological home of Vulcan, the Roman god of fire, this shallow volcanic crater is part of the large, multi-cratered Campi Flegrei volcanic area. Solfatara has many fumaroles, which emit jets of steam and sulphurous fumes, as well as hot mud pools, but the volcano itself has been dormant since 1198. Mollusc burrows found 7 m (22 ft) up three Roman columns show that the area has sunk below sea level and subsequently been uplifted again since the columns were erected. This is evidence of the uplifting and deflating of the caldera as a result of the filling and emptying of its magma chamber, a process called bradyseism.

## Vulcano

**Location** Off north coast of Sicily, Italy

As with other volcanoes in the region, activity on this small island has resulted from the collision of the African and Eurasian plates. The volcanic cones here are of three different ages. The oldest are three ancient composite volcanoes, the cones of which have largely collapsed into their caldera. The more recent Fossa cone last erupted in 1888–90. This event was well studied and involved the ejection of blocks of rocks, lava bombs, and ash but no lava flows;

it gave its name to Vulcanian-type eruptions. The most recent cone is Vulcanello, which formed in 183 BCE and was last active in 1550.

## The Dinaric Alps

**Location** From Italy to Kosovo, southeast Europe

Much of this range, which stretches 645 km (400 miles) along the eastern Adriatic coast, is made of limestone and dolomite, and these rocks display some of the most spectacular karst landscape features in the world. These include sinkholes, underground rivers, caves, and gorges where caves have collapsed. One of the biggest underground systems is at Škocjan Caves in Slovenia, through which a 34-km- (21-mile-) long underground section of the Reka River flows.

## The Pindus Mountains

**Location** From the Albanian border to the Peleponnese, Greece

Greece's longest range, the Pindus Mountains, are geologically, an extension of the Dinaric Alps (see above). Some peaks, including the highest – Mount Smólikas – are composed of metamorphic ophiolites, sections of ancient oceanic crust that have been uplifted high above sea level and subsequently eroded. Other

▲ MOUNT SMÓLIKAS IN THE PINDUS MOUNTAINS

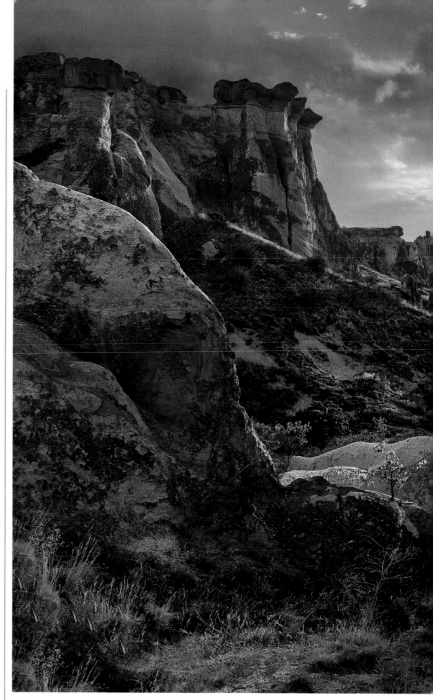

▲ HOODOOS AT CAPPADOCIA

areas are made up of limestone and dolomite, and have produced karst features. Vikos Gorge, in the far north of the range, is one of the deepest gorges in the world, plunging 900 m (2,950 ft) in places.

## Santorini

**Location** The Cyclades Islands in the Aegean Sea, Greece

This oval archipelago is all that remains above sea level of a giant volcanic caldera. The islands of Thera, Aspronisi, and Therasia surround a lagoon that is 12 km (7 miles) by 7 km (4 miles). In 1610 BCE, Santorini produced one of the largest eruptions of the past 5,000 years, blasting 100 cubic km (24 cubic miles) of rock particles and ash into the atmosphere and producing a tsunami that devastated the north coast of Crete.

## Cappadocia

**Location** Central Anatolia, Turkey

Ancient volcanic activity on this high plateau has given rise to one of the world's most extraordinary landscapes, the "fairy chimneys" (hoodoos) of Goreme National Park. Volcanoes deposited thick layers of ash, which later consolidated into a rock called tuff, then covered it with basalt. Runoff from heavy rainfall cut gullies through the resistant basalt before rapidly eroding the less resistant tuff beneath. Over thousands of years, this produced numerous tall, thin spires of tuff topped by protective basalt caps.

## The Manpupuner Rock Formations

**Location** The Komi republic, west of the northern Urals, Russia

These seven huge columns of metamorphic crystalline schist stand up like giants from the western slopes of the northern Urals. Their almost-human shapes have inspired a number of legends about their origin. The formations, which range in height from 30 to

42 m (100 to 138 ft), are flat-topped, almost vertical-sided rock towers, or buttes, created by freeze-thaw and wind erosion over many thousands of years.

### Other mountains and volcanoes in Europe

- The Alps » pp.138-39
- The Dolomites » p.136
- The Massif Central » p.135
- Mount Etna » pp. 140-41
- The Pyrenees » p.137
- The Scottish Highlands » p.134
- Strokkur Geyser » p.135
- Stromboli » p.144
- The Urals » p.144
- Vesuvius » p.145

# Africa

## Pico de Teide

**Location** Tenerife, the Canary Islands

Measured by its summit elevation above the ocean floor (7,500 m/24,600 ft), this composite volcano on Africa's continental shelf is the highest in the world outside the Hawaiian islands. It rises 3,718 m (12,198 ft) above sea level and is capped by the Las Canadas caldera. Pico de Teide and other vents have erupted several times since the island was settled in 1402, the most recent eruption being in 1909, when lava flows caused some damage.

# The Hoggar Mountains

**Location** Central Sahara Desert, Southern Algeria

This vast lunar landscape in Algeria formed from rocks about 2 billion years old – some of the oldest in Africa. Also known as the Ahaggar Mountains, this range is largely rocky desert, with scant vegetation, at an elevation above 900 m (2,950 ft). Mount Tahat is the highest summit, peaking at 2,908 m (9,541 ft). The bedrock largely comprises ancient metamorphic rocks, although several of the more dramatic peaks result from the erosion of extinct volcanic cones. The droppings of a critically endangered Saharan cheetah were found in the mountains in 2006, and there is a population of Dorcas gazelles in the area.

## Kapsiki Peak

**Location** The Mandara Mountains, northern Cameroon

Close to the village of Rhumski in the volcanic Mandara Mountains, a series of sentinel-like volcanic plugs rise sheer from the grassland. The tallest of them is Kapsiki Peak, which rises to 1,224 m (4,016 ft) above sea level. The plug formed when lava cooled and solidified as resistant rock in a volcano's vent. Subsequent erosion

removed the less-resistant volcanic cone, leaving behind just the hard plug.

## Lake Nyos

**Location** Northwest Cameroon

Situated on the flank of an extinct volcano, Lake Nyos is one of only three "exploding lakes" in the world. Its deep waters are saturated with carbon dioxide ($CO_2$) from a pocket of magma deep underground. On 21 August 1986, when the lake's waters were disturbed by a landslide, the gas rose into areas of lower pressure, bubbled out of solution, and

escaped. A massive cloud of $CO_2$ – about 1 cubic km (¼ cubic mile) – poured down the mountainside and suffocated 1,746 people in the valley below.

## Pico Cão Grande

**Location** Near the southern end of São Tomé island, São Tomé and Príncipe

Also known as Great Dog Peak, this moss-covered column of rock on the island of São Tomé rises 365 m (1,198 ft) above the surrounding landscape and 668 m (2,192 ft) above sea level. It is a volcanic plug, which formed when magma solidified in the vent of an active volcano. The volcano's cone

has subsequently been eroded, and volcanic activity has long since ceased in the area.

## The Ethiopian Highlands

**Location** The Horn of Africa, mostly central and northern Ethiopia

This range is called the roof of Africa for a good reason: it is the largest area above 1,500 m (4,900 ft) on the continent, and its highest peak, Ras Dashen, rises to 4,550 m (14,928 ft). The source of the Blue Nile, Lake Tana, is in the highlands. Bisected by the East African Rift Valley, they are also home to

▲ CLIFFS AND ESCARPMENTS IN THE ETHIOPIAN HIGHLANDS

several endemic species, including the critically endangered Ethiopian wolf and mountain nyala.

## Lake Bogoria

**Location** Western Kenya, in the East African Rift Valley

This shallow, very saline lake lies in a section of the down-faulted East African Rift Valley between the Maji Moto block to the west and the Bogoria Escarpment to the east. The water depth ranges from 11 to 14 m (36 to 46 ft), with the deepest area being the south basin, a relict volcanic crater. Many hot springs and geysers in and around the lake testify to continued geothermal activity. At times, 1.5 million flamingos feed in the alkaline waters.

## The Great Dike

**Location** Running south-north from east of Bulawayo to just south of the Mozambique border, Zimbabwe

Despite its name, this feature is not a dike but a loppolith, a mass of intruded igneous rocks that is more or less saucer-shaped in cross-section, fed by a vertical dike beneath. Its surface manifestation is a series of narrow ridges, running 550 km (340 miles) from south to north. It was injected into the surrounding rocks about 2.5 billion years ago and is rich in minerals including gold, silver, chromium, platinum, and nickel.

## Piton de la Fournaise

**Location** Réunion Island, in the Indian Ocean, off the east coast of Madagascar

In English, Piton de la Fournaise translates as "Furnace Peak". This is one of the biggest and most active volcanoes in the world, with more than 150 eruptions recorded since the 17th century, the most recent being in 2017. More than 530,000 years old, this shield volcano grew over the Réunion hotspot, which is thought to have been active for 66 million years. For most of its history, its lava flows have intermingled with those of Piton des Neiges. There are many craters within the caldera, which is 8 km (5 miles) across. The volcano's activity is closely monitored at an observatory on its slopes.

Other mountains and volcanoes in Africa

- **The Afar Depression** » p.183
- **The Atlas Mountains** » p.182
- **The Brandberg Massif** » p.188
- **The Drakensberg** » p.189
- **The Great Rift Valley** » pp. 184–87
- **Table Mountain** » p.188

▲ STEAM RISING FROM THE VALLEY OF GEYSERS

# Asia

## The Siberian Traps

**Location** Northwest Siberia, Russia

The traps (step-like hills typical of a basalt landscape) are the legacy of the largest known volcanic eruption in Earth's history. About 250 million years ago, numerous fissures and vents in western Siberia released an estimated 3 million cubic km (720,000 cubic miles) of flood basalts over a vast area. A mantle plume may have been responsible for the eruptions, but this is still debated by scientists. The greatest mass extinction in history followed, triggered by dramatic changes in the composition of the atmosphere.

## The Valley of Geysers

**Location** The Kamchatka Peninsula in eastern Siberia, Russia

The world's second-biggest concentration of geysers is in the valley of the fast-flowing Geysernaya River, which carries geothermal waters from a nearby stratovolcano. The valley contains at least 20 large geysers, the most spectacular of which, Velikan, can send out jets of water 40 m (130 ft) high. Some gush every few minutes, others only once every three hours or more. A massive mudslide in 2007 inundated the valley and shut off many of the geysers, but new ones have become active since.

## Azerbaijan Mud Volcanoes

**Location** The eastern Caucasus and Caspian coast, Azerbaijan

More than half the world's known mud volcanoes, possibly more than 400, are located in Azerbaijan. Mud volcanoes, also known as sedimentary volcanoes, occur where groundwater at up to 100°C (212°F) mixes with mineral deposits and is forced upwards. Reaching the surface through fissures, the mud-water mixture forms a small cone, usually no more than 4 m (12 ft) high. Some mud volcanoes also emit methane, as shown when the Lok-Batan mud cone spectacularly erupted in flames in 1977 and 2001.

## The Arabian Highlands

**Location** Yemen and southwest Saudi Arabia, Southwest Asia

At 3,666 m (12,028 ft), the rocky peak of Jebal Nabi Shu'ayb in the southern sector of this range is the highest point on the Arabian Peninsula. To the west, running parallel to the Red Sea as far north as Makkah (Mecca), an escarpment rises from the coastal plain of Tihamah. To the east, a high plateau slopes towards the desert interior of Saudi Arabia. In contrast, there are areas of moist deciduous forest, high-elevation agricultural terraces, and sparkling mountain streams within the highlands.

## The Deccan Traps

**Location** Mostly Maharashtra, Gujarat, and Madhya Pradesh provinces, India

The Deccan Traps form one of the largest volcanic features on Earth, a vast area of multiple step-like layers of solidified basalt released about 65 million years ago. In places, the basalt is more than 2,000 m (6,600 ft) thick and covers 500,000 square km (190,000 square miles). One theory proposes that the eruptions were caused by the Réunion hotspot, a mantle plume that once lay under South Asia.

## Sigiriya

**Location** Central Province, Sri Lanka

The near-vertical sides of Sigiriya, or Lion Rock, rise abruptly 180 m (590 ft) from the forest below, with the ruins of an ancient royal palace on its flattened summit. The rock is the solidified magma plug of a volcano whose cone has long since been eroded. As the magma in its vent cooled, the minerals crystallized at different rates, so the composition of Sigiriya changes with height.

## Seongsan Ilchulbong

**Location** Off the east coast of Jeju, South Korea

This World Heritage Site is a tuff cone – made of compacted volcanic rock – that rises from the sea close to the South Korean coast. The sides of the former volcano tower 180 m (590 ft) above the water. Seongsan Ilchulbong was created by a volcanic eruption on the seafloor about 5,000 years ago. The cliffs rise to a circular summit rim, which encloses a crater – an almost perfect bowl 90 m (295 ft) deep and 450 m (1,480 ft) across.

▶ SIGIRIYA'S SUMMIT

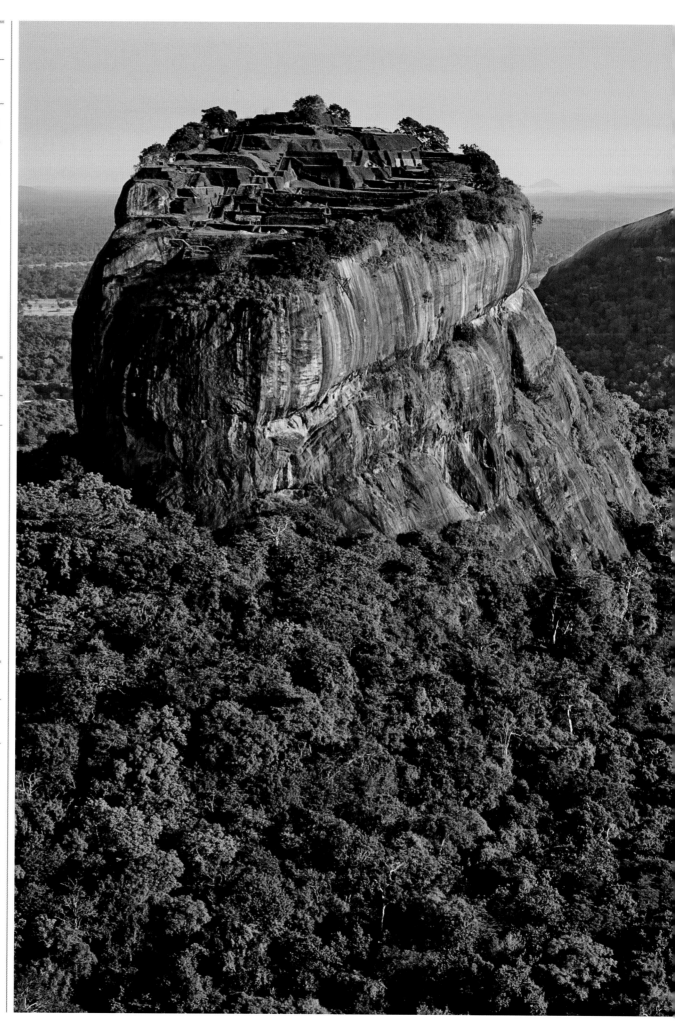

## Beppu Hot Springs

**Location** Kyushu island, Japan

Beppu has almost 3,000 geothermal springs, in eight major groups, the second-highest concentration in the world. The springs supply water at temperatures up to 99°C (210°F) for bathing pools (*onsen*). The pool water is a range of different colours, depending on the mineral content. For example, Umi-Jigoku *onsen* is cobalt-blue, while Chinoike-Jigoku is blood-red. The heat source for the geothermal activity is the nearby Tsurumi lava dome.

## Mount Unzen

**Location** Kyushu island, Japan

This volcano complex was responsible for Japan's worst volcanic disaster when, in 1792, the collapse of one of its lava domes triggered a huge tsunami that killed about 15,000 people. Its most recent period of activity was 1990–96, when one eruption generated a pyroclastic flow (a rapid surge of hot gas and lava) that killed 43 people, including three volcanologists. The highest of the Unzen peaks is Heisei-shinzan, at 1,486 m (4,875 ft).

## Sakurajima

**Location** Kyushu island, Japan

Prior to 1914, this composite volcano was an island, but enormous lava flows in an eruption of that year filled the main strait between the island and the mainland. Sakurajima is one of Japan's most active volcanoes,

▲ SANDSTONE COLUMNS IN THE TIANZI MOUNTAINS

erupting almost daily, and regularly sending ash 5,000 m (16,000 ft) into the sky. Its summit has three peaks, but since 2006 activity has centred on the Showa crater, east of the Minami-dake summit.

## The Tianzi Mountains

**Location** Hunan province, China

The Tianzi Mountain Nature Reserve is part of the spectacular Wulingyuan Scenic Area, whose main claim to fame is more than 3,000 quartzite sandstone columns, many of which are more than 200 m (660 ft) tall. There are also ravines, caves, fast-flowing streams, and waterfalls. Lush forest covers the area, which is often shrouded in mist and fog. The columns are the product of water eroding vertical joints in the 380-million-year-old sandstone.

## The Tibetan Plateau

**Location** Mostly in China (Tibet) and India, Central and East Asia

This is the world's largest and highest plateau, a vast area of flat or undulating steppe about five times the area of France and on average 4,500 m (14,800 ft) above sea level. Bordered to the south by the Himalayas, to the north by the Kunlun Mountains, and to the northeast by the Qilian Range, the plateau supports tens of thousands of glaciers and is the source of the several major rivers, including the Indus and Brahmaputra.

## Baishuitai Hot Springs and Terraces

**Location** Yunnan province, China

Also known as the White Water Terraces, Baishuitai is located high in the uplands of Yunnan, in the foothills of the Haba Snow Mountains. It has some of the world's most beautiful terraces of travertine (a form of calcium carbonate). Water from geothermal springs pours over hundreds of white semicircular steps and terraces, which cover an area 140 m (460 ft) long and 160 m (520 ft) wide. The high concentrations of calcium carbonate in the water precipitate out at lower temperatures to leave behind the travertine. The terraces are one of the sacred sites of the Dongba culture.

## The Huangshan Mountains

**Location** Anhui province, China

Celebrated for their beauty in art and literature for hundreds of years (for example, in the Shanshui, or "mountain and water", style of painting), the Huangshan Mountains feature oddly shaped pines, towering granite peaks and cliffs,

▲ VIOLENT ERUPTION ON THE SHOWA CRATER OF SAKURAJIMA

waterfalls, lakes, and forests, all sometimes draped in mist. There are 77 peaks exceeding 1,000 m (3,300 ft), with Lian Hua Feng (Lotus Flower Peak) reaching 1,864 m (6,115 ft). Many pines grow from fissures in the rock faces, and 19 plant species are endemic to the mountains. Also called The Yellow Mountains, the Huangshan Mountains were listed as a World Heritage Site by UNESCO in 1990.

## Mayon Volcano

**Location** In the south of the island of Luzon, the Philippines

This is the most active volcano in the Philippines, having erupted at least 49 times in the last 400 years. Rising 2,462 m (8,077 ft) above the Abay Gulf, this stratovolcano is considered to be one of the most perfectly symmetrical in the world. Mayon's most violent eruption, in 1814, killed over 1,200 people and ended up devastating several towns. Following another eruption in 2006, heavy typhoon rainfall produced lahars (mudflows) from the recently deposited ash, which killed at least 1,000 people.

▲ TAAL, WITH TAAL LAKE IN THE CALDERA

## Tambora

**Location** On the island of Sumbawa, Indonesia

Tambora's eruption in 1815 was the biggest in recorded history, as it killed about 12,000 people. Huge clouds of ash and pumice were emitted into the atmosphere and dimmed sunlight, lowering global temperatures and creating the "year without a summer" in 1816. Tambora lost 1,500m (4,900ft) of its height in the eruption, and the present giant caldera formed, 6km (4 miles) across and 1,110m (3,640ft) deep. At sea level, Tambora is 60km (40 miles) in diameter.

## Kelimutu

**Location** On the island of Flores, Indonesia

This complex volcano has three crater lakes, whose waters are different colours despite being at the summit of the same volcano. Different oxides and salts give the lakes their varying colours. Tiwu Ata Mbupu is usually blue, while Tiwu Ata Polo and Tiwu Nua Muri Kooh Tai, which share the same crater wall, are typically red and green, respectively. The lakes, which are fed by underwater fumaroles, periodically change colour.

## Taal Volcano

**Location** In the west of the island of Luzon, the Philippines

The Philippines' second most active volcano after Mayon (see p.355), Taal is known to have erupted 33 times since 1572, with particularly large events in 1754 and 1911. Recent eruptions have been concentrated on Volcano Island, which is in the middle of Taal Lake. The lake fills much of a massive, ancient caldera. Taal is one of a chain of volcanoes that were formed by the subduction of the Eurasian Plate under the Philippines.

## Lake Toba

**Location** In the north of Sumatra, Indonesia

The largest volcanic lake in the world, this was the site of the biggest known eruption of the last 25 million years. The blast, which occurred about 75,000 years ago, covered parts of Sumatra beneath 600m (2,000ft) of ash. The collapse of the volcano after the eruption formed a huge caldera, which subsequently filled with water. The lake is now 100km (60 miles) by 30km (20 miles) and plunges to 505m (1,657ft) deep.

## Mount Merapi

**Location** On the island of Java, Indonesia

This stratovolcano is the most energetic of Indonesia's 129 active volcanoes, and forms part of the Pacific Ring of Fire. It has erupted regularly in the last 500 years, causing many fatalities. Merapi rises 2,968m (9,738ft), several metres lower than before its 2010 eruption, which killed more than 300 people living on its slopes. The volcano is situated over a subduction zone where the Indo-Australian Plate is pushing under the Sunda Plate.

### Other mountains and volcanoes in Asia

- The Altai Mountains » p.224
- The Caucasus Mountains » p.221
- The Himalayas » pp.223-23
- The Karakoram Range » p.225
- Kliuchevskoi » p.229
- Mount Everest » pp.226-27
- Mount Fuji » p.229
- Mount Pinatubo » p.231
- Pamukkale Hot Springs » p.220
- The Tian Shan Mountains » p.224
- The Tengger Volcanic Complex » p.230
- The Zagros Mountains » p.220
- Zhangye Danxia » p.228

# Australia and New Zealand

## Rabaul

**Location** At the eastern end of New Britain island, Papua New Guinea

The large caldera at Rabaul was created about 1,400 years ago. Subsequently, a large breach on the eastern side flooded most of the caldera, which became Blanche Bay, a sheltered inlet of the Bismarck Sea. Three small volcanoes lie just outside the rim of the caldera, and in 1994 (after a long period of inactivity) two of these – Tavurvur and Vulcan – erupted violently, destroying much of Rabaul town, which had developed as a port within the caldera. Tavurvur remains the most active vent, continuously emitting ash.

## The Bungle Bungles

**Location** Northeast Western Australia, Australia

The horizontally striped, beehive-shaped towers of this range form the most impressive sandstone karst landscape in the world. Deposited 360 million years ago, before being uplifted, the sandstone has been worn away by the combined effects of heavy rainfall, chemical weathering, and erosion by the wind. These have cut gullies and gorges, such as Cathedral Gorge. The orange layers of sandstone contain oxidized iron compounds and do not retain water. The grey layers contain more clay and so retain more moisture, supporting cyanobacteria that protect the surface of the sandstone.

▼ STRIATED SANDSTONES AT THE BUNGLE BUNGLES

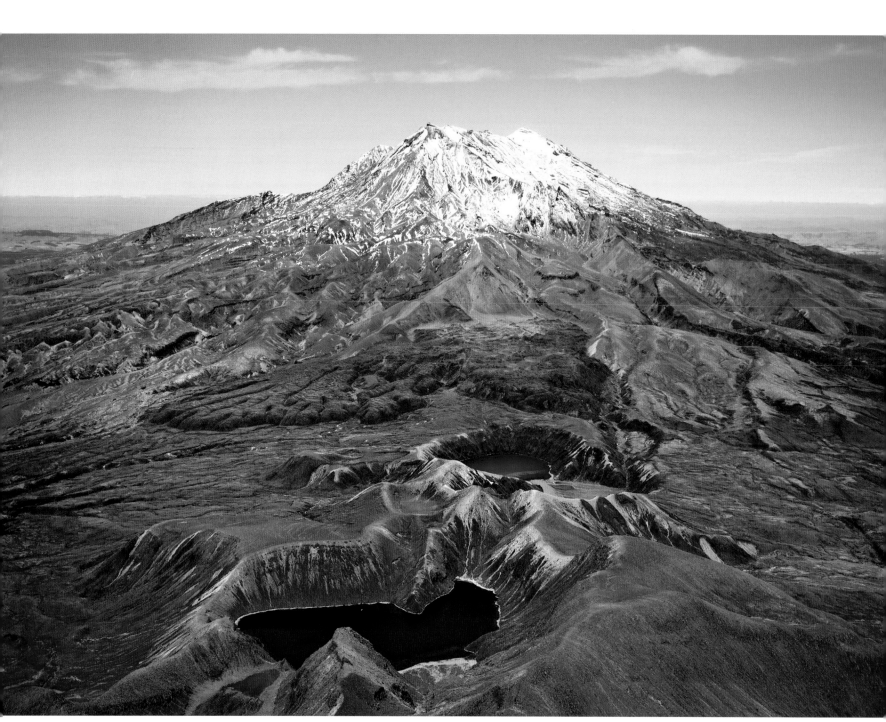

▲ CRATERS AND CRATER LAKES OF RUAPEHU

# Kata Tjuta

**Location** In the south of Northern Territory, Australia

This spectacular group of 36 domed rock formations towers high above the surrounding countryside and is visible from a great distance. The structures are composed of conglomerate, a mixture of rounded boulders in a sandstone matrix. Their orange–red hue is due to a thin cover of iron oxide. The biggest of the domes is Mount Olga, which stands 546 m (1,791 ft) above the plain.

# Waimangu

**Location** North Island, New Zealand

In June 1886, New Zealand's largest eruption in 700 years created the Waimangu Volcanic Rift Valley, which has been a hotspot of hydrothermal activity ever since and has a number of dramatic features. Acidic Frying Pan Lake is the largest hot spring on Earth. Nearby Inferno Crater Lake contains the largest geyser-like feature in the world – although it can't be seen because it jets into the bottom of the deep lake.

# Ruapehu

**Location** North Island, New Zealand

New Zealand's largest active volcano, and North Island's tallest mountain, Ruapehu has three peaks surrounding a crater, which becomes a crater lake between eruptions. One feature of Ruapehu eruptions has been the breaching of the lake dam and the creation of destructive lahars (mud flows) as water mixes with volcanic sediments and speeds downslope. A lahar in 1953 killed 151 people. The last major eruption was in 1995, although there have been smaller events since.

---

**Other mountains and volcanoes in Australia and New Zealand**

- The Great Dividing Range » p.274
- Hyden Rock » p.278
- Rotorua » pp.276-77
- The Southern Alps » p.279
- Tongariro Volcanic Park » p.275

# GLACIERS AND ICE-SHEETS

Most of Earth's glacial ice is concentrated in Antarctica and Greenland, but glaciers are found on every major landmass except Australia. Most exist either in the polar regions or in mountainous zones where the climate is cold enough, for at least a part of the year, for snow to accumulate and ultimately become consolidated as ice. The amount of snowfall a glacier receives is crucial for its survival; and regions with suitably cold winter temperatures but little snowfall, such as Siberia, do not have the ice accumulation necessary for glacier formation.

## Europe

### The Kongsvegen Glacier

**Location** Svalbard archipelago, on the western side of Spitsbergen

One of more than 1,500 glaciers on Spitsbergen, Kongsvegen shares its lower end (terminus) in the seawater of Isfjorden with another glacier, Kronebreen. Both calve ice blocks into the waters of the fiord. Although Kongsvegen's terminus is now slowly retreating, it is a surge glacier, meaning that periodically it advances rapidly. It last surged in 1948, pushing several kilometres along Isfjorden. It is a polythermal glacier, being warmer near the base than near the surface.

### The Rhône Glacier

**Location** The Bernese Alps, Switzerland

At the head of an impressive U-shaped valley, which runs roughly northeast to southwest below the giant peak of Finsteraarhorn, the glacier is the source of the River Rhône (see p.372) and one of the main providers of water for Lake Geneva. It is about 8 km (5 miles) long, although its snout has retreated 1 km (0.6 miles) since 1880. Every summer, sections of the ice are covered with white blankets in an attempt to reduce ice melt.

### The Pasterze Glacier

**Location** The eastern Alps, Austria

At just over 8 km (5 miles) from head to terminus, Pasterze is the longest glacier in the eastern Alps. It is also located in one of the most dramatic settings, with Austria's highest mountain, Grossglockner, towering above it. The glacier's head is Johannisberg, at 3,453 m (11,329 ft), and at its snout it gives rise to the Moll River. Pasterze is currently retreating at a rate of 10 m (33 ft) a year, and its volume has halved since the first measurements were taken in the mid-19th century.

> **Other glaciers and ice-sheets in Europe**
> - The Aletsch Glacier » p.151
> - The Jostedalsbreen Ice-field » p.149
> - Mer de Glace » p.150
> - The Monaco Glacier » p.148
> - Vatnajökull » pp.146–47

▲ ICE CAVE AND MELTWATER UNDERNEATH THE PASTERZE GLACIER

▲ GLACIATED PEAKS NEAR THE SIACHEN GLACIER

# Africa

## The Kilimanjaro Ice-cap

**Location** At the southern end of the Great Rift Valley's eastern fork, northeast Tanzania

Despite lying virtually on the equator, low temperatures at the top of Kilimanjaro (5,895 m/19,340 ft) ensure that snow accumulates, and, in the recent past, glacier ice has formed. Global climate change reduced the area of permanent ice by 85 per cent between 1912 and 2011. The northern ice field at the summit is the biggest area, but only 1.0 square km (0.4 square miles) of this remains, having shrunk by 29 per cent since 2000.

# Asia

## The Inylchek Glacier

**Location** The Tian Shan Mountains of Kyrgyzstan, Kazakhstan, and China, Central Asia

The world's sixth longest non-polar glacier has two branches – North and South. The glacier's source is in the western slopes of the Khan Tengri massif, part of the Tian Shan mountain range. At 7,439 m (24,406 ft), the range's highest peak, Jengish Chokusu, towers over the southern branch of the Inylchek Glacier, which is 60 km (40 miles) long, covers an area of 17 square km (7 square miles), and is up to 200 m (660 ft) thick. Inylchek is the longest and fastest-moving glacier in the Tian Shan, and it feeds the seasonal Lake Merzbacher – named after Austrian explorer Gottfried Merzbacher, who discovered the lower branch in 1903. The lake drains annually into the Inylchek River below when its ice dam melts.

## The Siachen Glacier

**Location** The Karakoram Mountains of Jammu and Kashmir state, India

This glacier, situated between the main Karakoram and Saltoro ranges, falls from 5,753 m (18,875 ft) to 3,620 m (11,877 ft) over the course of its 76-km (47-mile) length. It is the second longest non-polar glacier and, including tributary glaciers, covers an area of 700 square km (270 square miles), but it is retreating rapidly. The Siachen Glacier is supplied by annual snowfall of snow-covered mountains of the Zanskar Range. It is an immense mountain glacier 23 km (14 miles) long, with an average thickness of 150 m (490 ft), which feeds the Doda (or Stod) River, and so is important for irrigating the fertile farmland downslope.

## The Rongbuk Glacier

**Location** Southern Tibet, China

Two tributary glaciers (the East and West Rongbuk) converge to form one flow of ice that is 22 km (14 miles) long – a fine example of a valley glacier moving north from Mount Everest onto the Tibetan Plateau. Climbers have used East Rongbuk's impressive medial moraine as a way to access Everest's North Col, describing it as the "magic highway". Either side of the moraine, the ice has been churned into fields of seracs (towering, jagged ice pinnacles) – some up to 30 m (100 ft) high – with ice bridges and pillars adding to the chaotic impression.

> ### Other glaciers and ice-sheets in Asia
>
> - **The Baltoro Glacier** » p.233
> - **The Biafo Glacier** » p.234
> - **The Fedchenko Glacier** » p.232
> - **The Khumbu Glacier** » p.235
> - **The Yulong Glacier** » p.232

# Australia and New Zealand

## The Fox Glacier

**Location** The Southern Alps of South Island, New Zealand

Fed by four alpine glaciers, this heavily crevassed maritime glacier plunges down 2,600 m (8,500 ft) over the course of its 13-km (8-mile) flow from Mount Tasman to the coast of New Zealand's South Island. Its outflow, 300 m (980 ft) above sea level, is the Fox River, and it is one of the few glaciers to terminate in lush temperate rainforest. Its recent history of advance and retreat is fascinating. During the last ice age, it extended beyond the present coastline, since which time it has retreated a long way. Between 1985 and 2009, it advanced again, sometimes rapidly, but since then the retreat has continued.

> ### Other glaciers and ice-sheets in Australia and New Zealand
>
> - **The Franz Josef Glacier** » p.280
> - **The Tasman Glacier** » p.281

▼ CREVASSES IN THE FOX GLACIER

up to 10,000 mm (394 in) and avalanches in an environment where winter temperatures may plunge to -50°C (-58°F).

## The Kolahoi Glacier

**Location** Jammu and Kashmir state, India

Below the pyramidal rock mass of Kolahoi Peak, this glacier flows from an ice-field that also feeds three other glaciers. Kolahoi, which is the source of the Lidder River, has an average elevation of 4,700 m (15,400 ft). Between 1963 and 2005, its area contracted from 14 square km (5 square miles) to 11 square km (4 square miles), and it is still retreating at 3 m (10 ft) a year. The interior of the glacier is believed to be hollowing out.

## Durung Drung

**Location** Jammu and Kashmir state, India

One of the most beautiful Himalayan glaciers, Durung Drung is a gently meandering stream of ice flowing between the towering,

# RIVERS AND LAKES

Every continent – even Antarctica, where most water is locked up in ice – has rivers and freshwater lakes. Rivers are fed by groundwater springs, lakes, glaciers, and surface runoff. Measured by the amount of water flowing through its channels, the Amazon–Orinoco basin in South America is the largest, while the Nile, in Africa, closely followed by the Amazon, is the longest. The largest lake by volume is Lake Baikal in Russia, which formed as a result of Earth's crust being pulled apart. Lakes also form in basins with no outlet, in volcanic craters, and behind natural and artificial dams.

## North America

### Great Bear Lake

**Location** Northwest Territories, western Canada

The eighth largest lake in the world by surface area, the Great Bear Lake is also deep, plunging to 413 m (1,355 ft). It drains to the west, via the Great Bear River, to the mighty Mackenzie River. Its waters are cold and, being both far from oceanic influences and close to the Arctic Circle, it is frozen from November to July. It is the largest inland body of water to exist at such a high latitude. During the last glaciation, about 10,000 years ago, it was part of the even larger Lake McConnell, another glacial lake.

### Spotted Lake

**Location** British Columbia, western Canada

Also known as Lake Okanagan, this salty alkaline basin contains some of the most mineral-rich waters on Earth – and the most colourful – with high concentrations of magnesium sulphate, calcium, and sodium sulphate, as well as smaller amounts of silver and titanium. Each summer, as the lake's waters evaporate, about 300 small, shallow pools are all that remain, each coloured

▲ COLOURFUL MINERAL DEPOSITS ON SPOTTED LAKE

according to the minerals concentrated within – a mix of blue, green, yellow, and orange. The lake is a traditional healing site for the First Nations of Okanagan, who use its mud and water to treat aches, pains, and other medical problems.

## The Columbia River

**Location** From British Columbia to Oregon, western North America

This river has the largest volume of any river flowing into the Pacific Ocean from the Americas. It rises in the Rocky Mountains of British Columbia and flows 2,000 km (1,200 miles) through a drainage basin the size of France before entering the Pacific between Oregon and Washington. The river's steep

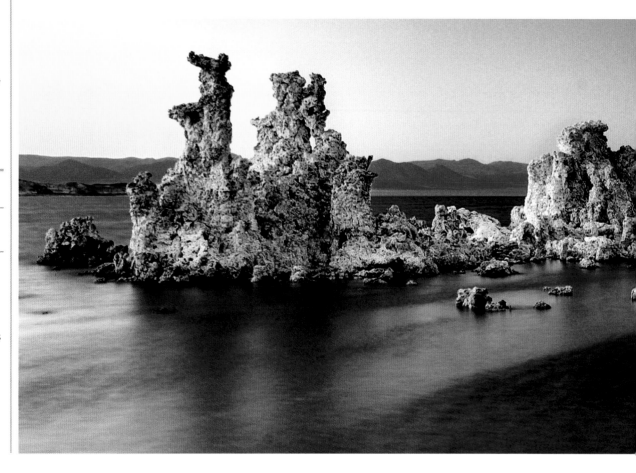

gradient and huge flow make it ideal for hydroelectric generation, and almost half the USA's hydroelectric power is generated in its basin.

## Manicouagan Lake

**Location** Quebec province, eastern Canada

This ring-shaped crater lake has been called the eye of Canada after its rough-edged, circular shape on a map. It was formed by the impact of a massive meteor, which measured 5 km (3 miles) across, about 214 million years ago. The original crater was about 100 km (60 miles) across, although the lake now has a diameter of 72 km (45 miles). Situated on the island in its centre, Mount Babel marks the crater's central peak, which formed after Earth's crust rebounded from the original impact. The lake drains into Manicouagan River to the south.

▼ TUFA TOWERS ON MONO LAKE

## Lake Seneca

**Location** New York state, eastern USA

This is the largest of New York's 11 freshwater finger lakes and has a reputation as a superb trout fishery. The lakes are in north-south valleys that were originally eroded by rivers but were later greatly deepened by south-flowing glaciers about 2 million years ago. The deepest point in Lake Seneca is 188 m (617 ft), well below sea level. It is fed by large underground springs, whose currents keep the lake mostly ice-free.

## The Hudson River

**Location** From the Adirondack Mountains to New York state, eastern USA

Although not especially long, at 507 km (315 miles), this is a river of great contrasts, rising in the sparsely populated Adirondack Mountains

and flowing south to reach one of the world's busiest ports and greatest cities, New York. A rise in sea level since the last ice age has "drowned" the lower reaches, meaning that almost half the Hudson's length is tidal. When sea levels were lower, its estuary was 320 km (200 miles) south of New York City, at the edge of the continental shelf.

## Mono Lake

**Location** California, western USA

Mono Lake is a shallow, salty water body with no outlet, lying in a basin just east of the Sierra Nevada. The basin is probably 3–4 million years old, and the lake is known to have existed for about 750,000 years. Although no more than 50 m (165 ft) deep now, it has been much deeper in the past. Along parts of the shoreline close to underwater

springs, towers of tufa (a kind of limestone) have formed. The lake's saline water supports a huge population of brine shrimps, which provide good feeding for up to 2 million water birds.

## The Rio Grande

**Location** Southwest USA and USA-Mexico border

This iconic river forms much of the USA–Mexico border. Near its source in the San Juan Mountains of Colorado, it flows as a clear, snow-fed stream before passing through a series of deep canyons near Big Bend, Texas, then sluggishly meandering across a broad floodplain to reach the Gulf of Mexico. At 3,100 km (1,900 miles) long, it is the 20th longest river on Earth, and its drainage basin is vast, but heavy water extraction for agriculture, industry, and domestic use means only 20 per cent of its natural discharge reaches the ocean. Nowhere is the river deep, and only small boats can navigate it.

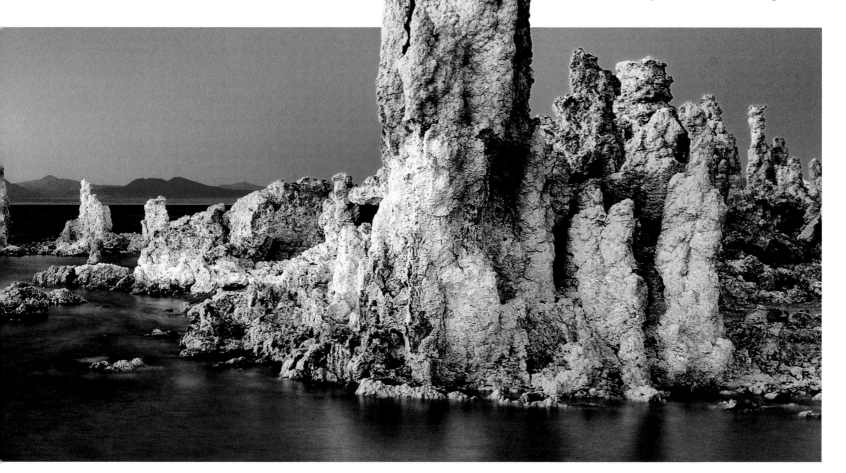

## The Arkansas River

**Location** From Colorado to Arkansas, southern USA

The second longest tributary of the Mississippi–Missouri system, the Arkansas River rises in the Rocky Mountains, and its headwaters descend 1,400 m (4,600 ft) in 193 km (120 miles) through many systems of rapids. As it flows out of the mountains and is joined by numerous tributaries, the river widens significantly before joining the Mississippi near Napolean, Arkansas, 2,364 km (1,469 miles) from its source. Water extraction has greatly reduced the flow of the river, but it is still an important waterway for barge transportation in eastern Oklahoma and Arkansas.

## The Red River of the South

**Location** From Colorado and Texas to Louisiana, southern USA

Apart from being one of North America's longest rivers, this river's main distinction is that the Prairie Dog Town Fork, one of the two major forks that form its headwaters, flows through the Palo Duro Canyon in northern Texas – the USA's second largest canyon after the Grand Canyon. Half the river's total length of 2,190 km (1,360 miles) forms the border between Texas and Oklahoma. The river once flowed into the Mississippi River but now joins the Atchafalaya River. Most of the river's traffic is concentrated in Louisiana.

## Mammoth Cave

**Location** The Appalachians of central Kentucky, eastern USA

The world's longest known cave network, Mammoth Cave consists of 652 km (405 miles) of tunnels, shafts, and underground caverns in limestone that is more than 320 million years old. A resistant cap of sandstone covers the limestone. But where water has eroded through this, sinkholes have formed and water has flowed underground, dissolving the limestone and creating stalactites, stalagmites, and a spectacular accumulation of calcium carbonate that has been named the Frozen Niagara as it looks like a waterfall.

## The Great Dismal Swamp

**Location** Virginia and North Carolina, eastern USA

Despite its discouraging name, the Great Dismal Swamp is a beautiful complex of wetland habitats. This poorly drained area of forest, with Lake Drummond at its centre, is one of the USA's largest swamps, with an area of more than 450 square km (170 square miles). Groundwater moves into the swamp from higher ground to the west but is held close to the surface by impermeable clay beneath. In winter and spring, the water table rises and the swamp floods, then in summer evaporation dries much of it out. Biodiversity is rich, with bald and Atlantic white cypress, maple, tupelo, and many other tree species, a diverse bird and reptile fauna, and populations of black bears and bobcats.

▼ THE HEADWATERS OF THE
ARKANSAS RIVER

## Ox Bel Ha

**Location** Quintana Roo state, southern Mexico

Meaning "three paths of water" in Mayan, this is the longest known undersea cave network in the world, comprising 270 km (170 miles) of tunnels close to and beneath the east coast of the Yucatan Peninsula. Created as water dissolved limestone, the caves later flooded as the sea level rose. They were inundated both by saltwater from the Caribbean Sea and fresh groundwater. The lighter fresh water flows on top of the heavier, near-static saltwater layer. The caves are accessed through several sinkholes.

## Sistema Huautla

**Location** Oaxaca state, southern Mexico

The ancient limestone highlands of the Sierra Mazateca mountains contain one of the world's finest collections of caves. First discovered in 1965, Sistema Huautla has subsequently been found to be the deepest in the Americas, extending to 1,545 m (5,069 ft) beneath the surface. Explorers have to dive through long submerged sections to reach the deepest point of the 64-km- (40-mile-) long system. When dye was put in the system's water, it was found to discharge at a spring in the Santa Domingo Canyon, at a much lower level, indicating that more passages are yet to be discovered. As the canyon has cut deeper, it has triggered changes throughout the cave system. Fresh stream passages have been formed under now-abandoned tunnels.

# Sac Actun

**Location** Quintana Roo state, southern Mexico

Translated from the Mayan language, Sac Actun means White Cave. With a total surveyed length of 331 km (205 miles), it is the longest cave system in Mexico and the second longest on Earth. While the upper parts of this coastal system are above water, most of the system is flooded with a mixture of fresh water from inland and seawater from the Caribbean. The cave can be reached through about 170 sinkholes. Some deep flooded shafts are yet to be explored, suggesting that there are more tunnels yet to be found.

### Other rivers and lakes in North America

- Lake Abraham » p.49
- Carlsbad Caverns » p.58
- The Everglades » pp.60–61
- The Grand Canyon » pp.54–57
- The Great Lakes » pp.50–51
- The Mississippi-Missouri » p.53
- Niagara Falls » p.52
- The Okefenokee Swamp » p.59
- The Yukon » p.48

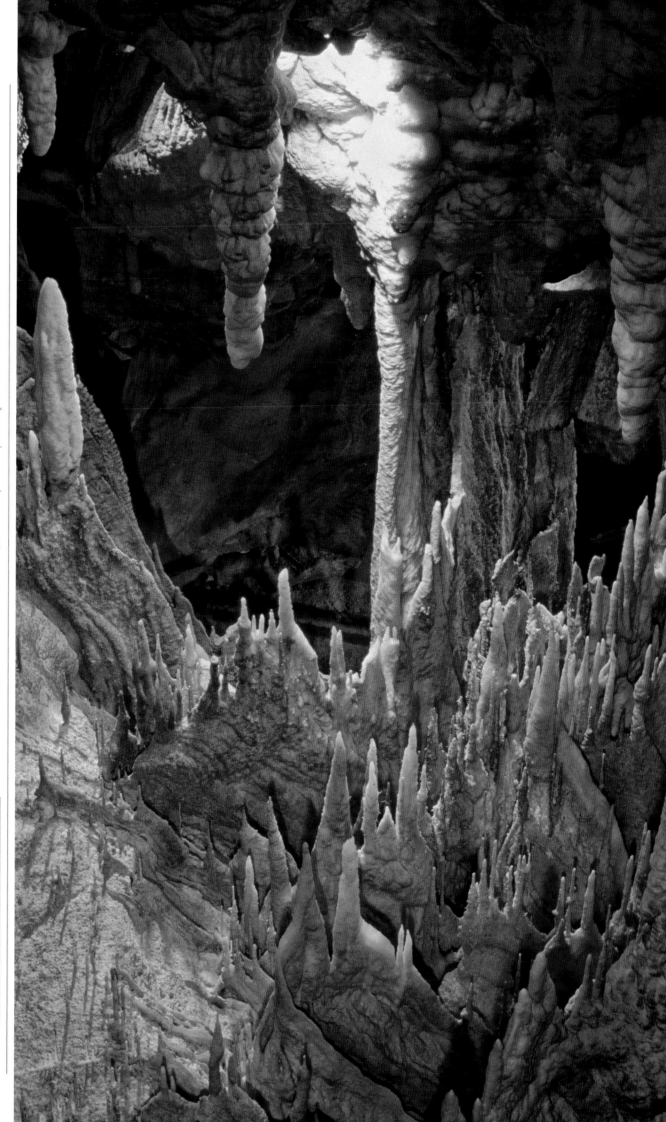

▶ STALACTITES AND STALAGMITES IN THE DRAPERY ROOM AT MAMMOTH CAVE

# Central and South America

## Windsor Caves

**Location** Jamaica, Caribbean Sea

These are some of the biggest underground caverns of the forested limestone hills of Jamaica's Cockpit Country. "Cockpits" refer to the steep-sided depressions up to 100 m (330 ft) deep that dominate the landscape of this area. Thick layers of the White Limestone Formation have been weathered and eroded by the action of rainwater and groundwater to form underground caverns and other karst features. Windsor Great Cave is 3 km (2 miles) long and has an underground river, stalactites, stalagmites, and some beautiful roof scalloping. It also has a high-level remnant of an ancient drainage system, now largely choked with calcite deposits, and houses the island's largest bat roost.

▼ THE LONG DROP OF KAIETEUR FALLS

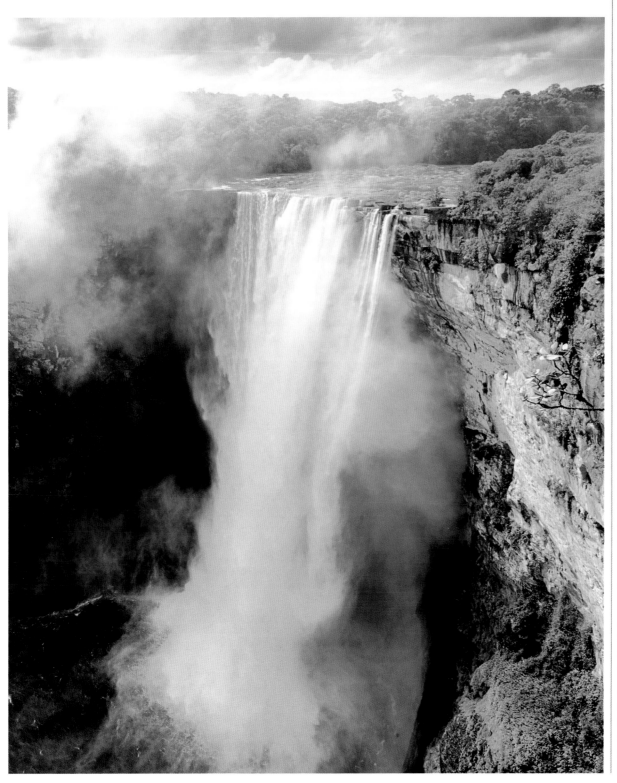

## Lake Nicaragua

**Location** The Pacific coast of Nicaragua

Along with its neighbour, Lake Managua, this giant freshwater lake formed in a part of the bottom of a graben – a faulted section of crust that slipped down between the land on either side. Lake Nicaragua is the largest lake in Central America, 177 km (110 miles) long and 58 km (36 miles) wide. Although only a thin strip of land separates it from the Pacific Ocean, the San Juan River drains the lake into the far more distant Caribbean Sea.

## The Orinoco

**Location** Venezuela and Colombia, from the Guiana Highlands to the Atlantic Ocean

Along with the Amazon, the Paraná, and the Tocantins, the Orinoco is one of the four great rivers of South America. It rises in the forested Guiana Highlands and flows through them in a great arc before crossing the seasonally flooded plains of the Llanos. Finally, it splits into many distributaries in a vast delta, its waters reaching the Atlantic Ocean 2,740 km (1,700 miles) from its source. The Orinoco's basin incorporates 80 per cent of Venezuela and 25 per cent of Colombia. The river's rate of flow varies dramatically – at Ciudad Bolívar, for example, its depth varies from 15 to 50 m (50 to 165 ft) according to the season.

## Kaieteur Falls

**Location** Mazaruni-Potaro Shield, Guyana

A combination of height, width, and water volume makes Kaieteur one of the world's largest single-drop waterfalls. In the pristine rainforest of the Guiana Highlands, the Potaro River

▲ THE TURQUOISE WATERS OF LAGUNA 69

plunges vertically 226 m (741 ft) over a resistant conglomerate lip atop a sandstone cliff. The water splashing at the bottom is eroding the less resistant sandstone, creating an overhang. The river then thunders down a further series of cascades into a 32-km- (20-mile-) long gorge. Although many waterfalls are higher, what makes Kaieteur remarkable is its volume of flow, averaging 663 cubic metres (23,400 cubic feet) per second.

## Laguna 69

**Location** Huascaran National Park, Ancash region, central Peru

Probably the most beautiful of Huascaran National Park's 400-plus lakes, what Laguna 69 (Laguna Sesentineuve) lacks in size it more than makes up for in its setting. The lake's turquoise colour is produced by fine sediment that has been eroded and transported by glaciers. It lies at an elevation of 4,600 m (15,100 ft) and is surrounded by snow-clad mountains. In summer, the lake is replenished by a waterfall plunging down the slopes of Chakraraju. In winter, it is frozen. Most of the national park's lakes did not have traditional names, hence the numeric designation.

## Tres Hermanas Falls

**Location** Otishi National Park, Junin, central Peru

After Angel Falls in Venezuela and Tugela Falls in South Africa, this is the third highest waterfall in the world and one of only three to attain a total drop of 915 m (3,000 ft). Tres Hermanas, or "Three Sisters", carries an unnamed Andean stream over a three-stepped drop in an area of undisturbed rainforest. Water plunges down the first two tiers into large, natural catch basins, before the final drop deposits water into the Cutivireni River.

## The Cotahuasi River

**Location** Arequipa region between the Altiplano and the Ocona River, southwest Peru

One of the best whitewater kayaking rivers on Earth, the Cotahuasi's modest source is high on the Altiplano. It descends quickly over rapids and waterfalls through the Cotahuasi Canyon – one of the world's deepest canyons, reaching a maximum depth of 3,501 m (11,486 ft) – before merging with the Maran River to form the Ocona, Peru's largest river, flowing into the Pacific. In 240 km (150 miles), the Cotahuasi descends 3,500 m (11,500 ft).

## The Colca River

**Location** Arequipa region between the Andean Altiplano and Pacific Ocean, southwest Peru

On the Colca river's 450-km (280-mile) journey from the high mountains of the western Andes to the Pacific coast, it descends 4,500 m (14,800 ft) and has three different names – Colca, Majes, and Camana. The Colca's main claim to fame, though, is flowing through the Colca Canyon, one of the world's deepest. At its deepest point, the river is at least 3,500 m (11,500 ft) below the summits on either side. The canyon is 120 km (70 miles) long and is famed for its whitewater rafting.

▲ THE ALGAE-RICH WATERS OF OF LAGUNA COLORADA

## Laguna Colorada

**Location** The Andean Altiplano, southwest Bolivia

Situated at an elevation of 4,278 m (14,035 ft) on the arid Altiplano of Bolivia, this hypersaline lake is also called the Red Lake because of the high concentrations of algae that colour it. White islands of the mineral borax dot its vast expanse. This endorheic lake covers 60 square km (23 square miles) but only averages 30 cm (12 in) in depth. It sits on ignimbrite, a volcanic rock. Large numbers of endangered James's flamingos feed in the lake's shallow water, along with smaller numbers of Chilean and Andean flamingos.

## The Paraguay River

**Location** From Brazil through Bolivia, Paraguay, and Argentina, southern South America

South America's fifth largest river has its source in the Mato Grosso uplands of Brazil and flows through a variety of landscapes for 2,621 km (1,629 miles) before its confluence with the Paraná River near Corrientes. For much of its length, seasonal flooding is characteristic. North of Corumba, its maximum discharge is in February, but to the south the greatest flow is in July. After leaving the Mato Grosso, the river runs south through the vast Pantanal wetland before dividing Paraguay into two, with alluvial savanna (the Gran Chaco) to the west and wetter, forested regions to the east. The river is navigable for a greater length than any other river on the continent apart from the Amazon. Fishes such as the salmon-like dorado, the piranha, and the bass-like pacu inhabit the Paraguay River.

## Gruta do Janelão

**Location** Minas Gerais state, eastern Brazil

This impressive limestone cave system in the valley of the Peruaçu River has many interesting features but is most famous for its rock art, which is more than 4,000 years old. The Peruaçu runs through part of the 4.7-km- (3-mile-) long system. Spectacular features include the tallest free-standing stalagmite in the world, which is 28 m (92 ft) from top to bottom, and one chamber that is 100 m (330 ft) from floor to ceiling. In places, sinkholes admit enough light through the cave roof for vegetation to grow within.

### Other rivers and lakes in Central and South America

- **The Amazon** » pp.106–109
- **Caño Cristales** » p.105
- **General Carrera Lake** » p.113
- **Iguaçu Falls** » pp.110–11
- **The Llanos** » p.104
- **The Pantanal** » p.112
- **Lake Titicaca** » p.105

# Europe

## Dettifoss

**Location** Vatnajökull National Park, northeast Iceland

Fed by meltwater from the Vatnajökull Glacier, Dettifoss is often described as the most powerful waterfall in Europe, based on its drop of 44 m (144 ft) and width of 100 m (330 ft). It is fed by the sediment-rich waters of the Jökulsá River, which flows from Vatnajökull across columnar basalt before thundering over the sheer cliff of Dettifoss. The water ranges from greyish white to brownish white, depending on the sediment load. The average flow is 175 cubic metres (6,180 cubic feet) per second.

## Seljalandsfoss

**Location** South Region, Iceland

This waterfall carries the waters of the Seljalands River 60 m (200 ft) over an ancient sea cliff into a pool and then to the coast. Unusually, it is possible to walk behind the fall as it pours over an overhang where resistant solidified lava lies on top of less resistant tillite (a glacial deposit). The river is fed by meltwater from the Eyjafjallajökull ice-cap, lying on the caldera of the active volcano of the same name (which last erupted in 2010).

## Skógafoss

**Location** South Region, Iceland

This is one of Iceland's most famous waterfalls, where the Skógá River thunders down a drop of 60 m (200 ft), producing clouds of spray – and rainbows on sunny days. Skógafoss is the most impressive of many falls that plunge down the basalt cliffs marking the position of the 3,000-year-old former coastline. This relic shoreline, now 5 km (3 miles) inland, dates from the last ice age, when Iceland was covered in ice; the coastline has subsequently fallen as the land has rebounded upwards following the melting of the ice.

▼ WATER CASCADING FROM BASALT CLIFFS AT SKÓGAFOSS

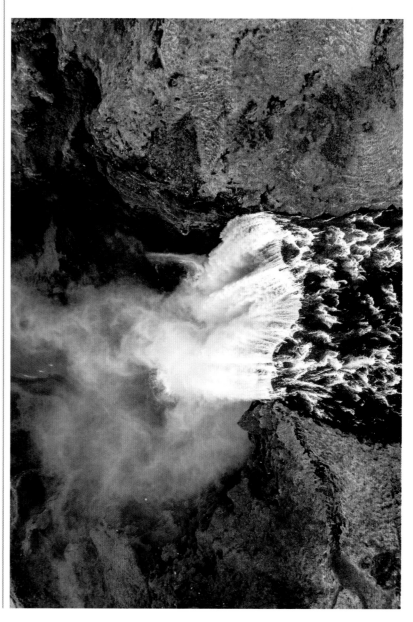

## The Jagala Waterfall

**Location** Jõelähtme municipality, Estonia

Estonia's largest waterfall, on the Jagala River not far from where it enters the Gulf of Finland, is 8 m (25 ft) high and more than 50 m (165 ft) wide. Downstream, the river runs through a small gorge 300 m (980 ft) long, charting the slow movement of the waterfall upstream as a result of erosion of the tough limestone bedrock. In winter, the waterfall freezes, creating a tunnel between it and the rock behind.

## Loch Ness

**Location** The Great Glen of Scotland, northern UK

Loch Ness is a deep, narrow freshwater lake that runs 37 km (23 miles) along the bottom of the Great Glen, a straight valley created by the Great Glen Fault and later scoured by glaciation. The greatest depth of Loch Ness is 230 m (750 ft). The Great Glen Fault is an ancient strike-slip fault, with crust to the north moving northeast relative to the south. It is now mostly inactive.

## Lough Derg

**Location** Clare, Galway, and Tipperary counties, Republic of Ireland

This long, thin lake is the second biggest in the Republic of Ireland. It is both fed and emptied by the River Shannon. The lough is 39 km (24 miles) long and covers an area of 130 square km (50 square miles). While its average depth is only 8 m (25 ft), the deepest point is 36 m (118 ft) – the result of the deepening of the Shannon valley by glaciers during the last ice age.

## The Severn

**Location** From the Cambrian Mountains to the Bristol Channel, UK

This is the longest river in the United Kingdom, at 354 km (220 miles). It also has the largest discharge, averaging 60 cubic metres (2,100 cubic feet) per second but occasionally many times more than this. Its source is the upland of Plynlimon, in Wales, and it discharges into a tidal estuary and the Bristol Channel. The latter has a large tidal range, and during the highest tides, rising water is funnelled up the lower reaches of the river, against the river current; this phenomenon is known as the Severn Bore.

## The Thames

**Location** From the Cotswold Hills to the North Sea, UK

The Thames is the second longest river in the United Kingdom, generally accepted as 346 km (215 miles) long. Its source is in the Cotswold Hills, and it flows roughly east, through London, to empty into a broad estuary and the North Sea. Huge quantities of water are removed for domestic and industrial use, so its discharge is significantly less than that of the Severn (see left). However, it poses a considerable threat. Heavy rains combined with exceptionally high tides have flooded the city on many occasions. The lowest 89-km- (55-mile-) long stretch of the river, to Teddington Lock, is tidal.

▲ FROZEN WATER AT THE JAGALA WATERFALL

# The Rhine

**Location** From the Alps to the North Sea, central Europe

Second only to the Danube as the greatest river of west and central Europe, the Rhine rises from two headstreams (the Vorderrhein and the Hinterrein) high in the Swiss Alps and flows for 1,230km (764 miles). Along the way, it passes through Lake Constance (see p.373), the Upper Rhine graben (a block of crust that has moved downwards between two faults), the Rhine Gorge, and across the North German Plain before reaching its delta near Rotterdam. It is one of the most important arteries of industrial transport in the world.

# The Seine

**Location** From Burgundy to the English Channel, northern France

France's second longest river has its source in the limestone hills of Burgundy, from where it flows 777km (483 miles) northwest through the agricultural heartland of the Île de France region, then through Paris to the English Channel between Le Havre and Honfleur. The Seine carries most of France's inland waterway traffic, and canals link it to the Rhine (see left), Loire (see right), and Rhône (see p.372) navigations. The estuary experiences a bore, called the *mascaret*, when the highest tides send a surge of water sweeping upriver.

▲ THE RHINE WINDING THROUGH THE RUINAULTA CANYON

# The Loire

**Location** From the Massif Central to the Bay of Biscay, France

The longest river in France rises from three springs near the base of Mont Gerbier de Jonc, a lava dome in the Massif Central. The Loire flows north to the Orleans region, where it swings to the west, eventually discharging via a tidal estuary into the Bay of Biscay. It is 1,012km (629 miles) long, and drains more than 20 per cent of France.

flows for 813 km (505 miles) to the Mediterranean west of Marseilles. Its course follows large glaciated valleys to Lake Geneva, then southwest to Lyon, where it converges with its biggest tributary, the Saône. The Rhône changes direction to the south and flows between the Alps and the Massif Central before splitting into several branches of its delta at Arles.

## Vercors Caves

**Location** Provence-Alpes region, southwest Alps, southeast France

This large limestone massif comprises a series of high plateaus divided by steep-sided valleys, including the Bourne and Furon gorges. Beneath the surface are many cave systems that quickly flood when the winter snow melts, or after heavy rain, and become treacherous. Highlights include the Grotte de Bournillon, which has an entrance 80 m (260 ft) high and 30 m (100 ft) wide – said to be the biggest in Europe. The Hall of the Thirteen, the main passage in the Gouffre Berger system, has massive stalagmites and delicate stalactites. Gouffre Berger was the first cave on Earth to be explored to a depth of more than 1,000 m (3,300 ft).

## Grotte Casteret

**Location** The Pyrenees of Aragon, Spain

At an elevation of more than 2,600 m (8,500 ft), the limestone cave system at Grotte Casteret contains a huge chamber called Grande Salle. The subterranean hall, which is 70 m (230 ft) long and 60 m (200 ft) wide, has an extraordinary frozen floor of clear ice covering 2,000 square metres (22,000 square feet); the deeper layers of ice are believed to be very old. There are also spectacular ice columns and a 20-m- (65-ft-) high ice wall. The cavern was discovered in 1926.

▲ LIMESTONE STALACTITES AT VERCORS CAVES

## Lascaux

**Location** Dordogne, southwest France

The walls of dozens of caves in the valley of the Vézère River, which flows through a limestone karst landscape, have ancient illustrations painted or etched on them. Those in the cave at Lascaux, a UNESCO World Heritage Site, include a huge depiction of fighting aurochs (an extinct species of wild cattle) and are believed to be about 17,000 years old. The main cavern at Lascaux is 20 m (65 ft) wide and 5 m (16 ft) high.

## The Rhône

**Location** From the Alps to the Mediterranean, Switzerland and southeast France

Meltwater from the Rhône Glacier (see p.359), high in the Swiss Alps, is the source of this river, which

## The Lakes of Covadonga

**Location** The Picos de Europa, Asturias, Spain

The permeable limestone landscape of the high Picos de Europa massif in the Cantabrian Mountains is renowned for its amazing karst features, including gorges, caves, sinkholes – and a virtual absence of standing surface water. Two exceptions are lakes Enol and Ercina, at an elevation of more than 1,100 m (3,600 ft), which retain water because of the presence of impermeable glacial deposits left behind after the last ice age.

## The Tagus

**Location** The Sierra de Albarracín to Lisbon, Spain and Portugal

The longest river of the Iberian Peninsula (1,007 km/626 miles long) drains the second most expansive drainage basin (after that of the Ebro River). Rising in the Sierra de Albarracín, in eastern Spain, the Tagus flows through limestone ravines and gorges before reaching the vast semiarid plains of Castille. There it has been dammed in several places to provide water and hydroelectric power before it reaches the lowlands of Portugal, discharging into the Atlantic Ocean near Lisbon.

## Doñana

**Location** Andalucía, Spain

This vast area at or close to sea level was given national park status for its rich tapestry of marshland, reedbeds, shallow streams, pools, and sand dunes – and the fauna that these habitats support, especially breeding and migrating birds. In the past, Doñana has been covered by both fresh water and salt water, but it is now slowly drying out, mainly as a result of the extraction of water from its aquifer for agriculture. It is protected from inundation by a wide barrier of sand dunes.

## Lake Constance

**Location** Western Austria, eastern Switzerland, and southern Germany

This freshwater body lies in a *zungenbecken*, or tongue basin, a long hollow scoured out and left behind by the Rhine Glacier as it retreated. The hollow filled with water and is now 63 km (39 miles) long and up to 252 m (827 ft) deep. The Rhine (see p.371) and other rivers are constantly depositing sediments in the lake, extending its coastline and reducing its depth.

▼ SHALLOW WATERS OF LAKE CONSTANCE

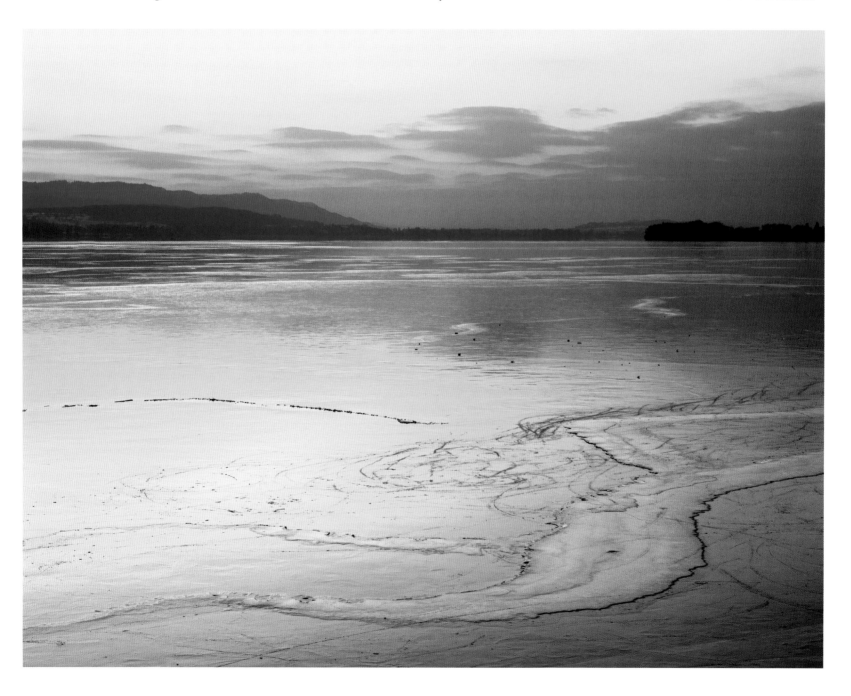

## The Po

**Location** From the Alps to the Adriatic Sea, northern Italy

The Po, Italy's longest river, at 652km (405 miles), takes water from the country's largest drainage basin, including its most fertile plain. It descends steeply from its source in the Cottian Alps, but from Turin follows a more gentle descent as it passes across a broad, flat plain. It is connected to Milan by a series of channels. Then it flows more or less due east to its delta, the most complex of any European river, with at least 14 mouths entering the Adriatic Sea.

## The Plitvice Lakes

**Location** The Velebit Mountains, Croatia

This stepped series of 16 lakes and interconnecting waterfalls cascades down the Korana River in the forested limestone landscape of the Velebit Mountains. Over the course of 8km (5 miles), the river falls in elevation by 133m (440ft). The waters of the river contain high concentrations of calcium carbonate, some of which is deposited to form travertine barriers, which dam the lakes. As water flows over the top of the dams, it forms waterfalls, the highest of which – Veliki Slap – drops 78m (256ft). The river runs through a gorge, and its valley includes other karst features, such as caves and underground streams.

▼ WATERFALLS AND LIMESTONE HILLS AT PLITVICE LAKES NATIONAL PARK

## Bigar Waterfall

**Location** The Anina Mountains, southwest Romania

Located in the Caraș-Severin County, Romania's most famous waterfall is neither spectacularly high nor thunderously powerful, but what it lacks in size it more than makes up for in beauty. The Bigar Waterfall is a stream flowing into the Miniș River gorge over an 8-m- (25-ft-) high domed rock overhanging the river below.

As the water flows over the moss-covered rock, it divides into myriad tiny rivulets to form a delicate fringe dropping into the underlying pool.

## Optimisticheskaya

**Location** Ternopil Oblast, western Ukraine

With 230 km (140 miles) of passageways, this is the longest gypsum cave system in the world – and the longest cave of any description in Eurasia. Despite this, it exists in layers of gypsum just 20 m (65 ft) thick. Gypsum is a soft sulphate mineral, which is more soluble than limestone and readily dissolves when in contact with slightly acidic rainwater to form underground features similar to those in limestone caves. Optimisticheskaya has an extraordinarily dense network on several levels, giving rise to its name of "the maze cave". It is accessible only through a single entrance, dug through the floor of a doline in 1966 by local cavers.

## The Black Sea

**Location** Bulgaria, Romania, Ukraine, Russia, Georgia, and Turkey

This saline water body is described as a trapped ocean basin. It is almost completely surrounded by land but underlain by oceanic crust that was trapped by tectonic plate collisions. To the southwest, it connects to the Mediterranean Sea through the Bosporus Strait, Sea of Marmora, and Dardanelles Strait. To the north, it connects to the Sea of Azov. Excluding the latter, the Black Sea has an area of 436,000 square km (168,300 square miles) and a maximum depth of 2,200 m (7,218 ft). Many rivers discharge into it, including the Danube and Dnieper.

### Other rivers and lakes in Europe

- The Biebrza Marshes » p.158
- The Camargue » p.154
- The Danube » pp.156-57
- Lake Geneva » p.155
- The Hortobágy Wetlands » p.158
- The Lake District » p.153
- Lake Ladoga » p.160
- Litlanesfoss » p.152
- Slovak Karst Caves » p.159
- The Verdon Gorge » p.155
- The Volga » p.161

▲ ACCUMULATION OF CYANOBACTERIA ON LAKE NATRON

# Africa

## Lake Chad

**Location** The Chad Basin, Nigeria, Niger, Chad, and Cameroon

This freshwater lake is situated in the Chad Basin, the second largest endorheic (closed) drainage basin on Earth. Between 1963 and 1998, it shrank from 26,000 square km (10,000 square miles) to 1,350 square km (520 square miles), although a small increase followed. Mostly fringed by marshes, it averages just 1.5 m (5 ft) in depth, although this varies from year to year. Lake Chad is fed by rivers such as the Ngadda and Komadugu Yobe and has no outlet. The lake is a vital source of water and fish for local communities.

## Lake Nakuru

**Location** The eastern arm of the Great Rift Valley, western Kenya

One of the ultra-alkaline soda lakes in the bottom of the down-faulted East African Rift Valley, this water body is famous for supporting hundreds of thousands of flamingos that come to feed on the lake's algae. Nakuru varies in size from 5 to 45 square km (2 to 17 square miles), but its maximum depth is just 3 m (10 ft). About 10,000 years ago, Nakuru and its neighbouring lakes (Elmenteita and Naivasha) formed a single deep freshwater lake, which shrank dramatically as the climate became drier, leaving behind three separate remnant lakes.

## Lake Natron

**Location** Arusha region, northern Tanzania

Fed by mineral-rich hot springs and the Ewaso Ng'iro River, Lake Natron is unusual for its deposits of the evaporite minerals natron and trona, its brightly coloured waters, and the huge flocks of flamingos that feed and breed there. With water temperatures as high as 60°C (140°F), this shallow soda lake at the bottom of the Gregory Rift (the eastern branch of the East African Rift Valley) is inhospitable for most organisms, but abundant cyanobacteria colour its waters deep red or orange.

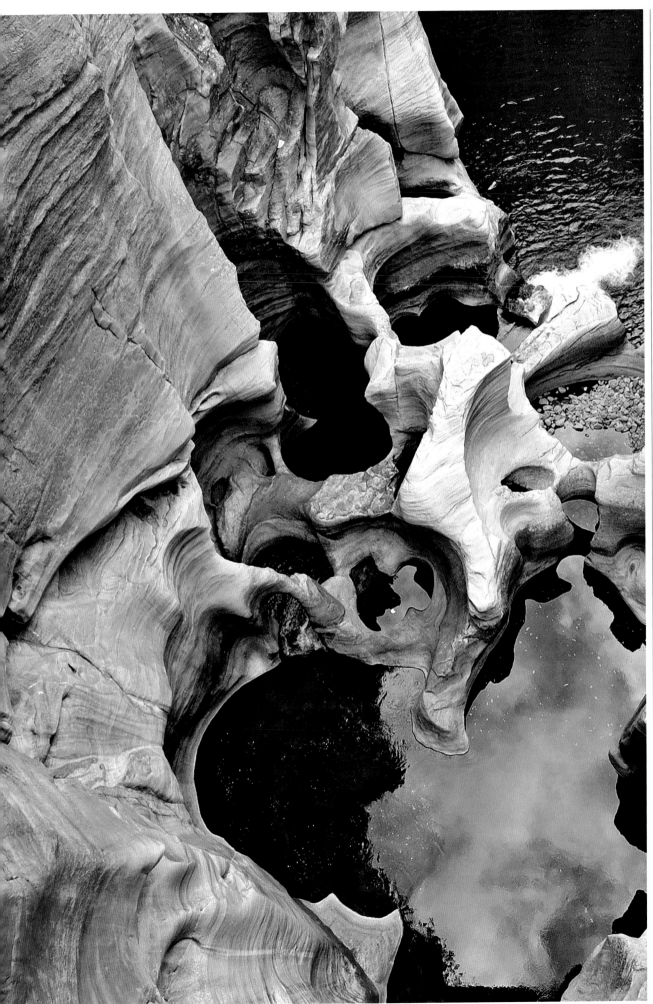

## The Zambezi

**Location** Northwest Zambia to the coast of Mozambique

This is the fourth longest river in Africa, after the Nile, Congo, and Niger, flowing 2,574 km (1,599 miles) through six countries from its source to the ocean. It rises in a marshy miombo woodland close to the watershed with the Congo Basin and discharges into the Indian Ocean at a large delta on the Mozambique coast. After plunging down the Victoria Falls, the river is dammed to form lakes Kariba and Cahora Bassa, in Zambia and Mozambique, respectively.

## Blyde River Canyon

**Location** Drakensberg Mountains of Mpumalanga, South Africa

The Blyde River Canyon is also called the Motlatse Canyon. Where the Blyde River meanders through the Drakensberg Mountains, it has carved one of the largest canyons on Earth. Some 25 km (15 miles) long and averaging 750 m (2,460 ft) deep, it features some remarkable rock formations, including three buttes capped by ultra-resistant Black Reef quartzite, known as the Three Rondavels (rondavels are traditional huts whose shape they mimic). The canyon is also the site of a quartzite column called the Pinnacle and, at the bottom of the canyon, the Kadishi waterfall, the second highest waterfall flowing over tufa.

## Cango Caves

**Location** Swartberg Mountains of Western Cape, South Africa

This limestone cave system in a ridge of the Swartberg Mountains is probably the best known in Africa. Although relatively small –

◀ BOURKE'S LUCK POTHOLES
AT BLYDE RIVER CANYON

▲ THE FRESH AND SALINE WATERS OF LAKE BALKHASH

about 4km (2 miles) of passageways and caverns are known – it is famous for cave paintings dating back to the Early Stone Age and for impressive screens of stalactites, large stalagmites, and other dripstone features. Some of the last bear a resemblance to various figures or structures and have been given individual names, including Madonna and Child, Cleopatra's Needle, Dried Tobacco Leaves, and Leaning Tower of Pisa.

> **Other rivers and lakes in Africa**
>
> - **The Congo** » p.192
> - **Lake Malawi** » p.193
> - **Lake Retba** » p.192
> - **Lake Victoria** » p.196
> - **The Nile** » pp.190-91
> - **The Okavango Delta** » p.197
> - **Victoria Falls** » pp.194-95

# Asia

## The Amur

**Location** From northeast China to the Sea of Okhotsk, eastern Asia

One of Asia's longest rivers, the Amur proper begins at the confluence of the Shilka and Argun (Ergun) rivers, then flows 2,824km (1,755 miles) to the Tartar Strait. If the headwaters are included, it is 4,444km (2,761 miles) long. Its remarkable drainage basin encompasses desert, steppe, taiga, and tundra, and covers an area of 1.8 million square km (700,000 square miles). Fed largely by monsoon rains in summer and autumn, the river's waters often rise 14m (46ft) above normal at this time, overflowing its banks in the marshy country downstream of Khabarovsk. For much of its course, the Amur forms the border between China and Russia.

## Lake Balkhash

**Location** Eastern Kazakhstan

The freshwater western section of this vast, shallow lake is separated from the saline eastern section by the Saryesik Peninsula, which almost divides Balkhash in two. The lake narrows and deepens to the east, and there is little mixing between the two parts. Most of the lake's input comes from the Ili River, one of several that drain the mostly arid Balkhash-Alakol Depression (a closed, or endorheic, basin) and discharge into the lake. Balkhash has a surface area of 16,400 square km (6,330 square miles).

## The Aral Sea

**Location** Northern Uzbekistan and southern Kazakhstan

In the 1960s, this endorheic lake was the fourth largest on Earth, with an area of 66,000 square km (25,500 square miles), but it shrank rapidly after its feeder rivers were diverted for irrigation projects. By 2007, its waters had become so shallow that it had split into four lakes, and its surface area was reduced to just one-tenth of what it had once been. Much of what was once water is now desert. Efforts to replenish the waters of the northern section (the North Aral Sea) have succeeded in increasing its surface area and depth, and fish populations are increasing.

## Krubera Caves

**Location** The Arabika Massif of the western Caucasus Mountains, Georgia

There are several major cave systems in the thick beds of folded limestone of the Arabika Massif, including Krubera – the world's deepest known cave. It descends through a series of vertical pits and steep meandering passageways to a level 2,197 m (7,208 ft) below its entrance in a glacial valley. For most of this depth, it is relatively dry, with no indication of major flooding, but the very deepest part, which was explored in 2012, is submerged.

## Jeita Grotto

**Location** Nahr al-Kalb valley, Lebanon

This is the longest cave system in the Middle East, stretching for 9 km (6 miles). It has two separate, but interconnected, sections, one 60 m (200 ft) below the other. The upper galleries contain the White Chamber and the Red Chamber. The former houses the world's longest stalactite, measuring 8 m (25 ft), while the limestone walls of the latter have been stained red by iron oxide. An underwater river flows through the lower galleries, and there is a subterranean lake. The chamber known as Grand Chaos Hall, is 500 m (1,640 ft) long.

## Sambhar Salt Lake

**Location** Rajasthan, northwest India

This is India's largest saline lake, with a surface area of 190–230 square km (75–90 square miles) and a depth varying from 3 m (10 ft) after the monsoon to 0.6 m (2 ft) at the end of the dry season. It is set in an endorheic basin and receives water mainly from the Mendha and Rupangarh rivers. When salinity levels reach a sufficient concentration, water is released through a dam into evaporation pools for commercial salt production. Many flamingos spend the winter at the lake.

## Lonar Lake

**Location** Maharashtra, western India

India's largest impact-crater lake is almost circular in shape, with an average diameter of 1.2 km (¾ mile) and a depth of about 6 m (20 ft). The lake's waters are saline and are fed by one stream. The lake is surrounded by a crater rim, created when a meteorite impacted the Deccan Traps plateau. The date of impact is debated – estimates range from 570,000 to 52,000 years ago.

## Borra Caves

**Location** The Eastern Ghats of Andhra Pradesh, eastern India

These are some of the deepest caves in India, containing formations such as stalactites, stalagmites, columns, and shapes reminiscent of mushrooms, a human brain, and a mother and child. The main chamber is 200 m (660 ft) long and 12 m (39 ft) high. Sulphurous springs discharge into the cave system, which is the source of the Gosthani River.

◀ THE LOWER CAVERN OF JEITA GROTTO

## Dudhsagar Falls

**Location** The Western Ghats of Karnataka-Goa, southwest India

One of India's highest waterfalls, this four-tiered cascade is produced by the Mandovi River as it plunges 310 m (1,020 ft) from the Deccan Traps (see p.353) in the Western Ghats. The falls are ordinary in the dry season, but their character changes completely during monsoons, between June and September. *Dudhsagar* means "sea of milk" in the local Konkani language, a reference to the creamy-white appearance of the water as it pours down the near-vertical rock face.

## Hogenakkal Falls

**Location** Karnataka-Tamil Nadu border, southern India

*Hogenakkal* means "smoking rocks", a reference to the clouds of spray that hang in the air above the falls. While the 20 m (65 ft) that the Kaveri River drops does not sound great, the width of the waterfall is. As the river approaches "India's Niagara", it forks into multiple streams that gush into a narrow gorge from both sides. Hogenakkal is at its most spectacular during monsoons (July–August).

## Yarlung Tsangpo Grand Canyon

**Location** The Eastern Himalayas, China (Tibet) and India

Arguably the world's deepest canyon, the difference in elevation between the Yarlung Tsangpo River and the mountain peaks on either side of it exceeds 5,000 m (16,000 ft) in places. Much of the river's course is in open valleys on the Tibetan Plateau, but in southeast Tibet it enters a deeply incised canyon through the eastern Himalayas, switching its course from northeast to southwest as it flows around Namcha Barwa. After 505 km (315 miles), it emerges into the broad Brahmaputra valley.

▼ MULTIPLE STREAMS OF HOGENAKKAL FALLS

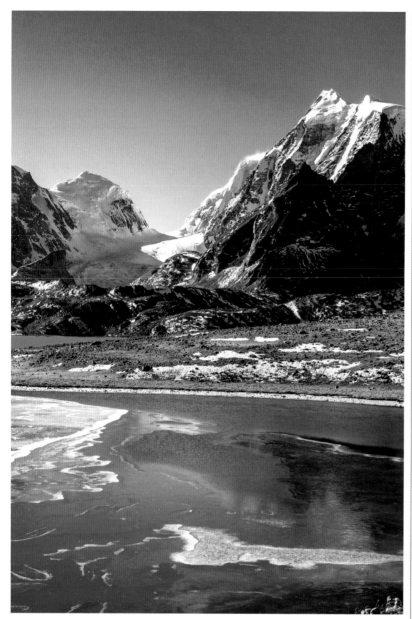

▲ THE SACRED WATERS OF GURUDONGMAR LAKE

## The Irrawaddy

**Location** From the eastern Himalayas to the Andaman Sea, Myanmar

This major river arises from the confluence of the N'mai and Mali rivers, both of which have their sources in Himalayan glaciers. As it bisects Myanmar (formerly Burma), the Irrawaddy flows 2,170 km (1,350 miles) through rainforest, dry forest, extensive areas of rice cultivation, and – in the delta zone – freshwater swamp forest and mangroves before entering the Andaman Sea. The river's massive sediment load extends the delta by 50 m (165 ft) each year. At peak times in the monsoon season, the river's discharge is more than 40,000 cubic metres (1.4 million cubic feet) per second.

## Gurudongmar Lake

**Location** The Himalayas of northern Sikkim, northeast India

Situated close to the summit of Mount Khangchen Gyao, this lake is one of the highest on Earth, at an elevation of 5,430 m (17,815 ft). Surrounded by snow-covered mountains, Gurudongmar is fed by glacial meltwater and drains into a small stream that later becomes the Teesta River. The lake is frozen in the winter months. It is considered sacred by both Buddhists and Sikhs.

## Loktak Lake

**Location** The Manipur Valley of Manipur, northeast India

The largest freshwater lake in northeast India is best known for its *phumdis*, islands of vegetation in various stages of decomposition, some of which are inhabited. The lake has a surface area of 287 square km (111 square miles), and the largest of the *phumdis* covers 40 square km (15 square miles). Water enters and leaves the lake through the Manipur River. Discharge from the lake is restricted by an 8 m (25 ft) rocky barrier called Sugnu Hump, which reduces the river's flow.

## Akiyoshi-do

**Location** Southwest Honshu, Japan

Akiyoshi-do is Japan's longest cave, a product of the action of water on the thick Akiyoshi Group reef limestone over hundreds of thousands of years. This cave is 9 km (6 miles) long, has a ceiling up to 80 m (260 ft) above the floor, and has a river flowing through it. There is also a beautiful complex of 500 terraced pools. Akiyoshi-do is one of more than 400 caves in the area, which is Japan's most important karst region.

## Red Beach

**Location** Liaoning, northeast China

Large parts of the saltmarsh at the Liaohe River delta near Panjin have been colonized by a succulent halophyte (a salt-tolerant plant) called seepweed, which thrives in salty alkaline soils. During the summer months, this plant is an unremarkable green colour, but it turns deep crimson from August to October, colouring a vast expanse of beach. Seepweed flats provide an important nesting location for the endangered red-crowned crane and Saunders' gull.

## Qinghai Hu

**Location** The Qilian Mountains of Qinghai, western China

This vast saline lake is the largest in China and the most expansive mountain lake without an outlet in Central Asia. It covers

4,317 square km (1,667 square miles) in a large depression in the Qilian Mountains and lies at 3,205 m (10,515 ft) above sea level. Most of the rivers that once discharged into Qinghai Hu now run dry before they reach it, mostly because water is diverted from them for irrigation. Consequently, the lake is shrinking.

## Jiuzhaigou Valley National Park

**Location** The Minshan Mountains of Sichuan, central China

This extraordinarily beautiful valley has been afforded World Heritage Site status for its superb landscape features. Notably, these include numerous blue, green, and turquoise lakes with clear, mineral-rich water, spectacular waterfalls, and limestone terraces set below high peaks. The valley also boasts classic limestone features such as caves and a variety of old-growth forest habitats, which are home to a population of giant pandas.

## Xiaozhai Tiankeng

**Location** Chongqing municipality, central China

Xiaozhai Tiankeng is the world's largest and deepest sinkhole. On the surface, it has a diameter of more than 500 m (1,640 ft), and apart from one sloping ledge, its walls descend vertically 662 m (2,172 ft) through the bedrock to Di Feng Cave, into which the limestone of the sinkhole dissolved and was washed away by the subterranean Migong River. Although local people have long known about Xiaozhai, naming it the Heavenly Pit, it was only discovered by western explorers in 1994.

▼ AUTUMN LANDSCAPE AT JIUZHAIGOU VALLEY NATIONAL PARK

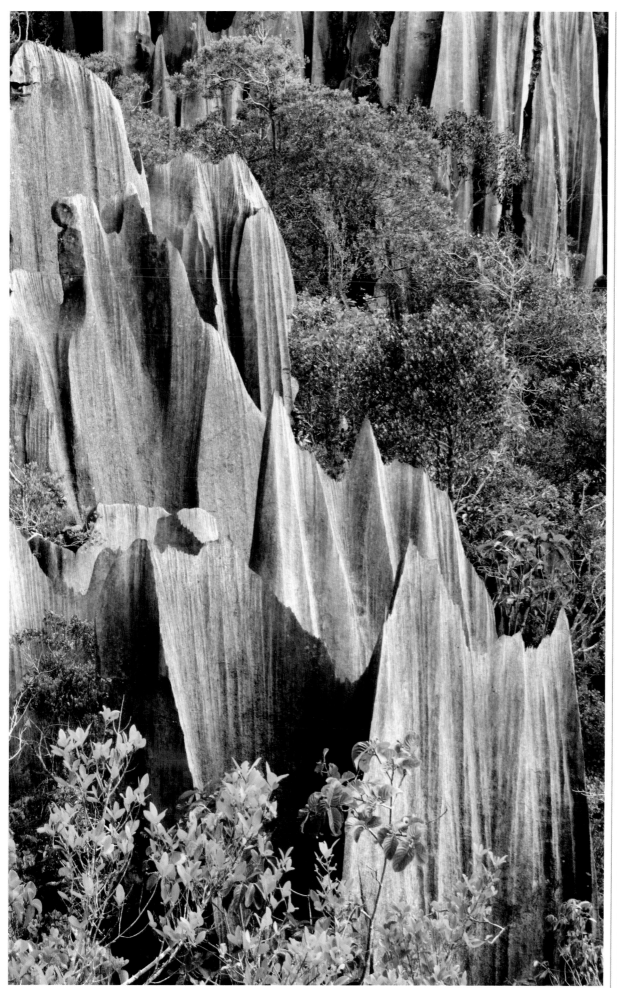

▲ LIMESTONE PINNACLES AT GUNUNG MULU NATIONAL PARK

## The Pearl River

**Location** Guangdong and Guangxi provinces, southern China

This name usually encompasses the three great rivers of Guangdong – the Xi, Bei, and Dong – that share a common delta. Measured from the source of the longest of these (the Xi) to the South China Sea, the Pearl River is the third longest in China at 2,197 km (1,365 miles). Its basin covers an area of 410,000 square km (158,300 square miles). The river is named after the pearl-coloured shells found on its bed in Guangzhou.

## Puerto Princesa Subterranean River National Park

**Location** The western coast of Palawan, Philippines

The most extraordinary feature of this World Heritage Site is its underground river, which is 8 km (5 miles) long, discharges directly into the South China Sea, and is tidal in its lower reaches. Underground chambers, waterfalls, stalactites, and stalagmites are other features of the cave system. Measured by volume, the 360-m (1,200-ft) Italian's Chamber is one of the largest in the world.

## Gunung Mulu

**Location** Sarawak, Borneo

Located beneath the rainforests of the Gunung Mulu National Park, everything about this cave system is on a massive scale. It contains the Sarawak Chamber, which at 600 m (2,000 ft) long, 415 m (1,360 ft) wide, and 80 m (260 ft) high is the largest on Earth. Gunung Mulu has a total of 295 km (183 miles) of explored caves, including Clearwater Cave, the longest in Asia. Dating of sediments in Clearwater Cave shows that the caves have evolved over a period of 2 million years. The surrounding landscape has many other features of a limestone karst environment, including deep canyons, underground rivers, and limestone pinnacles. Gunung Mulu is home to millions of bats and cave swiftlets.

### Other rivers and lakes in Asia

- Lake Baikal » pp.238-39
- The Dead Sea » p.236
- The Ganges » p.240
- Hang Son Doong » pp.244-45
- The Indus » p.240
- The Lena » p.237
- The Mekong » p.243
- The Ob'-Irtysh » p.237
- South China Karst » pp.246-49
- The Sundarbans » p.241
- The Yangtze » p.242
- The Yellow River » p.242

▲ LAKE EYRE, COLOURED RED BY BACTERIA

# Australia and New Zealand

## Koonalda Caves

**Location** The Nullarbor Plain, South Australia

The main significance of this cave in the Nullarbor Plain, a vast area of limestone, is its Aboriginal art. A sinkhole 30 m (100 ft) deep leads through relatively soft limestone to a steep passage that connects to a large chamber with a domed roof 45 m (150 ft) above the floor. Two more passages lead off from the chamber, one connecting to three underground lakes. The wall art, which is more than 20,000 years old, consists of concentric circles, parallel lines, and herringbone patterns. In the first chamber, archaeologists have found plenty of evidence that flint mining was practised here 24,000–14,000 years ago. Hearths and charcoal were presumably used to provide firelight by which to work.

## Lake Eyre

**Location** The Great Artesian Basin, South Australia

When it is full, this endorheic lake covers at least 9,500 square km (3,670 square miles) of the Great Artesian Basin. Most of its water comes from monsoon rains falling on Queensland, in the northeast part of the basin, and entering the lake via the Diamantina River and Coopers Creek. On average, a small flood occurs once every three years, but Lake Eyre fills completely much more rarely. Occasionally, such as in 1984 and 1989, local rainfall topped up the lake. Now officially known as Kati Thanda-Lake Eyre, it also has the lowest point in Australia – 15 m (50 ft) below sea level.

## Waituna Lagoon

**Location** The South coast of South Island, New Zealand

Important as a feeding area for a range of breeding and migratory waterbirds, especially shorebirds, Waituna Lagoon is a freshwater lake pooled behind a coastal gravel barrier. The lake covers an area of 36 square km (14 square miles) and is fed by three streams. It lies over gravel deposited by glacial meltwater during the last ice age and is bordered on its landward side by saltmarsh and bogs. The wide coastal barrier was created by longshore drift.

## Huka Falls

**Location** Waikato River, North Island, New Zealand

This series of waterfalls is on New Zealand's longest river, the Waikato, which drains Lake Taupo. Just upstream of the falls, the river narrows from 100 m (330 ft) wide to just 20 m (65 ft) as it passes through a canyon that has been incised into hard volcanic rock. The effect of this narrowing, and a rapid descent over a series of small falls, is to accelerate the water as it approaches the final 11-m (36-ft) drop. Consequently, the water thunders over the falls with immense power.

### Other rivers and lakes in Australia and New Zealand

- Jenolan Caves » p.283
- The Murray-Darling » p.282
- Waitomo Caves » p.283

# COASTS, ISLANDS, AND REEFS

The world's continents and islands are all surrounded by coastlines. North America has the longest coast of any continent with 310,000 km (193,000 miles), and Canada is the individual country with the longest at 210,000 km (130,500 miles). Coastal features include gentle shores with a broad intertidal zone, sand dunes, vertical cliffs, rock stacks and arches, mangrove swamps, salt marshes, and wide estuaries. Some are deeply indented, especially where sea-level rise has drowned coastal valleys. Coasts advance where deposition exceeds erosion, but recede where marine erosion is greater than deposition.

# North America

## Ilulissat Icefiord

**Location** Near Ilulissat, west coast of Greenland

This tidal fiord opening into the Davis Strait is the outlet for the Ilulissat Glacier (also known as Sermeq Kujalleq), the fastest-moving glacier outside Antarctica, pushing into the fiord at 20–35 m (65–115 ft) per day. As the glacier has retreated, the fiord's open water has extended further inland, and it is now about 50 km (30 miles) in length. It cuts through gneiss, granite, and mica schist 2.5–1.6 billion years old. The ice-carved fiord is up to 1,000 m (3,300 ft) deep, but much shallower at the iceberg bank, where it opens into the ocean; many of the larger icebergs run aground there.

## Puget Sound

**Location** Washington state, western USA

This huge complex of tidal basins and channels is the USA's second largest estuary, stretching 160 km (100 miles) from north to south. Its waters connect with those of the Strait of Juan de Fuca and the Pacific Ocean at Admiralty Inlet. Puget Sound's deep basins, which plunge to 280 m (920 ft) below sea level, were carved and scoured by glaciers. Shallower parts of the estuary mark where there are glacial deposits on the seafloor.

## Dungeness Spit

**Location** North coast of Washington state, western USA

Stretching 9 km (6 miles) into the Strait of Juan de Fuca, this is the longest natural spit in the USA. Prevailing westerly winds drive longshore drift, which has extended the sandy spit by an average of 4.5 m (15 ft) per year over the past 120 years. It is vulnerable to the destructive action of the ocean and was temporarily broken into three sections by a storm in 2001.

## Mount Desert Island

**Location** Maine, eastern USA

This is the second largest island on the USA's eastern seaboard. Much of it is granite, the remnants of an ancient igneous intrusion. More recently, it has been shaped by the erosive action of glaciers, which carved out Somes Sound, a deep inlet that almost cuts the island in two. This is the only fiord-like inlet on the east coast of the USA. Most of Mount Desert Island is in Acadia National Park.

## Long Island

**Location** New York state, eastern USA

The largest island on the east coast of the USA, Long Island stretches 190 km (120 miles) from east to west and is separated from the mainland by the East River and Long Island Sound. Long Island is also incredibly diverse, ranging from the New York City boroughs of Brooklyn and Queens to the wilderness of the Pine Barrens temperate coniferous forest. Much of the surface geology is made up of moraines and other glacial deposits.

## Lighthouse Reef

**Location** Belize coast, Caribbean Sea

Oblong in shape, 35 km (22 miles) from south to north, and generally less than 6 m (20 ft) deep, Lighthouse Reef is a large atoll, part of the Belize Barrier Reef. Divers consider it one of the best dive sites on Earth. Near the centre is the Great Blue Hole – a large sinkhole that plunges more than 100 m (330 ft) below sea level. It was formed during the last ice age when the sea level was much lower and the surrounding limestone was dry land. The atoll has an amazing variety of corals and fish, including sharks and sea turtles, which breed here.

---

### Other coasts, islands, and reefs in North America

- Acadia Coastline » p.65
- The Bay of Fundy » p.64
- Big Sur » p.63
- The Hawaiian Islands » pp.316-19
- Monterey Bay » p.62
- Thor's Well » p.62

▲ ICEBERGS AT THE ILULISSAT ICEFIORD

# Central and South America

## Les Pitons

**Location** Soufrière Bay, St Lucia, Lesser Antilles

Translated as "the Peaks", Les Pitons are two adjacent forest-clad volcanic lava domes that rise from the Atlantic Ocean and tower over St Lucia. Linked by Piton Mitan Ridge, steep-sided Gros Piton (770 m/2,526 ft high) and Petit Piton (at 743 m/2,438 ft) are cloaked in tropical and (at higher elevations) subtropical moist forest. Sulphurous fumaroles, hot springs, and deposits of ash, pumice, and lava flows are all indications of the volcanic origins of the peaks, which are the remains of a collapsed stratovolcano. The region was declared a World Heritage Site in 2004.

## The Paracas Peninsula

**Location** Pacific coast of Peru

Sometimes referred to as the Galápagos Islands of Peru, the Paracas Peninsula is part of Peru's only marine reserve. This arid peninsula extends into the Pacific Ocean towards the cold, nutrient-rich Humboldt Current, which is extremely rich in fish and other oceanic life-forms and attracts huge numbers of seabirds, including Humboldt penguins, as well as whales, dolphins, and sea turtles. A massive prehistoric design (or geoglyph), the Paracas Candelabra, has been carved into the north side of the peninsula; pottery found nearby dates from 200 BCE and could be of the same age.

◄ THE OBLONG-SHAPED ATOLL OF LIGHTHOUSE REEF

▲ THE LOFOTEN ISLANDS RISING FROM THE GREENISH BLUE WATERS OF THE NAPPSTRAUMEN STRAIT

## The Bahia Coastline

**Location** Bahia state coast, eastern Brazil

Stretching for about 1,000 km (620 miles), this is the longest coastline in Brazil, featuring sandy beaches, river estuaries, small deltas, large bays, and coral reefs. Much of the Bahia coast is backed by expanses of Atlantic forest. There are fringing reefs at Pintaunas, Praia do Forte, and Itacimirim. Just south of the idyllic sandy bay at Trancoso is Monte Pascoal National Park, an old-growth Atlantic forest with a population of the endangered red-browed Amazon parrot, along with scarce mammals such as jaguar, cougar, and giant armadillo.

Other coasts, islands, and reefs in Central and South America

- Cape Horn » p.117
- Chilean Fiordland » p.116
- The Galápagos Islands » pp.320-21
- The Great Blue Hole » p.114-15
- Lençóis Maranhenses » p.116

# Europe

## The Lofoten Islands

**Location** Nordland region, northern Norway

This archipelago of five large islands and many small ones stretches for 160 km (100 miles), and is characterized by steep-sided mountains of ancient metamorphic gneiss and quartzite, rising directly from the sea. The landscape of deep valleys and fiords was crafted by the action of glaciers. The Rost Reef, discovered in 2002, is the world's largest deepwater coral reef. Built by Lophelia corals, it is 35 km (22 miles) long. One of the world's most powerful tidal eddies occurs close to the island of Mosken.

## The North Jutland Dunes

**Location** Between Frederikshavn and Skagen, northern Denmark

For centuries, wind-blown sand caused problems for farmers in the Skagen area. Persistent southwesterly winds shifted dunes onto farmland and even buried buildings. Most of the dunes have been stabilized by the planting of vegetation, but one – the Rabjerg Mile – has been left unrestrained. Up to 40 m (130 ft) high and containing an estimated 4 million cubic metres (140 million cubic feet) of sand, this huge dune migrates up to 18 m (60 ft) northeast every year. It began its movement from the coast of Skagerrak, a strait separating Denmark from southwest Sweden, 300 years ago.

## The Devon Ria Coast

**Location** Plymouth and Exmouth, Devon, UK

The coast of south Devon has several drowned valleys, or rias, partially submerged by a rise in sea level of up to 25 m (80 ft)

layers of white and pale grey chalk are 96–86 million years old. They form vertical cliffs up to 102 m (335 ft) high. Marine erosion has produced three natural arches and a spike-like structure called L'Aiguille, which rises 70 m (230 ft) from the sea.

## Cap Ferret

**Location** Gironde, southwest France

This shingle-and-sand barrier spit almost completely separates the Bay of Biscay from the Arcachon Lagoon, into which the Leyre River discharges. The spit formed as a result of the longshore drift of sediments from the north. Cap Ferret is known to have built seawards and southwards since the early 18th century, pushing the entrance to the lagoon south as it did so. However, it has undergone erosion at its southern tip since 1970.

## Cathedral Beach

**Location** Cantabrian coast, northern Spain

Described as one of the most beautiful beaches in the world, Cathedral Beach has a series of rock arches that are reminiscent of a cathedral's nave. The dramatic caves, arches, and stacks are best appreciated at low tide. The incessant wave action has exploited faults in the metamorphic quartzite and slate strata to open up caves, which later form arches and then stacks when their roofs collapse.

## The Galician Coast

**Location** Northwest coast of Spain

This dramatic coastline, facing the Atlantic Ocean to the west and the Bay of Biscay to the north, is about 1,500 km (900 miles)

long and very irregular, with dozens of inlets and headlands, and at least 316 islands and islets. The cliffs at Vixia Herbeira rise 621 m (2,037 ft) above the sea and are among the highest in Europe. Drowned estuaries, or rias, such as those of the rivers Arousa, Pontevedra, and Vigo, are the result of rising sea level since the last glaciation.

## Cape de Creus

**Location** Catalonian coast, northeast Spain

Many of the classic features of crustal folding, faulting, shearing, and deformation are exhibited in the cliffs of this headland as well as anywhere on Earth. Located in the northeastern part of the Iberian Peninsula, Cape de Creus marks the point where the Pyrenees reach the editerranean. The rock structures displayed here demonstrate the massive pressures and strains applied to the crust during the formation of these mountains. Folded layers of schist and other metamorphic rocks have been intruded into the rocks by dikes of pegmatite, an igneous rock with large crystals. The Cape de Creus peninsula and its marine environments were designated a National Park by the Catalan government in 1998.

▼ AN ARCH IN THE ÉTRETAT CLIFFS

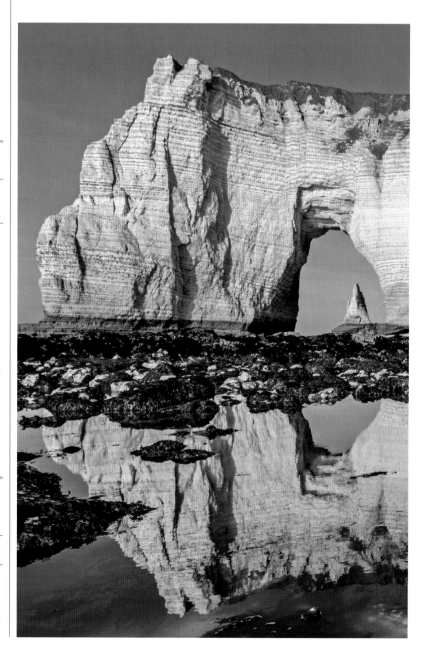

since the last ice age. The most important rias are the valleys of the Tamar–Lynher, Kingsbridge, Dart, Teign, and Exe. The drowned section of the Kingsbridge valley is a broad estuary that, despite being tidal 8 km (5 miles) inland, is fed only by small streams. Longshore drift has built bars across parts of the mouths of the Teign and Exe estuaries, and sedimentation occurs behind the bars.

## The Étretat Cliffs

**Location** Normandy coast, northern France

These are part of a 130-km (80-mile) stretch of chalk cliffs on the English Channel coast of Normandy. Étretat's horizontal

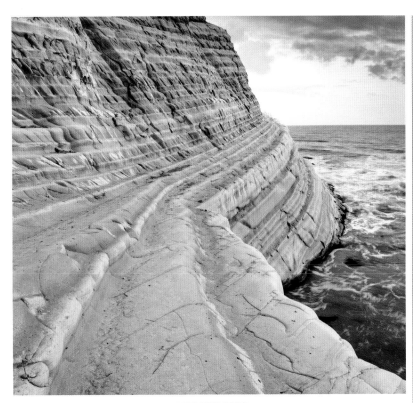

▲ DAZZLING WHITE TRUBI MARL OF THE TURKISH STEPS

## The Amalfi Coast

**Location** Sorrento Peninsula,
west coast of Italy

The southern coastline of the
Sorrento Peninsula is one of
the most beautiful in the world.
Mountains of limestone and
volcanic deposits plunge into
the sea in a stunning scene of
precipitous crags, with narrow
valleys descending steeply through
them to the shore. The seismic
activity associated with Vesuvius,
an active volcano not far to
the north, produces frequent
earthquakes, and the rocks of the
peninsula are extensively faulted.
The Amalfi Coast experiences
more flash floods and landslides
than any other region of Italy.

## The Turkish Steps

**Location** Southern Sicily, Italy

These cliffs of slightly inclined
strata of white Trubi marl have been
eroded into a series of long, parallel
steps. Trubi marl is a relatively soft
calcareous sedimentary rock, which
was deposited on the ocean floor
about 5 million years ago. It
contains the tracks of animals,
probably worms that lived within
it while it was being deposited.

> **Other coasts, islands, and
> reefs in Europe**
>
> ● **The Algarve Coast** » p.170
> ● **The Cliffs of Dover** » p.167
> ● **The Cliffs of Moher** » p.166
> ● **The Dalmatian Coast** » p.171
> ● **Dune du Pilat** » p.167
> ● **Fingal's Cave** » p.166
> ● **The Giant's
>   Causeway** » pp.164–65
> ● **The Great Chagos Bank** » p.315
> ● **The Jurassic Coast** » pp.168–69
> ● **Norwegian Fiords** » pp.162–63

# Africa

## Legzira

**Location** Atlantic coast of
southern Morocco

Until 2016, there were two massive
arches of red sandstone on the
beach at Legzira. The layers of
sandstone are 100–150 million
years old and were deposited
on an unconformity over much
older granite. They derive their
coloration from iron oxide
in the natural cement binding
the grains in the rock. The arches
were eroded by the action of
waves, with one of them collapsing
onto the beach in September 2016.

## Banc d'Arguin

**Location** Golfe d'Arguin, west coast
of Mauritania

A national park covers 12,000
square km (4,630 square miles)
of this coastline – which includes
sand dunes, seagrass flats,
intertidal mud, sand, and shell
fragments – as well as the shallow
Golfe d'Arguin stretching up to
60 km (40 miles) from the shore.
An offshore upwelling zone
generates huge quantities of
invertebrate food for fish and birds,
with up to 50,000 pairs of colonial
birds breeding and more than two
million wintering in the park.

## Socotra

**Location** Arabian Sea

One of Earth's most isolated
islands, which originated as part
of a continent rather than as new
oceanic crust, Socotra is 240 km
(150 miles) from the African
mainland. It is also a major centre
of endemic plants and animals.
Socotra has an area of 3,665
square km (1,415 square miles),
divided between coastal plains,
a limestone plateau with caves
and other karst features, and a
mountainous core. The climate
is arid or semiarid, apart from
high elevations, which experience
more rainfall. Once part of the
supercontinent of Gondwanaland,
it became separated when the
Gulf of Aden opened up. More
than one-third of its plants are
found nowhere else, including the
bizarre-looking dragon's blood
tree, and there are also endemic
birds, reptiles, and invertebrates.

## Coffee Bay

**Location** Eastern Cape province,
South Africa

Near this coastal settlement is
the spectacular Hole in the Rock
formation, where wave action has
punched a hole through an offshore
cliff of shale and sandstone. Erosion
has created a rock arch, rather than
dividing the cliff into two stacks,
because a thick intrusion of hard
volcanic dolerite lies above the
less resistant shale and sandstone,
which is 260 million years old.
The arch is large enough for a
yacht to sail through at high tide.

▶ DRAGON'S BLOOD TREES
AT SOCOTRA

## Jeffreys Bay

**Location** Eastern Cape province, South Africa

One of the best surfing locations in the world, Jeffreys Bay has ideal wave conditions, which are attributed to a combination of swell size, direction, and frequency; wind direction; tides; and the shape of the seabed. Just offshore, the warm Agulhas Current passes west before turning back on itself at the eastern edge of the Agulhas Bank, where it meets cold water from the west. Here, upwelling has produced a huge marine ecosystem.

## Cape Point

**Location** Western Cape province, South Africa

This headland lies just east of the Cape of Good Hope and forms the western entrance to False Bay. Hard quartz sandstones form cliffs that rise more than 200 m (660 ft) from the sea. There are two lighthouses, although only one is operational. The headland is part of the Cape Floral Region, an area with an extraordinary variety of plants that grow nowhere else in the world. Offshore are the cold, nutrient-rich waters of the Benguela Current, which fuel high rates of phytoplankton growth and sustain a very productive marine ecosystem.

**Other coasts, islands, and reefs in Africa**
- **Los Gigantes Cliffs** » p.198
- **The Red Sea Coast** » p.199
- **The Seychelles** » p.314

# Asia

## The Yaeyama Islands

**Location** East China Sea, southwest of Okinawa, Japan

This archipelago of 32 islands and numerous islets has volcanic origins. More than 90 per cent of Iriomote, the largest island, is covered by tropical rainforest, with mangroves around the coast. It is home to the critically endangered and endemic Iriomote cat. Ishigaki is the second biggest island. Its offshore coral reefs provide prime habitat for dolphins, sea turtles, manta rays, and whale sharks, although dugongs are now virtually extinct here.

▲ BRILLIANT WHITE SILICA SANDS OF WHITEHAVEN BEACH

## El Nido

**Location** Northern Palawan Island, Philippines

Spanish for "the nest", El Nido is the gateway to the famous Bacuit Archipelago. El Nido comprises the northern tip of Palawan Island as well as 45 islands and islets. Erosion of thick layers of limestone has produced a spectacular landscape of coastal karst, with cliffs towering vertically from the sea, combined with caves, white sand beaches, and coral reefs. The Secret Lagoon on Miniloc Island is a sea-filled cavern that can only be accessed via a narrow entrance. Experienced divers access a submarine tunnel beneath Dilumacad Island through an entrance that is 12 m (39 ft) below the water surface. The fauna of the offshore reefs includes pipefish, pufferfish, scorpion fish, lionfish, and many more beautiful creatures.

### Other coasts, islands, and reefs in Asia

- Andaman Sea Reefs » p.253
- Dragon Hole » p.251
- Ha Long Bay » p.250-51
- The Krabi Coastline » p.252-53
- The Maldives » p.315
- Nusa Tenggara » p.253
- Shiraho Reef » p.251

# Australia and New Zealand

## Horizontal Falls

**Location** Talbot Bay, Buccaneer Archipelago, western Australia

These tidal waterfalls are produced as ocean water pours through narrow gorges cutting through the two parallel rock barriers of the McLarty Ranges. On a rising tide, water builds up on one side of the barriers more quickly than it can push through the gorges. The resultant difference in water level reaches 5 m (16 ft) and produces the waterfall.

## Whitehaven Beach

**Location** Whitsunday Island, Queensland, Australia

The origin of Whitehaven's 7 km (4 miles) of fine, bright white sand has long been debated. Whitsunday Island is part of an ancient volcanic caldera, but its sand – which is 98 per cent quartz – could not have come from the island's igneous rhyolite and dacite rocks. In fact, it is derived from eroded granite on the Queensland mainland and was deposited by ancient longshore drift. This conveyor belt of coastal movement stopped at least 6,500 years ago, but the sand remains.

## Ninety Mile Beach

**Location** West coast of Aupouri Peninsula, North Island, New Zealand

Despite its name, this strip of sand, which faces the Tasman Sea, actually stretches 88 km (55 miles) between the rocky headlands of Scott Point and Ahipara. Even more remarkable are the sand dunes – probably the largest dunes in the southern hemisphere and a renowned sandboarding site – that lie behind the beach. The dunes increase in height from south to north, culminating in summits more than 140 m (460 ft) high.

### Other coasts, islands, and reefs in Australia and New Zealand

- The Great Barrier Reef » pp.284-87
- Moeraki Beach » p.290
- New Zealand Fiords » p.291
- Shark Bay » pp.288-89
- The Twelve Apostles » p.290

# FORESTS

About 30 per cent of Earth's land surface is covered by forest, which provides some of the richest of all habitats. At high latitudes, between 53 and 67°N, are the huge boreal forests of North America and Eurasia, which are dominated by coniferous trees. Between 10°S and 10°N are the world's great tropical rainforests, including the Amazon and Congo, regions of great biodiversity. In between are a wide variety of subtropical and temperate forests, some of the latter being rainforests. Russia has the greatest area of forest and Suriname the greatest proportion of cover.

# North America

## The Tongass National Forest

**Location** Alaska, northwest USA

Covering most of southeast Alaska, this is the largest national forest in the USA, at 69,000 square km (26,640 square miles). It is part of the Pacific Northwest Rainforest, Earth's largest temperate rainforest. The dominant trees are red cedar, sitka spruce, and western hemlock. There is logging of old-growth forest, but it is strictly controlled. Brown and black bears live in the forest. The region also supports large populations of breeding birds.

## The Redwood National Forest

**Location** California, western USA

Half of the world's remaining old-growth coast redwood trees – the tallest on Earth – are protected within this forest reserve. One, Hyperion, is 116m (380ft) tall. Coast redwoods are native to coastal northern California and southern Oregon. Black bears, cougars, beavers, flying squirrels, and more than 400 species of bird have been recorded in the forest, which covers 158 square km (61 square miles).

## Californian Chaparral and Woodland

**Location** California and Baja California, western USA and northern Mexico

Many coastal areas of central and southern California, the Central Valley, and Baja California are characterized by this scrub-dominated environment. Scrub oaks, sages, and *Ceanothus* are typical plants, and the flora is well adapted to recover from frequent wildfires. The climate is a Mediterranean type, with warm, wet winters and hot, dry summers.

## The Sierra Madre Pine-Oak Forest

**Location** Between Jalisco state, northwest Mexico, and New Mexico, southwest USA

Large areas of these subtropical mixed forests survive at high altitudes in the dramatic mountains and steep-sided valleys of the Sierra Madre Occidental, although much has also been logged. Typical trees in the region include several species of pine, Douglas fir, Emory oak, and Mexican blue oak. The forests are home to black bears, jaguars, and more than 300 breeding birds, including some that are found nowhere else on Earth – the thick-billed parrot, eared quetzal, military macaw, Sinaloa martin, golden eagle, and tufted jay.

▶ SNOW-COVERED TREES AT THE TONGASS NATIONAL FOREST

▲ CARACOL FALLS SURROUNDED BY RESTINGA FOREST

## The Jalisco Dry Forest

**Location** Nayarit, Jalisco, and Colima states, west to central Mexico

Although relatively small, this ecoregion along the Pacific coast of central Mexico is one of the most biodiverse on Earth. Of 1,200 plant species, 16 per cent are found nowhere else, and of 733 vertebrates, 29 per cent are endemic, including Merriam's desert shrew, the Mexican parrotlet, and the San Blas jay. The Jalisco Dry Forest also serves as an important stop for migrating birds from North America on their route towards South America. A wet season lasts from June to September, but the region is dry for most of the year.

### Other forests in North America

- **Cherokee National Forest** » p.70
- **The Giant Forest** » pp.68-69
- **Green Mountain National Forest** » p.71
- **North American Boreal Forest** » p.66
- **Pando Aspen Grove** » p.70
- **Pacific Northwest Rainforest** » p.67

# Central and South America

## Chocoan Rainforest

**Location** Eastern Panama and northern Ecuador, northwest South America

This biome of ultra-wet tropical rainforest is one of the most species-rich lowland areas on Earth. More than 11,000 types of plant have been recorded, about a quarter of them endemic, and there are 650 types of bird, many of them found nowhere else on Earth. This rain forest covers about 180,000 square km (70,000 square miles) of the zone between the Pacific coast and an elevation of about 1,000 m (3,300 ft) in the western Andes. The annual rainfall here exceeds 13,000 mm (512 in).

## Atlantic Coast Restinga Forest

**Location** Three enclaves in Atlantic Forest, eastern Brazil

This is now restricted to three relatively small fragments of tropical forest in the north and, further south, subtropical forest on sandy, nutrient-poor soils that have often formed on stabilized sand dunes. The forest typically comprises closed-canopy broadleaf trees of medium height, 5–15 m (16–50 ft) tall, but there is also an open-canopy variant, more like savanna. Several animals that are dependent on restinga are threatened with extinction. Much of the biome has been destroyed for urban development, and this still poses a major threat.

## South American Dry Forest

**Location** Western Paraguay, southern Bolivia and Brazil, and northern Argentina

Occupying a huge lowland area of sandy and silty sediments west of the Paraguay River, south of the Pantanal, and east of the Andes, this forest type generally grades from park-like savanna in the east through palm savanna to thorn scrub in the west. The change is related to rainfall, which generally decreases moving west, and may be less than 50 mm (2 in) in the west. The plants are tolerant of low water levels. This forest is one of the last refuges of the flightless rhea and other animals, including jaguars, ocelots, pumas, tapirs, and spiny anteaters.

## Chilean Matorral

**Location** Between the Pacific Ocean and the Andes, central Chile

This biome of shrubland and forest covers 148,000 square km (57,100 square miles) of the central Chilean coastal strip. It is transitional between the arid Atacama Desert to the north and the wetter Valdivian forest to the south, with a Mediterranean climate of rainy winters and dry summers. It has an enormous variety of plant species, about 95 per cent of which are endemic to Chile, including many sclerophytic shrubs (with tough evergreen leaves to reduce water loss), cacti, and fuschias.

### Other forests in Central and South America

- **The Amazon Rainforest** » pp.120-21
- **The Andean Yungas** » p.119
- **The Monteverde Cloud Forest** » pp.118-19
- **The Valdivian Temperate Forest** » p.119

# Europe

## The Caledonian Forest

**Location** The Scottish Highlands, UK

Scots pine forest colonized the United Kingdom after the last ice age, 9,000 years ago. Wetter and windier conditions later reduced its extent, and grazing deer and sheep have also contributed to the Caledonian Forest being reduced to its present 180 square km (70 square miles), divided into 35 remnants. In addition to the pines, there is birch, rowan, aspen, juniper, and oak. The UK's only endemic vertebrate species, the Scottish crossbill, feeds on pine cones in the forest.

## The Crooked Forest

**Location** Western Pomerania, Poland

This is a bizarre-looking group of 400 pine trees that are bent at a 90-degree angle close to the ground and have subsequently readjusted to grow vertically again. Even stranger, the trees are surrounded by a forest of normal, upright trees. Several theories have been advanced to explain the phenomenon, but the most likely one is that the farmers who planted the trees in the 1930s were responsible. They may have wanted trees with curved trunks to use for shipbuilding or the manufacture of furniture.

## Hainich National Park

**Location** Thuringia, central Germany

This national park contains part of the largest continuous area of deciduous forest in Germany, Hainich Forest, which is dominated by European beech. The park is being managed with the intention of allowing the forest, which was formerly used for military training, to return to its natural state. In spring, before the thick canopy closes, an understorey includes masses of wild garlic. Later, 16 species of orchid grow here, which provide a habitat for wildcats, bats, woodpeckers, and thousands of species of invertebrates.

## The Black Forest

**Location** Baden-Württemberg, southwest Germany

This large upland area is called the Black Forest because the spruce trees here grow so close together that parts of it are very dark. The area is an uplifted fault block (horst) of sandstone, gneiss, and granite, which is bounded by the River Rhine to the south and west. The forest is roughly 160 km (100 miles) by 60 km (40 miles). The original cover of mixed forest was largely cleared in the 19th century to be replaced by spruce monoculture for forestry, but recent years have seen an increase in deciduous tree cover once more. Spruce and silver fir mingle with stands of beech in the hills and valleys of the forest. Germany's two largest nature parks are in the Black Forest, and there is also a national park.

▼ PINE TREES IN THE CROOKED FOREST

▲ FROST-COVERED TREES IN THE CARPATHIAN MOUNTAIN FORESTS

## Kalkalpen National Park

**Location** The Northern Limestone Alps, northern Austria

More than 80 per cent of this protected park is forest. Cloaking hills, ridges, and valleys on limestone, this is central Europe's largest continuous tract of forest. Within the forest are several enclaves of virgin spruce, fir, and beech forest, thought not to have been cleared since the end of the last ice age. Larger areas are now being left unmanaged, so dead wood is left to rot – providing habitats for many invertebrates and woodpeckers. Almost 1,000 species of plant have been recorded. Lynx, brown bear, and endangered bats inhabit the region.

## Carpathian Mountain Forests

**Location** The Carpathian Mountains, western Ukraine and eastern Slovakia

Ten areas of primeval beech forest along a 185-km (115-mile) stretch of the Carpathians were designated a UNESCO World Heritage Site in 2007 for the quality of the woodland and their biodiversity. About 70 per cent of the 780 square km (300 square miles) of forest, which includes oak and hornbeam in addition to beech, is in Ukraine, the remainder in Slovakia. The site includes old-growth and virgin forest and supports 64 species of mammals, including brown bear, lynx, wolf, and wild boar.

## The Ardennes

**Location** The Ardennes Mountains of Belgium, Luxembourg, Germany, and France

This upland region comprises a landscape of ridges up to 694 m (2,277 ft) above sea level and steep-sided valleys on sandstone, quartzite, slate, and limestone bedrock. At least half of its area is covered by forest, with a rich mix of oak, beech, ash, hazel, maple, aspen, spruce, and pine. Generally scarce bird species – such as hazel grouse, hawfinch, Tengmalm's owl, black and grey-headed woodpeckers, and nutcrackers – breed in the forest.

## Mediterranean Evergreen Forest

**Location** The Mediterranean Basin, including Spain, southern France, Italy, Greece, Morocco, Tunisia, and Algeria

This broadleaf evergreen forest, which is transitional between Mediterranean pine forest and

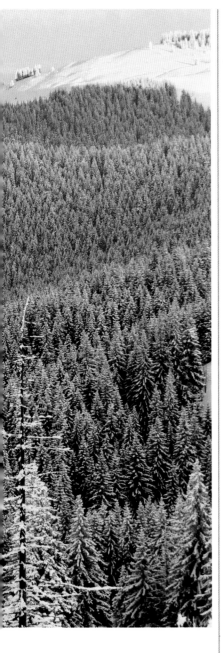

# Africa

## Cameroonian Highlands Forest

**Location** Eastern Nigeria and western Cameroon, West Africa

This moist broadleaf forest grows at elevations above 900 m (2,950 ft) on the fertile soils of a range of extinct volcanoes in an area roughly 625 km (390 miles) by 180 km (110 miles). Because of its altitude, the climate is cooler than most of tropical Africa, and parts of the forest experience heavy rainfall. Afromontane trees, including brittle-wood, yellow-wood, African cherry, and Cape beech, are plentiful. The forest is home to numerous endemic plants, reptiles, birds, and mammals, and there are also several species of endangered primates, including western gorillas and chimpanzees.

## East African Coastal Forests

**Location** Coastal regions from southern Mozambique to southern Somalia, East Africa

The tropical and subtropical moist forest extends over 112,000 square km (43,200 square miles), stretching in a relatively narrow strip just inland from the Indian Ocean. Its mosaic of closed-canopy woodland, more open savanna, grassland, and wetland varies from south to north. Much has been cleared for farmland, but it remains home to more than 600 species of resident and seasonal migrant birds.

scrubland and cool-temperate deciduous forest, grows to an elevation of 1,400 m (4,600 ft). The forest is dominated by small-stature holm oaks, usually 5–12 m (16–39 ft) tall. These trees are slow-growing, cast a deep shadow, and have tough, drought-resistant leaves. Although the forest is much reduced in area, some large tracts survive, for example in the lower Pyrenees of Aragon and the Sierra Morena of Andalucía, both in Spain.

> **Other forests in Europe**
>
> ● **The Bavarian Forest** » p.173
>
> ● **The Hallerbos** » p.172

> **Other forests in Africa**
>
> ● **The Congo Rainforest** » p.200
>
> ● **Madagascan Dry Forest** » pp.202-203
>
> ● **Madagascan Rainforest** » p.201

# Asia

## Russian Far East Temperate Forest

**Location** The Sikhote-Alin mountains of Primorsky Krai and Khararovsky Krai, Russian Far East

After escaping glaciation during the last ice age, this region of mixed broadleaf and conifer forest covering the Sikhote-Alin mountain range became a refuge for many animal species. Today, it is an area of extraordinary biodiversity, with a small population of Amur tigers and Amur leopards. This is the only forest on Earth where tigers, leopards, and brown bears co-exist. Plant diversity is varied, with at least 2,500 species known. Broadleaf trees are common at lower elevations but conifers become dominant as the altitude increases. Winters are bitterly cold, summers are mild, and rainfall can be heavy during summer and autumn.

## The Forest of the Cedars of God

**Location** Mount Makmel, northern Lebanon

Protected since 1876, this is the most famous surviving remnant of the cedar forest that once covered much of the Lebanese mountains. For thousands of years, a succession of civilizations used cedar wood for shipbuilding and the construction of important buildings, so the forest is now only a fraction of its former size. The 375 trees of the forest grow at an elevation of more than 1,900 m (6,200 ft) on Mount Makmel, whose slopes are blanketed by snow in winter. Four of the cedars are more than 35 m (115 ft) tall.

▼ WOODED SLOPES IN THE FOREST OF THE CEDARS OF GOD

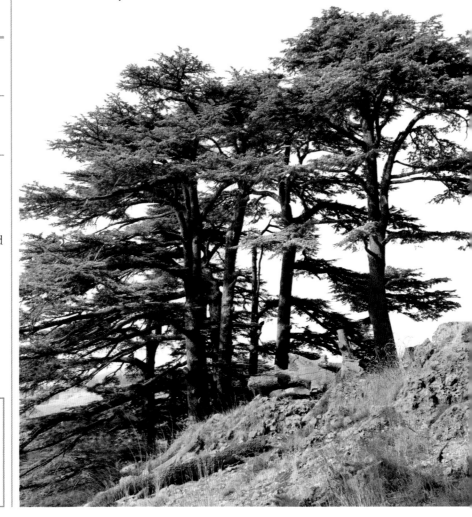

## Western Himalayan Temperate Forest

**Location** Nepal and far northern Pakistan and India, on the southern slopes of the western Himalayas

Comprising broadleaf forest at an altitude of about 600–2,600 m (2,000–8,500 ft) and subalpine conifer forest at higher elevations up to some 3,800 m (12,500 ft), this biome is more fragmented than its eastern Himalayan counterpart and receives less moisture from the Bay of Bengal monsoon. Birch and rhododendron are common broadleaf trees, and conifers such as blue pine, spruce, yew, and fir are plentiful. One area of forest, the Palas Valley, has the richest flora in Pakistan. The forest is home to endangered mammals, including the Himalayan tahr, and birds such as the western tragopan and Himalayan monal.

▼ THE SOUTHWESTERN GHATS MOIST FOREST

## The Mudumalai Forest

**Location** The Western Ghats mountains in Tamil Nadu, southern India

Up to 80 tigers live in the protected section of this forest, which is a 320-square-km (120-square-mile) national park. Other endangered species include the Indian leopard, sloth bear, and gaur. More than 200 bird species live here, including the endemic black-and-orange flycatcher. The forest can be divided into tropical moist deciduous, tropical dry deciduous, and tropical dry thorn forest – primarily reflecting differences in rainfall.

## Southwestern Ghats Moist Forest

**Location** The Western Ghats mountains in southwest India

This ecoregion is one of extremely rich biodiversity, including more than 4,000 types of flowering plant. Many animals are endemic to these tropical and subtropical moist broadleaf forests, including lion-tailed macaque, Nilgiri tahr, 90 species of reptiles, and 85 species of amphibians. This forest grows at elevations of 250–1,000 m (820–3,300 ft), between the Malabar coast moist forest below and montane forest at higher altitudes. It is characterized by high rainfall, most of which falls during the southwest monsoon. Protected areas include the Indira Gandhi and Periyar national parks.

## Sri Lankan Moist Forest

**Location** Southwest Sri Lanka

Covering about 15,500 square km (6,000 square miles), this tropical and subtropical moist broadleaf forest is one of the world's hotspots of endemism, with many animals and plants found nowhere else on Earth. The nature of the forest changes with elevation. At lower levels, more than 300 tree species are known, with some individuals in protected areas such as Sinharaja Forest towering to 50 m (165 ft). The region is wet – with annual rainfall exceeding 5,000 mm (197 in) in places – conducive to the greatest diversity of amphibians on Earth.

## The Sagano Bamboo Forest

**Location** Near Kyoto, the Japanese island of Honshu

Established by a monk in the 14th century, this forest now covers 16 square km (6 square miles). It is harvested regularly, but the bamboo regenerates rapidly, growing up to 25 m (80 ft) tall. A walk through this small area of *Phyllostachys* (moso) bamboo is an amazing experience aurally as well as visually as even the slightest breeze knocks the stems, or culms, together.

► TALL STALKS OF BAMBOO AT SAGANO BAMBOO FOREST

# Nansei Shoto Archipelago Forest

**Location** Ryukyu archipelago, southwest Japan

This archipelago consists of a long chain of larger volcanic and smaller coral islands between southern Japan and Taiwan and China. It has many tracts of subtropical moist broadleaf forest, which benefit from warm winters, hot summers, and high rainfall. The forest has a unique mix of fauna and flora, including many species that are found nowhere else on Earth. The Iriomote Islands contain the world's only habitat for the rare and endangered Iriomote cat. The Amami rabbit is another species endemic to the archipelago. The largest tract of forest is on Okinawa and is home to a woodpecker and a rail, both endemic to the island and bearing its name.

# Xishuangbanna Rainforest

**Location** Near Jinghong, Yunnan, China

This is the largest area of tropical China still covered by native forest. Despite the forest's high elevation (above 500 m/1,640 ft), it has most of the characteristics of the lowland rainforest found in southeast Asia. This is partly because it is protected from cold northerly winds in winter by the Hengduan Mountains. An extraordinary variety of more than 3,300 species of trees and other plants grow at Xishuangbanna. Some trees grow to a height of more than 80 m (260 ft).

### Other forests in Asia

- **Bornean Rainforest** » pp.258-59
- **Eastern Himalayan Forest** » p.256
- **Taiheiyo Montane Forest** » p.256
- **Siberian Taiga** » pp.254-55
- **Upper Yangtze Forests** » p.257

# GRASSLANDS AND TUNDRA

This varied biome includes tropical savannas and temperate prairies and steppes. Steppe is virtually treeless, but savanna can have an open woodland canopy. Despite their variety, grasslands are united in receiving too little rainfall to support forest but not so little that a desert forms. Rainfall may be seasonal or spread through the year. Other factors influencing the distribution and nature of grasslands include: seasonal flooding, which prevents forest growth; geology, which heavily influences soil type and hence vegetation; and heavy grazing that has created a grassland from a former forest.

## Central and South America

### The Cerrado

**Location** Between the Pantanal, Amazon Rainforest, and Atlantic Forest, central Brazil

This vast savanna stretches across more than 20 per cent of Brazil's total area. It has a semihumid tropical climate, with a wet season and a dry season, and comprises different types of savanna crossed by closed-canopy gallery forest along watercourses. More than 10,000 species of plants, including 800 trees, have been noted here. Mammals include the giant anteater, giant armadillo, jaguar, maned wolf, and marsh deer.

▼ MORICHE PALM TREES SPREAD ACROSS THE BRAZILIAN CERRADO

> **Other grasslands and tundra in Central and South America**
>
> ● The Pampas » pp.122-23

## Africa

### The Central and Eastern Miombo Woodlands

**Location** From southern Zambia to northern Tanzania, and from eastern Angola to northern Mozambique

Broadly divided into central Zambezian miombo woodland and eastern miombo woodland, these tropical and subtropical grasslands, savannas, and shrublands cover an area of about 1,930,000 square km (745,000 square miles). Although there are regional variations, shared features include nutrient-poor soils, the dominance of many types of miombo tree, and a long, hot dry season. Kafue National Park, Zambia's largest protected area, and the Niassa Reserve in Mozambique are examples of these ecoregions, which have a high diversity of large mammals.

> **Other grasslands and tundra in Africa**
>
> ● The Cape Floral Region » p.205
> ● Ethiopian Montane Grassland » p.204
> ● The Serengeti » pp.206-07
> ● The Sudanian Savannas » p.204

## Asia

### The Mongolian-Manchurian Steppe

**Location** Extending in a crescent north of the Gobi Desert, Mongolia to northeast China

Extensive temperate grasslands spread inland from northeastern China's coastal hills towards the boreal forests of southern Siberia, covering almost 900,000 square km (350,000 square miles). The rolling grasslands experience warm summers and bitterly cold, windswept winters, with most rain falling in the weak summer monsoon. Feathergrasses are dominant in many areas, with drought-resistant species being more common closer to the Gobi Desert. There are also small, spiny shrubs. Native wild mammals include Mongolian gazelles and Asiatic wild asses, and Przewalski's horses have been reintroduced. The brown eared-pheasant is the only endemic bird of this ecoregion; it uses the grassland and shrub habitat as winter refuge.

> **Other grasslands and tundra in Asia**
>
> ● The Eastern Steppe » p.260
> ● Siberian Tundra » p.261
> ● The Terai-Duar Savannas » p.260

# DESERTS

About one-third of Earth's land surface is desert, receiving under 25 cm (10 in) of precipitation annually. Deserts can be hot, cold, or polar. The Sahara is the largest hot desert, lying under a vast area of sinking, warming air, meaning rainfall is rare. The Gobi is the largest cold desert; hot in summer, cold in winter, it receives little precipitation because it lies in the rain shadow of the Himalayas. The world's biggest desert is polar – the Antarctic, where the air is too cold to hold much water vapour for rain or snow.

## North America

### The Colorado Plateau

**Location** Arizona, Utah, New Mexico, and Colorado states, southwest USA

This high, arid plateau made up of sedimentary rock layers boasts some of the most spectacular scenery in the world. Ranging in elevation from 600 m (2,000 ft) at the base of the Grand Canyon, which has been cut by the Colorado River, to 3,870 m (12,700 ft) in the La Sal Mountains, it owes its character to remarkable geological stability. There are many deep canyons, flat-topped mesas, towering buttes, and natural rock arches within the plateau's 337,000-square-km (130,000-square-mile) expanse. The walls of the formations reveal rocks ranging in age from billions to just a few hundred years old.

#### Other deserts in North America

- Antelope Canyon » p.80
- Bryce Canyon » p.78
- The Chihuahuan Desert » p.81
- The Great Basin Desert » p.74
- Mesa Arch » p.79
- Meteor Crater » p.79
- The Mojave Desert » p.75
- Monument Valley » pp.76–77
- The Sonoran Desert » p.76

## Central and South America

### La Guajira Desert

**Location** Guajira Peninsula, northeast Colombia

This arid zone of xerophytic (drought-resistant) plants lies in the rain shadow of the Serrania de Macuira, a range of low mountains at the eastern end of the Guajira Peninsula, which intercepts the northeast trade winds. La Guajira is a land of thorny bushes, cacti, and other succulents. Large numbers of American flamingos breed at the Parque Natural los Flamencos, and limited-range bird species include the vermilion cardinal, white-whiskered spinetail, and chestnut piculet.

### The Sechura Desert

**Location** Piura and Lambayeque provinces, northern Peru

Covering 190,000 square km (73,350 square miles) of low-lying coastal Peru, the Sechura Desert extends up to 100 km (60 miles) inland to the secondary ridges of the Andes mountains. To the north, it transitions into tropical dry forest while to the south it becomes the Peruvian Coastal Desert. Rain falls for only a few days each year. It owes its aridity to the atmospheric subsidence caused by the upwelling of cold water offshore. However, during the 1998 El Niño event, much higher than usual rainfall inland swelled rivers flowing through the desert and flooded a large part of the region, creating, for a time, Peru's second largest lake in the Bayovar Depression.

▼ THE GLEN CANYON REGION OF THE COLORADO PLATEAU

## The Caatinga

**Location** Northeast Brazil

This vast ecoregion covers almost 10 per cent of Brazil's territory, some 850,000 square km (328,000 square miles) of drought-resistant shrubland and thorn forest. The Caatinga experiences two seasons: a very hot dry season, when soil temperatures may reach 60°C (140°F) and plants drop their leaves to reduce transpiration; and a shorter, hot rainy season. After the first rain falls, the cerrado transforms in just a few days from a dead-looking greyish landscape to one that is bright green and full of life.

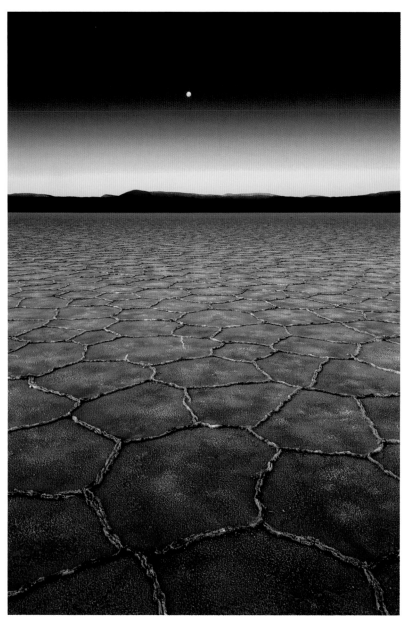

▲ HEXAGONAL PATTERNS AT SALINAS GRANDES SALT DESERT

## Salar de Arizaro

**Location** Puna de Atacama, northwest Argentina

The world's sixth largest salt flat covers 1,600 square km (600 square miles) and lies at an elevation of about 3,460 m (11,350 ft) in the Andes. Formed as a result of the evaporation of surface water and the deposition of its constituent salts, Salar de Arizaro also features yardangs (ridges of rock aligned with the prevailing northwesterly winds), rocks that have been weathered by the repeated growth of salt crystals, relic lake shores, and the small volcanic cone of Cono de Arita, which rises about 120 m (390 ft) above the salt.

## Talampaya National Park

**Location** La Rioja province, northwest Argentina

This World Heritage Site protects 2,150 square km (830 square miles) of arid scrub desert in the rain shadow of the Andes mountains. The Talampaya National Park includes dramatic landscapes shaped by water and wind, including the 143-m- (469-ft-) deep Talampaya Gorge and many red sandstone cliffs. The area's main claim to fame is its fossils – the most complete known range of life-forms dating from 250–200 million years ago, including the early dinosaur *Eoraptor* and mammals, fish, amphibians, and plants.

## Salinas Grandes

**Location** Jujuy and Salta provinces, Puna de Atacama, northwest Argentina

This large salt flat lies at the bottom of an endorheic basin (one with no outlet) at an elevation of 3,350 m (10,990 ft) on the arid Andean puna. Its dazzling white salt deposits, which cover more than 200 km (120 miles), were precipitated from evaporated runoff from the surrounding mountains. Rectangular pools of water mark places where salt extraction is taking place, and local artisans carve salt figures to demonstrate their skills.

> **Other deserts in Central and South America**
>
> ● **The Atacama Desert** » pp.124-25
> ● **The Patagonian Desert** » p.127
> ● **The Salar de Uyuni** » p.126
> ● **The Valle de la Luna** » p.127

▼ LARGE DUNES AT THE ERG CHEBBI

# Europe

## The Tabernas Desert

**Location** Almeria, Andalucía, southern Spain

Protected from humid winds coming from the Mediterranean by the Sierra de Alhamilla, this area of 280 square km (110 square miles) is semiarid rather than a true desert. Annual rainfall of 150–220 mm (6–9 in) is restricted to a few days. Occasional torrential rain has cut steep-sided gullies, whose river beds are usually dry, and natural underground pipes, through which water permeates below the surface at the top of a slope, passes underground, and emerges lower down the slope. The drought-resistant flora includes oleanders, tamarisks, and the endangered sea lavender.

## The Stone Desert

**Location** Varna depression, northeast Bulgaria

This area of low rainfall and sparse vegetation is Bulgaria's only desert. It is best known for the Pobiti Kamani (hammered stones) – 18 groups of upright limestone columns up to 7 m (22 ft) tall and 3 m (10 ft) in diameter. The most popular theory that explains this phenomenon is that methane-bearing fluids seeping up shafts to an ancient ocean floor were oxidized by bacteria to form cylinders of calcium carbonate. Subsequent uplift of the region and erosion of the softer, sandy sediments around the cylinders has left them standing prominently.

# Africa

## The Erg Chebbi Sand Dunes

**Location** Erfoud, Sahara Desert, eastern Morocco

Erg Chebbi is one of two large ergs – seas of dunes composed of wind-blown sand with little or no vegetation – in the Moroccan Sahara. Its biggest dunes rise 150 m (490 ft) above the surrounding desert. Erg Chebbi extends about 28 km (17 miles) from south to north and 7 km (4 miles) from west to east. Daytime temperatures soar to 40°C (104°F) in July, but at night in winter they may fall to 3°C (37°F). The wettest month of the year is November, with just 10 mm (½ in) of rain. Despite the aridity, reptiles and nocturnal mammals such as jerboas and fennec foxes live here.

## The Nyiri Desert

**Location** South of Nairobi, southern Kenya

Also known as the Nyika Desert and the Tarudesert, the Nyiri Desert lies north of Mount Kilimanjaro and encompasses Amboseli National Park, including the northern half of Lake Amboseli. The annual rainfall averages 350 mm (14 in), but outside of the brief rainy season in April and May, water is scarce in this region, being restricted to a few springs and riverbeds. Vegetation comprises a mixture of grassland and thorn scrub (including some poisonous plants), with scattered baobab trees – some more than 2,000 years old, their circular trunks often as much as 3 m (10 ft) in diameter. Large mammals include giraffes, lions, leopards, and the lesser kudu and impala. More than 400 species of birds are residents or migrants, including 47 different types of raptor.

## The Makgadikgadi Pan

**Location** Kalahari Desert, northeast Botswana

This is actually a group of enormous salt pans, together making up one of the largest in the world, with an area of 16,000 square km (6,200 square miles). The Sua, Nwetwe, Nxai, and other pans are relics of the formerly huge Lake Makgadikgadi, which dried up several thousand years ago. Almost lifeless dry, salty crusts for most of the year, they are fed fresh water seasonally by the Nata, Boteti, and other rivers. This encourages the growth of vegetation around their margins and attracts migrant ducks, flamingos, and other birds. The Makgadikgadi pan is home to one of only two breeding populations of greater flamingos in southern Africa.

### Other deserts in Africa

- The Adrar Plateau » p.210
- The Kalahari Desert » p.211
- The Karoo » p.211
- The Namib Desert » pp.212-13
- The Sahara Desert » pp.208-209
- The White Desert » p.210

# Asia

## The Kyzylkum Desert

**Location** Between the Amu Darya and Syr Darya rivers in Uzbekistan, Kazakhstan, and Turkmenistan

Sloping gently northwest towards the Aral Sea between the Amu Darya and Syr Darya rivers, this arid region of 300,000 square km (116,000 square miles) of sand dunes is occasionally punctuated by highlands, such as the Bukantau, Tamdytau, and Auminzatau mountains. Another feature of the desert is the presence of takirs: shallow hollows that are submerged after seasonal rains and dry out to form cracked crusts when the water evaporates. Annual rainfall averages just 150 mm (6 in), while temperatures fluctuate greatly from 30°C (86°F) or more in summer to -9°C (16°F) in winter.

## The Syrian Desert

**Location** Southern Syria, western Iraq, northeast Jordan, and northern Saudi Arabia

Covering a vast area of true desert and steppe, the Syrian Desert comprises an elevated plateau in southern Syria and eastern Jordan, and a plain that gently slopes northeast towards the Euphrates River. The whole area is dissected by wadis (dry valleys). Annual rainfall is as low as 100 mm (4 in) in some areas. Daytime temperatures reach 45°C (113°F) in summer, and a hot, dry wind called the khamsin frequently sweeps across the desert from the south or southeast, blowing up duststorms. The southern part of the desert is inhabited by several nomadic tribes and breeders of Arabian horses.

## The Dasht-e-Kavir Desert

**Location** Southeast of Tehran, in northern Iran

Extending 390 km (240 miles) from the Alborz mountains southeast to the Dasht-e-Lut desert, this extremely arid region is surrounded on all sides by mountains. Although it receives virtually no rainfall, runoff from the mountains flows into seasonal lakes and the salt marshes of Kavir Buzorg, where evaporation leads to the formation of a salty surface crust. Rig Jenn is a vast area of sand dunes where summer temperatures can reach 50°C (122°F). The Kavir

National Park has a small population of the critically endangered Asiatic cheetah.

## The Indus Valley Desert

**Location** Between the Chenub and Indus rivers, northwest Punjab province, Pakistan

This inhospitable plain, which covers 20,000 square km (7,700 square miles), bakes in temperatures up to 45°C (113°F) during the day in summer, but can be freezing in winter. Annual rainfall ranges from 600 mm to 800 mm (24 in to 31 in), and a variety of vegetation reflects this. Drought-resistant *Prosopis* shrubs and trees are plentiful. Five large mammal species are native to the desert: the Indian wolf, striped hyena, caracal, Indian leopard, and urial (a form of wild sheep).

### Other deserts in Asia

- The Arabian Desert » pp.262-63
- The Dasht-e-Lut » p.263
- The Gobi Desert » pp.266-67
- The Karakum Desert » p.264
- The Taklamakan Desert » p.264
- The Thar Desert » p.265

# Australia and New Zealand

## The Little Sandy Desert

**Location** Between Newman and Wiluna in Western Australia

Covering an area of 110,000 square km (42,500 square miles), this undulating country of red dune fields is vegetated by acacia and drought-resistant spinifex grass. The landscape is sometimes broken by escarpments of ancient sandstone, such as the Calvert Range. In the north of the desert is Lake Disappointment, an endorheic salt lake, which holds water only in very wet periods.

### Other deserts in Australia and New Zealand

- The Gibson Desert » p.298
- The Great Sandy Desert » p.295
- The Great Victoria Desert » p.299
- The Pinnacles » p.298
- The Simpson Desert » pp.296-97

◀ THE MAKGADIKGADI PAN

▲ SALT MARSHES OF THE DASHT-E-KAVIR DESERT

# GLOSSARY

## A

**A'A LAVA** Basaltic lava with a rough surface of broken lava blocks or clinker. See also *block lava, pahoehoe lava.*

**ABLATION** ▼ The loss of ice from a glacier due to melting, evaporation, sublimation, calving, or erosion by the wind. The ablation area is the region of a glacier where there is a net loss of ice due to these processes. See also *accumulation area.*

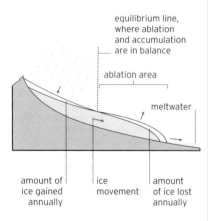

equilibrium line, where ablation and accumulation are in balance

ablation area

meltwater

amount of ice gained annually | ice movement | amount of ice lost annually

**ABYSSAL** Referring to the deep ocean floor and its environment, beyond the continental slopes. The abyssal zone is the part of the water column between a depth of 2,000 m (6,600 ft) and the abyssal plain. It is deeper than the bathyal zone but not as deep as the deep-sea trenches. See also *bathyal, deep-sea trench.*

**ABYSSAL PLAIN** The relatively flat, sediment-covered plain that forms the bed of most of the oceans, beyond the continental slopes. It lies at a depth of 4,000–6,000 m (13,000–20,000 ft). See also *continental slope.*

**ACCUMULATION AREA** ▼ The part of a glacier where snowfall exceeds losses by melting, evaporation, sublimation, calving, and wind erosion. See also *ablation.*

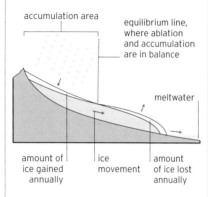

accumulation area

equilibrium line, where ablation and accumulation are in balance

meltwater

amount of ice gained annually | ice movement | amount of ice lost annually

**ACID RAIN** Precipitation (rain or snow) containing dissolved acids. Since it contains carbon dioxide, most rain is slightly acidic, but atmospheric pollution or the gases released by volcanic eruptions can make rain more acidic.

**ADVECTION FOG** A kind of fog that forms as moist air passes over cooler surfaces. It can form over land or the ocean. See also *radiation fog.*

**AEROSOL** A tiny particle suspended in the air. It may be dust or liquid, and is usually no larger than one millionth of a millimetre in diameter.

**AIR MASS** A body of air with relatively uniform characteristics derived from the surface region where it forms and distinct from surrounding air masses. Examples include polar maritime and tropical maritime air masses.

**ALBEDO** The proportion of the Sun's incoming radiation reflected by Earth's surface. For example, fresh snow may reflect up to 90 per cent, and its albedo will then be expressed as 0.9.

**ALLUVIAL DEPOSIT** River deposits that include sand, silt, mud, gravel, and organic matter. They may also be rich in minerals. See also *alluvium.*

**ALLUVIAL FAN** A cone-shaped deposit of sediment built up by rivers or streams. Fans are typically found where steep-flowing watercourses, rich in deposits, emerge from mountains onto a plain. See also *alluvium.*

**ALLUVIUM** Any sedimentary material deposited by rivers, including sand, silt, mud, gravel, and organic matter. Any accumulation of alluvium is called an alluvial deposit. See also *alluvial deposit, alluvial fan.*

**ANTICLINE** A fold of stratified rocks in which the oldest strata (layers) are at the core. Anticlines result from massive horizontal compression. See also *fold* (panel) *syncline.*

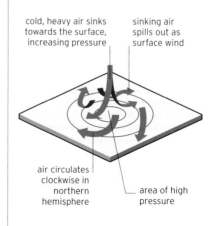

cold, heavy air sinks towards the surface, increasing pressure

sinking air spills out as surface wind

air circulates clockwise in northern hemisphere

area of high pressure

**ANTICYCLONE** ▲ A weather system in which atmospheric pressure is highest in the centre. Wind circulates around the system in a clockwise direction in the northern hemisphere, anticlockwise in the southern hemisphere. See also *cyclone, pressure system.*

**AQUIFER** ▼ An underground layer of permeable rock from which groundwater can be extracted. See also *groundwater.*

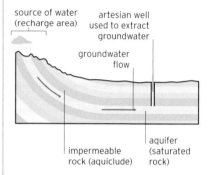

source of water (recharge area)

artesian well used to extract groundwater

groundwater flow

impermeable rock (aquiclude)

aquifer (saturated rock)

**ARCHIPELAGO** A chain or cluster of islands. See also *island arc.*

**ARÊTE** ▼ A sharp-crested ridge separating two adjacent cirques. See also *cirque.*

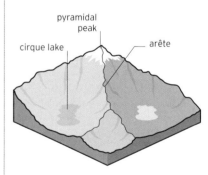

pyramidal peak

cirque lake

arête

**ASTHENOSPHERE** The viscous layer of the upper mantle immediately below the rigid lithosphere. Rock in this layer moves slowly in a solid state and is key to the movement of tectonic plates. See also *lithosphere* (panel), *mantle, tectonic plate.*

**ASYMMETRICAL FOLD** A fold whose axial plane is inclined, dipping in the same direction as the more gently dipping flank of the fold. See also *anticline, fold (panel), syncline*.

**ATMOSPHERE** The gaseous layers that surround Earth. The atmosphere is made up of four distinct layers – the troposphere, stratosphere, mesosphere, and ionosphere. See also *stratosphere, troposphere*.

**ATOLL** ▼ A ring-shaped coral island, or a ring of small coral islands, surrounding a lagoon. Atolls are created as coral reefs build up in shallow water around subsiding volcanic islands. See also *coral reef, fringing reef*.

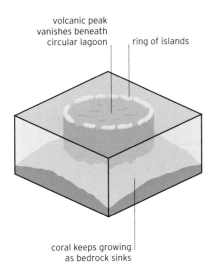

volcanic peak vanishes beneath circular lagoon

ring of islands

coral keeps growing as bedrock sinks

**AUREOLE** An area of rock around an igneous intrusion whose composition, structure, or texture has been thermally altered or metamorphosed. See also *country rock, intrusive igneous rock, metamorphism*.

**AURORA** Coloured streamers of light visible in the night sky at certain latitudes and at certain times of the year. They are caused by the interaction between high-energy particles from the Sun and Earth's magnetic field. In the northern hemisphere, displays are called the aurora borealis; the equivalent in the southern hemisphere is the aurora australis.

# B

**BACK-ARC BASIN** An undersea basin formed behind an island arc near where two tectonic plates meet. See also *island arc*.

**BACKSHORE** An area of the shore that is above the average high-water mark. This region is covered by water only during storms and the highest tides. See also *foreshore*.

**BACKWASH** The flow of water down a beach after a wave has broken on the shore. See also *swash*.

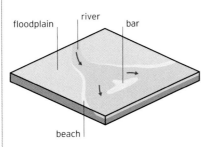

floodplain · river · bar

beach

**BAR** ▲ A linear deposit of gravel or sand deposited by a river or the sea. Bars may be exposed at low tide or permanently submerged. See also *barrier bar, longshore bar, point bar*.

**BARCHAN** A crescent-shaped dune formed by wind action in a sandy desert. The steeper slope, or slip face, is on the concave side. It is also called a barkhan dune or crescentic dune. See also *dune* (panel).

**BARRIER BAR** A longshore bar of gravel or sand roughly parallel to the coastline, whose surface is below mean sea level. See also *barrier island, longshore bar*.

**BARRIER ISLAND** A long, thin island of sediment roughly parallel to a coastline, whose surface is usually exposed. See also *barrier bar, longshore bar*.

**BASALT** A common, fine-grained or glassy volcanic rock that usually originates from solidified lava. The oceanic crust is composed of basalt, and basaltic lava can also erupt on continents. See also *extrusive igneous rock, flood basalt*.

**BASIN** A low-lying area or depression on land or in an ocean where sediment often accumulates. See also *river basin*.

**BATHOLITH** A very large igneous intrusion, 100 km (60 miles) or more across, which originates deep underground. Although they form deep underground, batholiths may be exposed by erosion. See also *igneous intrusion* (panel).

**BATHYAL** Referring to the zone of the ocean between about 200 and 2,000 m (660 and 6,600 ft) deep. The bathyal zone is not as deep as the abyssal zone. See also *abyssal*.

**BAYMOUTH BAR** A spit or bar that extends completely across the mouth of a bay. See also *barrier bar, spit*.

**BEACH** See panel, below.

**BEACH FACE** The sloping section of a beach, which is exposed to the wash of the waves. Also called the foreshore. See also *beach* (panel), *swash*.

**BEDDING** The sequence in which layers of sedimentary rock were originally laid down, one on top of another. See also *bedding plane, sedimentary rock*.

## BEACH

A beach is an accumulation of sand, gravel, shell debris, or other deposits along the shoreline of an ocean, lake, or river. Ocean beaches are constantly changing as tides and wave action deposit and carry away sediment. Every beach has a distinct profile above and below the water line. The intertidal zone lies between the low- and high-water marks.

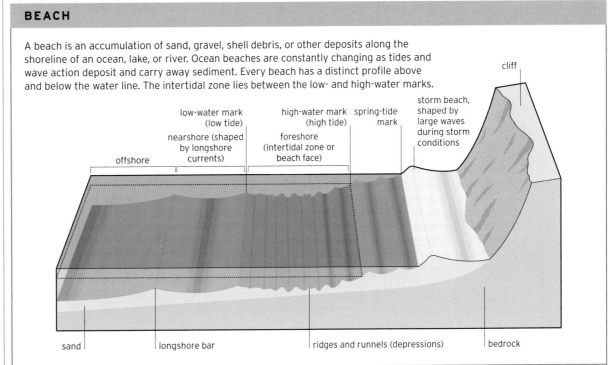

cliff

low-water mark (low tide)

high-water mark (high tide)

spring-tide mark

storm beach, shaped by large waves during storm conditions

nearshore (shaped by longshore currents)

foreshore (intertidal zone or beach face)

offshore

sand

longshore bar

ridges and runnels (depressions)

bedrock

**BEDDING PLANE** The surface separating a layer of a sedimentary rock from the layer above or below. See also *bedding, sedimentary rock.*

**BERGSCHRUND** A deep crack or crevasse that forms where the moving ice of a glacier pulls away from its headwall. See also *crevasse* (panel), *glacier.*

**BERM** A ridge of shingle on the upper part of a beach, above the beach face, usually marking the point of the highest high tides. See also *beach face.*

**BIODIVERSITY** The variety of living plants and animals in a given area or habitat.

**BIOLUMINESCENCE** The production of light by living life-forms. Bioluminescent organisms include some bacteria, fungi, cephalopods, jellyfish, and fish.

**BIOME** A large-scale biological community defined primarily by its vegetation. Examples include tropical rainforest, temperate grassland, taiga, and tundra.

**BLACK SMOKER** A hydrothermal vent in which the emerging, very hot water is coloured black with

dark minerals – mostly iron sulphide. See also *hydrothermal vent, white smoker.*

**BLOCK LAVA** Volcanic lava with a surface of smooth-sided angular fragments. This lava usually solidifies to form the rock andesite.

**BLOCKFIELD** An area of broken rocks, usually in a mountainous region. Also called felsenmeer.

**BOG** A peat-accumulating wetland. Bogs are acidic, with low levels of nutrients, and their soils are composed almost entirely of dead plant matter. See also *fen, marsh.*

**BORE** See *tidal bore.*

**BOREAL** A term describing parts of the northern hemisphere between the Arctic and temperate zones – for example, the boreal forest of Eurasia.

**BOSS** An igneous intrusion that is roughly circular in horizontal cross-section. See also *igneous intrusion* (panel).

**BOUNDARY** See panel, below.

**BRACKISH** Water that has a salt content less than that of ocean water but greater than that of fresh water.

**BRAIDED STREAM** A watercourse where water flows in several shallow channels that constantly separate and rejoin. This system is common downstream of many glaciers. See also *glacier, meltwater.*

**BREAKER ZONE** The part of a beach where waves break. See also *beach* (panel), *beach face, swash.*

**BRECCIA** A sedimentary rock made up of angular fragments of minerals and other rocks.

**BROADLEAF** A term describing trees that have broad leaves, as distinct from needle-like leaves. Examples of broadleaved trees include chestnuts, maples, and oaks. See also *coniferous, deciduous, evergreen.*

# C

**CAATINGA** A kind of thorny forest found in semi-arid areas of northeast Brazil.

**CALCITE** A common transparent or opaque mineral form of calcium carbonate.

**CALDERA** A large, bowl-shaped volcanic depression, typically more than 1 km (0.6 miles) in diameter. It forms when a volcano collapses into its magma chamber after it has been emptied by a volcanic explosion. See also *magma chamber, volcano* (panel).

**CALVING** The process of ice blocks shedding from a glacier or ice-sheet and falling into a lake or the ocean. See also *glacier, ice-sheet.*

**CANOPY** The crowns of the tallest forest trees, which receive full sunlight. The canopy may be closed, with crowns touching each other to form a continuous layer, or open. See also *rainforest.*

**CANYON** A deep, steep-sided, relatively narrow valley. See also *gorge.*

**CATCHMENT AREA** The total area drained by one river and its tributaries. See also *drainage basin, watershed.*

**CERRADO** A type of savanna dotted with small trees and shrubs, found mostly in central Brazil. See also *savanna.*

---

## BOUNDARY

The margin between two or more tectonic plates is called a boundary. At a divergent (constructive) boundary, two plates move apart, creating a rift. Magma rises from the mantle beneath the rift, pouring out through long fissures. A convergent (destructive) boundary forms where two plates collide, often where oceanic crust is pushing against continental crust and is being forced

beneath it, a process known as subduction. The continental crust often buckles up into a mountain range, while the subducting oceanic crust forms a deep seafloor trench. Volcanoes form above the area of subduction. A transform plate boundary involves two plates pushing past each other. Although no magma reaches the surface, earthquakes are regular features of transform fault locations.

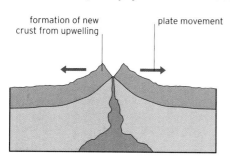

DIVERGENT (CONSTRUCTIVE) BOUNDARY

CONVERGENT (DESTRUCTIVE) BOUNDARY

TRANSFORM (CONSERVATIVE) BOUNDARY

**CINDER CONE** A steep-sided volcano made up of ash, cinders, or clinker (rough, irregular lava fragments). Also called a scoria cone. See also *volcano* (panel).

**CIRQUE** ▼ A steep-sided hollow at the head of a valley or on a mountainside carved out by a glacier. Many glaciers originate in mountain cirques, from which they flow to lower ground. See also *glacier*.

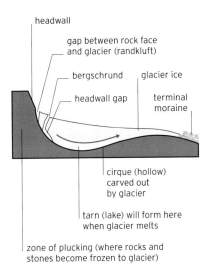

headwall
gap between rock face and glacier (randkluft)
bergschrund
glacier ice
headwall gap
terminal moraine
cirque (hollow) carved out by glacier
tarn (lake) will form here when glacier melts
zone of plucking (where rocks and stones become frozen to glacier)

**CIRQUE GLACIER** A glacier that originates in, and is mostly restricted to, a cirque. If a cirque glacier advances far enough, it may become a valley glacier. See also *cirque, glacier, valley glacier*.

**CLAST** A fragment of gravel, sand, or other sediment incorporated into another, newer rock. For example, sandstone is made up of many clasts of sand. See also *grain*.

**CLAY** An earthy material with mineral particles smaller than about 0.002 mm (0.00008 in) in diameter. Clay is porous, but water flows through it only slowly.

**CLOUDFOREST** A damp forest that is often shrouded in cloud. Cloudforest is generally found high in mountains.

**COLD DESERT** A desert that becomes very cold for at least part of the year, due either to high elevation or high latitude. Examples of cold deserts are Antarctica and the Gobi Desert.

**CONDENSATION** The conversion of a substance from a vapour to a liquid. Clouds form when water vapour condenses to form tiny droplets of water.

**CONFLUENCE** The place where two streams, rivers, or glaciers join.

**CONIFEROUS** Cone-bearing. Firs and pines are examples of coniferous trees.

**CONSERVATIVE BOUNDARY** A margin along which two tectonic plates slide past each other in opposite directions or at different speeds. See also *boundary* (panel).

**CONSTRUCTIVE BOUNDARY** See *divergent boundary*.

**CONTACT METAMORPHISM** The process by which very hot intrusive igneous rocks alter the country rocks around them. See also *country rock, igneous intrusion* (panel), *metamorphic rock*.

**CONTINENTAL CRUST** The layer of mainly sedimentary and metamorphic rocks that forms the continents and the relatively shallow oceans close to their shores. See also *continental shelf, crust, oceanic crust*.

**CONTINENTAL DIVIDE** A continent's drainage divide. Water either side of the divide drains into different oceans or seas. See also *watershed*.

**CONTINENTAL MARGIN** The zone of the ocean floor that separates thin oceanic crust from thick continental crust. The continental shelf, continental slope, and continental rise together make up the continental margin. See also *continental shelf, continental slope*.

**CONTINENTAL RISE** The lowest part of the continental slope, adjacent to the abyssal plain. See also *abyssal plain*.

**CONTINENTAL SHELF** The submerged, gently sloping portion of continental crust between the coast and the continental slope. See also *crust*.

**CONTINENTAL SLOPE** The sloping ocean floor between the continental rise and the continental shelf. See also *continental rise, continental shelf*.

**CONVECTION** The movement of gases, liquids, and molten rocks in response to differences in temperature. One example is when air is warmed as it passes over hot ground and rises high in the atmosphere.

**CONVERGENT BOUNDARY** The line along which two or more tectonic plates move towards each other and collide. The result is either a subduction zone or a continental collision. See also *boundary* (panel), *subduction, tectonic plate*.

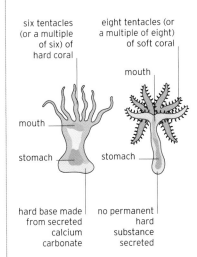

six tentacles (or a multiple of six) of hard coral
eight tentacles (or a multiple of eight) of soft coral
mouth
mouth
stomach
stomach
hard base made from secreted calcium carbonate
no permanent hard substance secreted

**CORAL** ▲ An ocean-dwelling invertebrate. Corals often live in huge colonies and can secrete calcium carbonate skeletons to support themselves. The accumulation of these corals eventually produces a coral reef. See also *coral reef*.

**CORAL BLEACHING** ▼ The process whereby corals expel the colour-giving algae (zooxanthellae) from their tissues and consequently turn white. One of the main causes is an increase in the temperature of seawater. Bleached corals are not necessarily dead, but bleaching can lead to death.

polyps    zooxanthellae

**HEALTHY CORAL**

zooxanthellae expelled

**BLEACHED CORAL**

**CORAL REEF** ▼ A structure built up over many years from the skeletons of corals. Examples include fringing reefs, barrier reefs, and atolls.

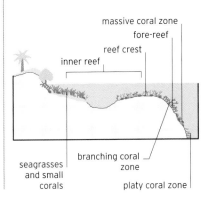

massive coral zone
fore-reef
reef crest
inner reef
seagrasses and small corals
branching coral zone
platy coral zone

**CORE** Earth's innermost layer. It consists of a solid inner core and a liquid outer core, both made of nickel-iron.

**CORIOLIS EFFECT** The tendency for the direction of winds and currents are deflected. Earth's rotation means that winds are deflected to the right in the northern hemisphere and to the left in the southern hemisphere. Without the Coriolis effect, air would simply flow directly from areas of high pressure to low pressure. See also *anticyclone, cyclone*.

**CORROSION** A form of chemical erosion in which water dissolves minerals in rocks and washes them away.

**COUNTRY ROCK** Existing rock into which an igneous intrusion invades. See also *igneous intrusion* (panel).

**CRATER** ▼ A bowl-shaped depression through which a volcano erupts gases, lava, ash, or volcanic bombs. Another form of crater is caused by a large meteorite impact.

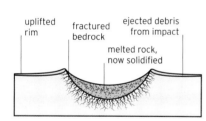

uplifted rim · fractured bedrock · ejected debris from impact · melted rock, now solidified

**CRATON** See *shield*.

**CRESCENTIC DUNE** See *barchan*.

**CREVASSE** See panel (below).

**CRUST** Earth's rocky outermost layer. The continents and their margins are made of thicker but less dense continental crust, while thinner, denser oceanic crust underlies the floor of the deep oceans. See also *continental crust, oceanic crust*.

**CRYOTURBATION** The churning up of different layers of soil by repeated freezing and thawing. Also called frost-churning.

**CRYSTAL** Any solid in which the individual molecules are arranged in a regular geometrical pattern. Calcite and quartz, for example, form crystals.

**CURRENT** The flow of water in an ocean, lake, or river.

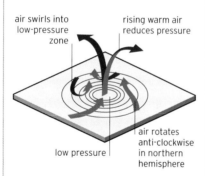

air swirls into low-pressure zone · rising warm air reduces pressure · low pressure · air rotates anti-clockwise in northern hemisphere

**CYCLONE** ▲ A weather system in which winds circulate around an area of low pressure. A tropical cyclone is another name for a hurricane or typhoon. See also *anticyclone, depression, hurricane, pressure system*.

# D

**DECIDUOUS** Describing trees and shrubs with leaves that die and fall at a particular time of year. See also *coniferous, evergreen*.

**DEEP-SEA FAN** A fan-shaped deposit of sediment on the ocean floor, usually at the bottom of the continental slope, deposited by turbidity currents. It is also known as an abyssal fan. See also *turbidity current*.

**DEEP-SEA TRENCH** A canyon-like depression in the ocean floor. Such trenches occur where one tectonic plate is pushing under another at a subduction zone. They are the deepest regions of the oceans. See also *boundary* (panel), *subduction*.

**DEFORMATION** The change in shape of rocks caused by the pressure of geological movements.

**DELTA** The area of gently sloping silt, sand, and other sediments built up by a river as it enters the ocean or a lake.

**DENDRITIC DRAINAGE** The pattern of streams and rivers in a drainage basin resembling the twigs and branches of a tree. See also *drainage basin*.

**DEPOSITION** The laying down of silt, sand, gravel, and other sediments by rivers, ocean currents, moving ice, or the wind.

**DEPOSITIONAL COAST** A coast where there is net deposition of sediments. This may result from longshore drift or the deposition of sediments in estuaries and deltas. See also *longshore drift*.

**DEPRESSION** A weather system where winds circulate around an area of low pressure. Also called a cyclone. See also *cyclone, pressure system*.

**DESERT PAVEMENT** ▼ A surface layer of closely packed rock fragments in a desert.

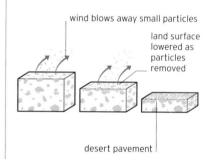

wind blows away small particles · land surface lowered as particles removed · desert pavement

**DESERT VARNISH** A glossy orange, brown, or black coating sometimes found on the surface of desert rocks that have been exposed to the atmosphere for a long time.

**DESERTIFICATION** The conversion of a formerly fertile region into desert.

**DESTRUCTIVE BOUNDARY** See *convergent boundary*.

**DEW POINT** The temperature at which, under given conditions, liquid water in the atmosphere begins to condense. See *condensation*.

## CREVASSE

A crevasse is a deep fissure in a glacier or other ice body. Crevasses usually form when some areas of ice move faster than others, consequently splitting the ice. This happens when glaciers change direction or pass over uneven topography, such as a rock ridge. Crevasses are usually wider at the top than at the bottom.

direction of stretching · crevasse is at right angle to flow · direction of flow

**TRANSVERSE CREVASSES**

crevasse forms along direction of flow · direction of flow

**LONGITUDINAL CREVASSES**

serac (ridge) forms between crevasses

**SERACS**

**DIKE** A sheet of intrusive igneous rock inclined vertically or at a steep angle to the surface. A large number of dikes in an area are referred to as a dike swarm. Also known as a dyke. See also *basalt, igneous intrusion* (panel), *sill*.

**DISTRIBUTARY** A branch of a river that flows away from the main watercourse and does not reconnect with it later, for example in a delta.

**DIVERGENT BOUNDARY** The line along which two or more tectonic plates are moving away from each other, for example at a mid-ocean ridge. See also *boundary* (panel), *mid-ocean ridge* (panel).

**DOLINE** See *sinkhole*.

**DOME VOLCANO** A mound-shaped, steep-sided volcano. The shape is determined by the viscous nature of the lava, which cannot flow far from the vent. See also *volcano* (panel).

**DRAINAGE BASIN** The area within which all surface water from rain, melting snow, or ice is drained by a single river system. See also *catchment area, watershed*.

**DROWNED COAST** A stretch of coastline that has been inundated by a rise in sea level. Features of such coasts include drowned valleys (or rias). See also *isostasy, ria*.

**DRUMLIN** A streamlined mound of sediments left behind by a glacier after it has retreated. See also *glacier, moraine* (panel).

**DRY FOREST** Woodland growing in an area with a primarily dry climate.

**DUNE** See panel, below.

**DYNAMIC METAMORPHISM** The transformation of rocks brought about mainly by directed pressure or stress, rather than heat or chemical change.

# E

**ECOREGION** A large area containing a distinctive group of species, communities of species, and environmental conditions.

**ECOSYSTEM** A community of interacting life forms and their environment. An ecosystem could be as small as a rotting log or as large as Earth.

**EDDY** A circular motion of any size and speed, in water or air. See also *vortex*.

**EKMAN SPIRAL** The tendency of deep ocean currents to change direction as they approach the surface, and of winds to change direction as they get closer to the ground. See also *Coriolis effect*.

**EL NIÑO** The phenomenon in which warm water from the western Pacific flows eastwards, making the eastern Pacific warmer than usual and triggering worldwide changes to weather patterns. See also *La Niña*.

**EMERGENT** Describing forest trees that grow taller than the height of the canopy. See also *canopy*.

**EMERGENT COAST** A stretch of coastline along which formerly submerged features (for example, beaches and wave-cut platforms) have been exposed by a fall in sea level.

**ENDEMIC** A term for animals and plants native to one region and found nowhere else.

**ENDORHEIC** Describes a watershed or a lake that has no outflow to a larger body of water.

**EPHEMERAL** Temporary, as in a water body present only in the wet season.

**EPIPHYTE** A non-parasitic plant that grows on another plant.

**ERG** A large expanse of vegetation-free sand in a desert.

**EROSION** The processes by which water, moving ice, and wind, as well as the sand grains and other particles they carry, scrape and wear away rocks or soil from a land surface. See also *weathering*.

**EROSIONAL COAST** A coastline where there is net removal of sand, gravel, and other sediments.

**ERRATIC** A rock that has been transported, usually by a glacier, from its original location. See also *glacier*.

---

## DUNE

A dune is a mound of sand in a desert, on a river bed, or close to the shore of a sea or lake, formed by the action of the wind or the flow of water currents. Dunes occur in a range of shapes and sizes, depending on how they have formed. With desert dunes, if the prevailing wind is usually from one direction, crescentic barchan or parabolic dunes will form. If the wind is variable, linear or star dunes will result. Dunes may migrate many kilometres or become stabilized by vegetation.

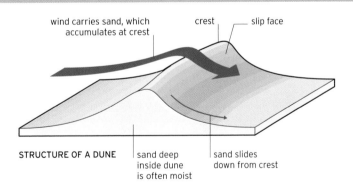

wind carries sand, which accumulates at crest — crest — slip face

**STRUCTURE OF A DUNE**

sand deep inside dune is often moist

sand slides down from crest

**DUNE TYPES**

prevailing wind direction

slip face

sand avalanches down slip face

**BARCHAN DUNE**

on vegetated sand, dune faces in opposite direction to barchan

prevailing wind direction

slip face

horns face into wind

**PARABOLIC DUNE**

longitudinal, parallel dunes

sparse sand

wind direction varies

**LINEAR (SEIF) DUNE**

variable winds

complex dune shape

**STAR DUNE**

## ERUPTION

The discharge of gas, lava, ash, or volcanic bombs from a volcano is called an eruption. Different types of volcanic eruption include Hawaiian, Plinian, Strombolian, Surtseyan, and Vulcanian. Depending on the type of eruption, lava, dust, ash, or pyroclastic bombs are ejected from a main vent, while secondary vents, or fumaroles, may produce smoke and steam.

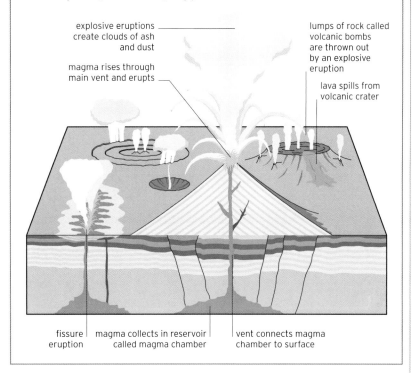

explosive eruptions create clouds of ash and dust

magma rises through main vent and erupts

lumps of rock called volcanic bombs are thrown out by an explosive eruption

lava spills from volcanic crater

fissure eruption

magma collects in reservoir called magma chamber

vent connects magma chamber to surface

**ERUPTION** See panel, above.

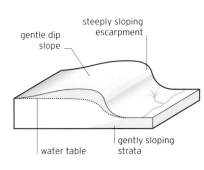

steeply sloping escarpment

gentle dip slope

water table

gently sloping strata

**ESCARPMENT ▲** The steep slope at the edge of a plateau, or at the edge of an area of exposed, gently sloping strata, separating two relatively level areas of different elevation. Also called a scarp.

**ESKER** A long, typically winding, ridge of gravel or sand left behind by a glacier. Eskers mark the position of former meltwater channels beneath or within the ice. See also *glacier, meltwater channel*.

**ESTUARY** The wide lower course of a river from where it flows into the ocean. Estuaries experience tidal flows and are often zones of deposition.

**EUSTASY** A global change in sea level produced by a change in the amount of water in the oceans. This change usually results from a change in the volume of Earth's ice-sheets.

**EUTROPHICATION** The process by which lakes and other water bodies are enriched with nutrients from fertilizers.

**EVAPORITE** A sediment formed when a salty lake or oceanic lagoon evaporates. Examples of evaporites include calcite, gypsum, and halite.

**EVERGREEN** Describing trees bearing leaves all year round. See also *deciduous*.

**EXFOLIATION** A weathering process that involves layers of rock peeling off in sheets instead of grain by grain.

**EXFOLIATION DOME** A large, dome-shaped mass or rock, such as granite, whose form has been produced by exfoliation. See also *exfoliation*.

**EXTRUSIVE IGNEOUS ROCK** Rock that forms when magma reaches Earth's surface and quickly cools and solidifies. For example, basalt and rhyolite. See also *intrusive igneous rock*.

# F

**FAST ICE** Frozen water that is secured to the coast. If it rises more than 2 m (6 ft) above sea level, it is called an ice-shelf. See also *ice-shelf*.

**FAULT** See panel, below.

**FEN** A wetland that receives water mainly from groundwater seepage and where the soils are composed almost entirely of dead plant matter, such as peat.

**FETCH** The distance across open water that a wave or the wind has travelled.

**FIORD** A former glacial valley on a coast that has been submerged by a rise in sea level and has become an inlet of the sea.

**FISSURE ERUPTION** The emission of lava through a linear vent, which may be many kilometres long. See also *eruption* (panel).

**FLASH POINT** The temperature at which superheated water changes to steam.

**FLASH-FLOOD** A sudden, often destructive flood occurring after heavy rainfall.

## FAULT

A fracture where the rocks on either side have moved relative to one another and often extend through the crust for many miles is called a fault. In normal and reverse faults, one section of the crust slips down, or rises up, in relation to the other. In strike-slip faults, one section slides past another, with little or no vertical movement. Great forces are involved in the formation of thrust faults, where older rocks are thrust over younger strata.

fault plane

surface lengthened by faulting

movement along fault plane

down-faulted block

NORMAL DIP-SLIP FAULT

surface shortened by faulting

upthrust, overhanging block

REVERSE THRUST DIP-SLIP FAULT

blocks move in opposing directions

STRIKE-SLIP FAULT

vertical and horizontal movement create diagonal movement

OBLIQUE-SLIP FAULT

**FLOOD BASALT** The result of a massive eruption that covers an extensive area with basalt. See also *basalt*.

**FLOODPLAIN** A flat area alongside a river that is liable to be covered with water when river volumes are particularly great. See also *alluvium*.

**FLOWSTONE** A coating of a carbonate mineral such as calcite on the walls or floor of a cave. The mineral precipitates out of flowing water. See also *calcite, precipitate*.

**FOG** A cloud of water droplets (or ice crystals) suspended in the air close to Earth's surface. See also *radiation fog*.

**FOLD** See panel, below.

**FOLIATION** Repetitive layering in metamorphic rocks where minerals become arranged in parallel bands. See also *lamination*.

**FORESHORE** The part of a shore lying between the low-water level and the average high-water level. See also *beach* (panel), *beach face*.

**FORKED LIGHTNING** A cloud-to-ground electrical discharge with a branched path in a storm. See also *sheet lightning, thunderstorm*.

**FOSSIL** The impression or remains of an ancient animal or plant buried in rock and preserved in petrified form.

**FOSSIL FUEL** An energy source such as coal, natural gas, or oil that is derived from the remains of long-dead plants and animals buried beneath the ground.

**FOSSILIZATION** The process of being transformed into a fossil.

**FRINGING REEF** A coral reef where either there is no lagoon between the reef and the shore or, if there is, it is entirely shallow. See also *atoll, coral reef*.

**FRONT ▼** In meteorology, the forward-moving edge of an air mass. See also *air mass*.

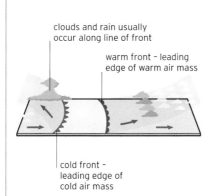

clouds and rain usually occur along line of front

warm front - leading edge of warm air mass

cold front - leading edge of cold air mass

**FUMAROLE** A small opening in the ground in volcanic regions through which steam and hot gases escape.

**FUNNEL** A tube of rapidly rotating air descending from a cloud. If its lower end touches the ground, it becomes a tornado. See also *tornado*.

# G

**GEODE** A hollow cavity in a rock that is lined with crystals.

**GEOTHERMAL** Relating to the heat energy that is generated in Earth's interior. Geothermal energy is mainly produced by natural radioactive decay of uranium, thorium, and potassium.

geyser outlet

constriction creates build-up in pressure

steam

chamber

rising hot water

**GEYSER ▲** A jet of steam and boiling water that periodically erupts from the ground. Geysers are powered by hot rocks heating groundwater. See also *hot spring*.

**GLACIER** A slowly moving mass of ice formed by the accumulation and compaction of snow over a long period. There are many different types of glacier. See also *cirque glacier, piedmont glacier, surge-type glacier, valley glacier*.

**GORGE** A deep, narrow valley bounded by vertical or near-vertical cliffs. See also *canyon*.

**GRABEN** A block of Earth's crust lying between parallel faults and depressed relative to the blocks on either side. See also *rift valley*.

**GRAIN** The texture of a rock. For example, clay is made up of fine particles so is described as fine-grained, whereas conglomerate is coarse-grained.

**GRANITE** A coarse-grained igneous rock made up of the minerals quartz, feldspar, and mica.

**GREASE ICE** An early stage of sea-ice formation when tiny ice crystals called frazil join together to give the water surface the appearance of an oil slick. See also *sea-ice*.

**GREENHOUSE EFFECT** The tendency of the atmosphere to absorb a proportion of the Sun's radiation after it has been re-radiated by Earth. As the amount of carbon dioxide in the atmosphere has increased, the greenhouse effect has become more marked. See also *greenhouse gas*.

**GREENHOUSE GAS** Any gas that promotes the greenhouse effect. The major greenhouse gases are carbon dioxide, methane, and water vapour. See also *greenhouse effect*.

---

## FOLD

A geological structure in which once-horizontal layers of rock have been flexed and bent is called a fold. Folds develop when rocks are subjected to compression or tension. If the layers bend upwards to form a ridge, the fold is an anticline; if they bend downwards to form a trough, the trough is a syncline. Folds can be symmetrical or asymmetrical. Overfolds and recumbents are asymmetrical folds where the layers have folded beyond 90°. In an overfold, the layers have folded back on themselves. A fold's axis is its angle of inclination to Earth's surface.

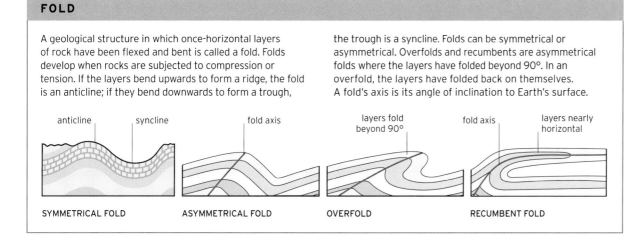

anticline    syncline    fold axis    layers fold beyond 90°    fold axis    layers nearly horizontal

SYMMETRICAL FOLD    ASYMMETRICAL FOLD    OVERFOLD    RECUMBENT FOLD

**GROUNDMASS** The fine-grained material in which larger crystals or fragments are embedded in igneous (and some sedimentary) rocks. See also *matrix*.

**GROUNDWATER ▼** Water held in spaces (interstices) between the grains of rocks underground. The water table represents the top of the groundwater zone. See also *aquifer, water table*.

surface water found where water table intersects surface

groundwater zone

water table

**GUYOT** A flat-topped, undersea volcanic mountain, or seamount. The summit of a guyot is more than 200 m (660 ft) below sea level. See also *seamount*.

**GYRE** The large-scale rotation of an ocean current. See also *current*.

# H

**HABITAT** Any area that can support a particular group of plants or animals.

**HALO** A hazy ring around the Moon or Sun, which is caused by the refraction of light as it passes through high clouds.

**HAMADA** A flat, rocky type of desert terrain, mainly devoid of sand. See also *desert pavement*.

**HAMMOCK** A stand of trees growing on raised ground above a wetland. Hammocks are common in the Florida Everglades.

**HANGING VALLEY ▼** A tributary valley that joins a main valley high above the floor of the latter. Hanging valleys usually occur where a glacier has deepened the larger valley.

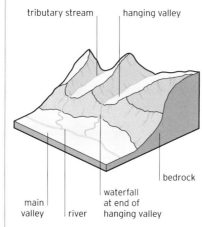

tributary stream    hanging valley

main valley    river    waterfall at end of hanging valley    bedrock

**HAWAIIAN ERUPTION** The least explosive type of volcanic eruption, with a relatively gentle flow of basaltic lava from a vent.

**HEADLAND** A narrow area of land projecting from a coastline. Headlands often exist because they are made of harder rocks than the coast on either side.

**HEADWATER** The upper portion of any stream or river, close to its source.

**HOT SPRING** The emergence of hot water and steam from the ground. Hot springs occur when hot rocks beneath the surface have heated groundwater. See also *geyser*.

**HOTSPOT** An area of volcanic activity thought to originate deep in Earth's mantle. Tectonic plates passing over hotspots are marked by lines of volcanoes that become older with increased distance from the hotspot. See also *mantle plume*.

**HUMUS** A dark-coloured component of soil made up mainly of decomposed plant material.

**HURRICANE** A large weather system that develops and moves over tropical and subtropical oceans. Hurricanes feature destructive winds circulating rapidly around an area of low pressure, torrential rain, and storm surges. Also called a tropical cyclone or a typhoon. See also *cyclone, latent heat*.

**HYDROTHERMAL VEIN** Any concentration of minerals formed by the precipitation of solids from hot, mineral-laden water. See also *mineral, vein*.

**HYDROTHERMAL VENT** An outlet for water that has been heated by molten or hot rocks. Hydrothermal vents that emit hot, dark water from the ocean floor are called black smokers. See also *black smoker, geyser, hot spring, white smoker*.

# I

**ICE AGE** A period of long-term reduction in global temperatures that results in the expansion of polar and continental ice-sheets and glaciers. See also *glacier, ice-sheet*.

**ICE LEAD** A large fracture in an expanse of sea-ice. See also *sea-ice*.

**ICE SHELF** A large area of floating ice that is permanently attached to land. The Ross Ice Shelf, attached to Antarctica, is the world's largest.

**ICE STREAM** Part of an ice-sheet that is moving faster than the surrounding ice. Ice streams are common in Antarctica. See also *ice-sheet*.

**ICE WEDGE** A vertical, wedge-shaped mass of ice in the soil in a periglacial area. See also *ice-heave, periglaciation*.

**ICE-CAP** An ice mass covering a large area, but less than 50,000 square km (19,000 square miles). An ice-cap's form overrides underlying landscape features. See also *glacier, ice-field, ice-sheet*.

**ICE-FIELD** A large expanse of ice whose surface shape and extent is determined by underlying landscape features. Ice-fields are often interconnected valley glaciers from which higher mountain peaks rise. See also *glacier, ice-cap, nunatak*.

**ICE-HEAVE** Disruption of the land surface due to the growth of ice within the soil. Ice-heave is one of the features of periglacial regions. See also *ice wedge, periglaciation, pingo*.

**ICE-SHEET** A layer of ice covering an area greater than 50,000 square km (19,000 square miles) for a prolonged period. See also *ice-cap, ice-field*.

**ICEBERG ▼** A floating mass of ice that has calved from a glacier, ice-sheet, or ice shelf. See also *calving, glacier, ice shelf, ice-sheet*.

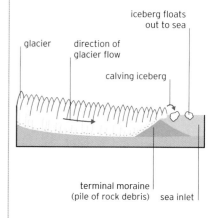

iceberg floats out to sea

glacier    direction of glacier flow

calving iceberg

terminal moraine (pile of rock debris)    sea inlet

**IGNEOUS INTRUSION** See panel, right.

## IGNEOUS INTRUSION

An igneous intrusion is a mass of magma that forces its way into Earth's crust and cools and solidifies before reaching the surface. There are three main types: batholiths, dikes, and sills. Batholiths are large igneous bodies that may force the crust up into a dome. Dikes and sills are tabular sheets of igneous rock that have been intruded between layers of sedimentary, metamorphic, or older volcanic rocks. Dikes are vertical or have a high angle of elevation, whereas sills are horizontal or have a low angle of elevation. Granite, gabbro, diorite, and pegmatite are common intrusive igneous rocks. Over time, the erosion of softer surrounding rocks (referred to as country rock) may expose an intrusion at the surface.

BATHOLITH

layered sedimentary rock, into which batholith was intruded

aureole of metamorphic rock altered by heat and pressure

batholith

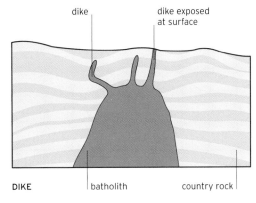

DIKE

dike

dike exposed at surface

batholith

country rock

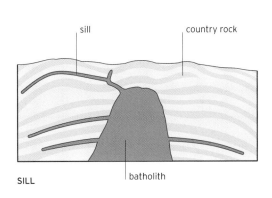

SILL

sill

country rock

batholith

**IGNEOUS ROCK** Rock formed from the solidification of molten magma.

**INFRARED** Invisible radiation with a long wavelength, which we feel as heat. See also *ultraviolet*.

**INSELBERG** An isolated, steep-sided hill or small mountain rising up out of a flat plain. The Sugarloaf Mountain in Rio de Janeiro is an inselberg.

**INTRUSIVE IGNEOUS ROCK** A rock that formed due to the slow cooling of magma beneath Earth's surface. Examples include diorite and granite. See also *granite, igneous intrusion* (panel).

**ISLAND ARC** A chain of volcanic islands close to a subduction zone. Typically, there is a deep ocean trench on one side. See also *back arc, subduction, volcanic arc*.

**ISOBAR** An imaginary line connecting points with the same air pressure.

**ISOSTASY** The concept that continents rise high above the sea floor because continental crust is less dense than oceanic crust.

# J

**JET STREAM** A ribbon of very strong, high-altitude winds, which move weather systems around Earth.

vertical joint

stream

bedding plane between layers of rock

**JOINT** ▲ A fracture dividing rock into sections. The rocks on either side of a joint have not moved relative to each other, unlike the rocks either side of a fault. See also *fault* (panel).

# K

**KARST** A landscape, usually found in limestone regions, characterized by caves, sinkholes, and underground watercourses. See also *limestone, sinkhole*.

**KETTLE LAKE** A water body occupying a kettle hole, a depression left after an ice block from a former glacier has melted.

# L

**LA NIÑA** The phenomenon in which the waters of the eastern Pacific Ocean become unusually cold. This is the opposite of the El Niño effect. See also *El Niño*.

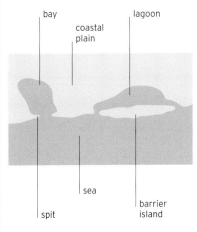

bay

coastal plain

lagoon

sea

spit

barrier island

**LAGOON** ▲ An expanse of seawater that is almost cut off from the open ocean and consequently relatively sheltered. See also *atoll*.

**LAHAR** A flow of mud, water, ash, and other debris down the slopes of a volcano. See also *mass movement*.

**LAMINATION** Fine and generally parallel layers within rocks. It is usually less distinct than bedding. See also *bedding*.

**LATENT HEAT** The heat that is released, for example, when water vapour condenses into water droplets. This heat release is an important factor in the growth of storms. See also *condensation*.

**LATERAL MORAINE** The ridges of silt, sand, and gravel that build up along the sides of a glacier. They indicate the extent of ice before the glacier retreated. See also *moraine* (panel).

**LAVA** Molten rock that has reached Earth's surface as a result of an eruption. See also *basalt, pillow lava*.

**LEACHING** The process of rainwater washing minerals and nutrients through topsoil. Leached materials tend to be lost from the topsoil and deposited in the subsoil.

**LEVEE** ▼ A ridge of sediment deposited alongside a river when in flood, or an artificial embankment built to stop the river inundating its floodplain. See also *floodplain*.

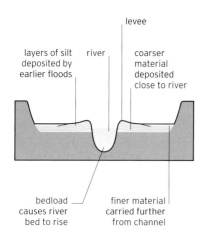

levee

layers of silt deposited by earlier floods

river

coarser material deposited close to river

bedload causes river bed to rise

finer material carried further from channel

**LIGHTNING** The visible flash that accompanies electrical discharges from storm clouds. See also *forked lightning, sheet lightning*.

**LIMESTONE** A sedimentary rock composed mainly from the skeletons of marine life-forms such as foraminifera and corals. Chemically, limestones are mostly calcium carbonate. See also *karst*.

**LIMESTONE PAVEMENT** An uneven, blocky surface layer of limestone rock with crevices or fissures (known as grikes) separated by blocks (or clints), created by rainwater dissolving the porous limestone.

**LINEAR DUNE** A long, narrow desert dune that is aligned parallel to the direction of the prevailing wind. These dunes usually form where wind from one direction dominates, and they often form parallel ridges many kilometres long. See also *dune* (panel).

**LITHOSPHERE** See panel, below.

**LITTORAL** Relating to the shoreline, especially between the low- and high-water marks. See also *beach* (panel), *tide*.

**LONGSHORE BAR** A ridge of sand, mud, or gravel parallel to the shore in the intertidal zone or just seaward of it. See also *beach* (panel).

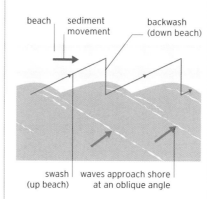

beach

sediment movement

backwash (down beach)

swash (up beach)

waves approach shore at an oblique angle

**LONGSHORE DRIFT** ▲ The transport of mud, sand, and gravel along a coastline by a current that flows roughly parallel to it. See also *current*.

# M

**MAAR** A shallow but steep-sided volcanic crater formed when magma comes into contact with groundwater to produce a violent steam explosion.

**MAGMA** Molten, or semi-molten, rock that rises from parts of the mantle into the crust. It can cool and solidify beneath the surface or at the surface. See also *intrusive igneous rock, lava, mantle*.

**MAGMA CHAMBER** A large underground reservoir of molten rock. Sometimes, pressure forces the magma contained within to push its way to the surface in an eruption. See also *caldera, plug*.

**MANGROVE SWAMP** ▼ An intertidal area where salt-tolerant trees (halophytes) thrive. Mangrove swamps are most common on shallow coasts in the tropics and subtropics. See also *swamp*.

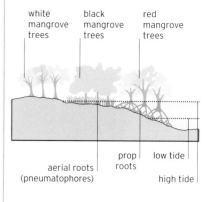

white mangrove trees

black mangrove trees

red mangrove trees

aerial roots (pneumatophores)

prop roots

low tide

high tide

**MANTLE** The region of Earth's interior lying between the core and the crust. It makes up 84 per cent of Earth's volume. See also *core, crust*.

**MANTLE PLUME** A column of hot rock that rises through the mantle and into the crust, giving rise to a hotspot, which is likely to lead to volcanic activity at the surface. See also *hotspot*.

## LITHOSPHERE

The solid, outer part of Earth, comprising the crust and the uppermost part of the mantle, is called the lithosphere. It lies above the asthenosphere, in which rocks are hot and viscous, and below the atmosphere and hydrosphere. The oceanic lithosphere is associated with oceanic crust and is slightly denser than the continental lithosphere, which is associated with continental crust and may extend to a depth of 200 km (120 miles).

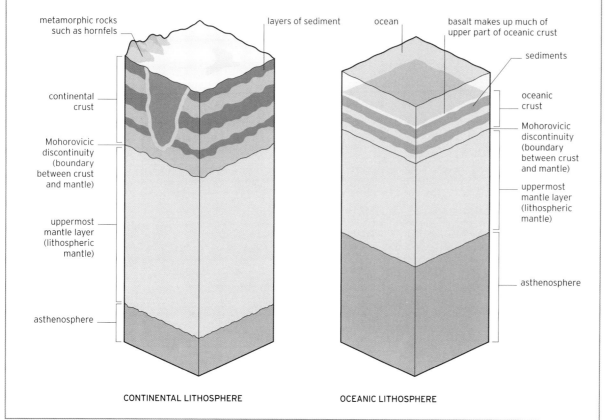

metamorphic rocks such as hornfels

layers of sediment

ocean

basalt makes up much of upper part of oceanic crust

continental crust

Mohorovicic discontinuity (boundary between crust and mantle)

uppermost mantle layer (lithospheric mantle)

asthenosphere

sediments

oceanic crust

Mohorovicic discontinuity (boundary between crust and mantle)

uppermost mantle layer (lithospheric mantle)

asthenosphere

CONTINENTAL LITHOSPHERE

OCEANIC LITHOSPHERE

**MARBLE** Metamorphosed limestone that is very hard and can be polished for use in construction and sculpture. See also *limestone, thermal metamorphism*.

**MARSH** A wetland dominated by plants such as grasses and reeds. Marshes are characterized by regular inundations, which can be of seawater or fresh water. See also *salt marsh* (panel), *swamp*.

**MASS MOVEMENT** The downslope movement of mud, soil, or rocks, as in a landslip. This process often takes place when soil and other surface debris are saturated with water. See also *transport*.

**MASSIF** A well-defined mountain mass or group of connected mountains whose rocks and landforms tend to be similar.

**MASSIVE** A term describing a rock without any obvious structures (such as bedding planes) or a mineral that is not visibly crystalline.

**MATRIX** The fine-grained material in which larger grains (in sedimentary rocks) or crystals (in igneous rocks) are embedded. See also *groundmass*.

**MEANDER** A loop in the course of a river. Meandering rivers gradually change their routes, as erosion takes place on the outside of the loop and deposition on the inside. See also *river valley*.

**MEDIAL MORAINE** A ridge of sediment running along the centre of a glacial valley, and parallel with its sides. Medial moraines often form when two glaciers converge and two of their lateral moraines merge. See also *confluence, lateral moraine, moraine* (panel).

**MELTWATER** Water, from melted snow and ice, that flows within or from the snout of a glacier.

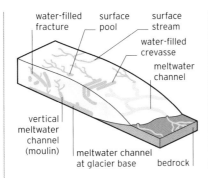

water-filled fracture · surface pool · surface stream · water-filled crevasse · meltwater channel · vertical meltwater channel (moulin) · meltwater channel at glacier base · bedrock

**MELTWATER CHANNEL ▲** A channel cut by melted snow or ice flowing beneath, through, or near a glacier. The channel and its deposits usually remain after the glacier has retreated. See also *meltwater*.

**MESA ▼** A "table-topped", steep-sided hill. Mesas are characterized by horizontal strata, capped by resistant top layers.

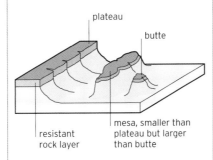

plateau · butte · mesa, smaller than plateau but larger than butte · resistant rock layer

**MESOSPHERE** The layer of Earth's atmosphere between the stratosphere and thermosphere. It ranges from about 50 to 80km (30 to 50 miles). See also *stratosphere, thermosphere*.

**METAMORPHIC ROCK** A rock that, under the influence of heat, pressure, or a combination of both, has had its texture or mineral make-up altered.

**METAMORPHISM** The processes – mainly heating and extreme pressure – by which rocks are transformed under ground. See also *contact metamorphism, dynamic metamorphism, regional metamorphism, thermal metamorphism*.

**MICROCLIMATE** The distinctive long-term weather conditions of a particular small place, such as a hilltop or a small valley.

**MID-OCEAN RIDGE** See panel, above.

**MINERAL** A naturally occurring inorganic material that has a well-defined chemical composition. Most rocks are a mixture of several different minerals.

**MIRE ▼** An area of swampy ground that retains wet, peaty soil. A mire often forms in a depression in impermeable rock.

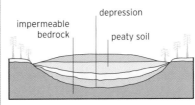

depression · impermeable bedrock · peaty soil

**MID-OCEAN RIDGE**

An undersea mountain range running across the floor of an ocean is called a mid-ocean ridge. It forms where two tectonic plates move apart and magma rises from the mantle to form new oceanic crust. See *oceanic crust, spreading ridge, tectonic plate*.

crest of mid-ocean ridge · crust · upper mantle · rising magma

**MONSOON** Refers to a pattern of seasonal winds, especially in southern Asia. The term is also used to refer to the heavy rains these winds bring during the rainy, or monsoon, season. The southwest monsoon brings heavy rain to India, usually between June and September. These winds blow from one direction for about half the year, and from the opposite direction for the other half.

**MONTANE** Relating to mountains – for example, montane climate or montane fauna.

**MORAINE** See panel, below.

**MORAINE**

Ridges of silt, sand, gravel, and rock debris resulting from glacial action are called moraines. These features often remain after a glacier retreats. Medial moraines form where two glaciers converge, lateral moraines form along the sides of a glacier, and terminal moraines form at the furthest extent of the ice.

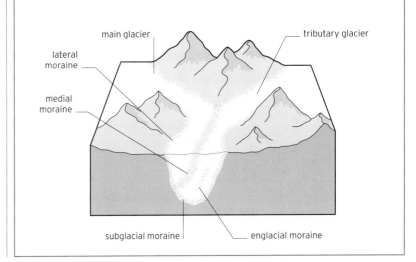

main glacier · tributary glacier · lateral moraine · medial moraine · subglacial moraine · englacial moraine

# N

**NATIVE** Describing a plant or animal indigenous to a particular area. See also *endemic*.

**NATIVE ELEMENT** A chemical element that is found in its pure state in nature. Copper, gold, sulphur, and tin can all be found in their native form. See also *mineral*.

**NEAP TIDE** The tide with the least difference between low and high water. Neap tides occur twice a month when the gravitational pull of the Moon and Sun on Earth's oceans are working against each other. See also *spring tide*.

**NEEDLE-LEAF** A needle-shaped leaf typical of coniferous trees. See also *coniferous*.

**NOCTILUCENT CLOUD** The highest clouds in Earth's atmosphere, illuminated at night when the Sun is below the horizon and lower levels of the atmosphere are in shade. These ice clouds are usually at an altitude of 76 to 85 km (47 to 53 miles). See also *mesosphere*.

**NORMAL FAULT** An angled fault where one side (the hanging wall) has slipped down relative to the other side (the footwall). See also *fault* (panel).

**NUNATAK** A mountain peak rising out of an ice-field or ice-sheet. See *ice-field, ice-sheet*.

# O

**OASIS** An isolated fertile area within a desert, usually fed by water from a spring. See also *aquifer*.

**OCCLUSION** In meteorology, the situation where a cold front overtakes a warm front, pushing a mass of warm air up and away from Earth's surface. There are two types: cold occlusions and warm occlusions. See also *front*.

**OCEAN TRENCH** A long, deep section of the ocean floor, usually lying above a subduction zone. See also *deep-sea trench, subduction*.

**OCEANIC CRUST** The outermost layer of Earth's lithosphere beneath the oceans. It is formed at spreading centres at oceanic ridges and is mainly made of basalt. See also *continental crust, crust, seafloor spreading, spreading ridge*.

**OFFSHORE BAR** A submerged or partly exposed ridge of sand or gravel away from and roughly parallel to the coast. Offshore bars are built up by waves and currents. See also *bar*.

**OOZE** The fine sediment covering a huge part of the deep ocean floor. Much of it is made up of the skeletal remains of microscopic marine life-forms, such as foraminiferans and radiolarians.

**ORE** A rock from which a metal or valuable mineral can be profitably mined.

**OROGENESIS** The process of mountain-building that results from massive horizontal pressures as tectonic plates push together.

**OROGENY** A mountain-building event resulting from the collision of tectonic plates, when sections of Earth's crust are compressed, folded, and faulted. See also *tectonic plate*.

**OUTLET GLACIER** A mass of slow-moving ice that flows from an ice-sheet, ice-cap, or ice-field before following a valley. See also *valley glacier*.

**OXBOW LAKE** ▼ A body of water that forms when a meander is completely cut off from the main channel of a river. See also *meander*.

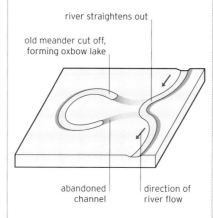

river straightens out

old meander cut off, forming oxbow lake

abandoned channel

direction of river flow

**OVERTURNED FOLD** A fold in which both limbs dip in the same direction. Overturned folds occur in areas of intense deformation of the rock layers. See also *fold* (panel).

# P

**PACK ICE** A mass of ice floating on the sea, formed from many smaller pieces that have frozen together. See also *sea-ice*.

**PAHOEHOE LAVA** A type of basalt lava with a smooth or gently undulating surface. It often changes to a'a lava as it flows downslope. See also *a'a lava, block lava*.

**PANCAKE ICE** Small, flat, roughly circular areas of floating ice with curled margins where the "pancakes" have bumped into each other. Pancake ice is an early stage in the development of sea-ice. See also *sea-ice*.

**PARABOLIC DUNE** A U-shaped or V-shaped mound of fine to medium sand with long "arms" trailing upwind. Unlike barchans,

the steepest side (slip face) is on the convex side. See also *barchan, dune* (panel).

**PARTIAL MELTING** The process by which some minerals (in a rock being heated to great temperatures) melt, while others remain solid. This happens because some minerals have a lower melting point than others. See also *igneous rock*.

**PEAT** Partly decomposed plant material that accumulates in bogs, fens, and mires. Peat has a low mineral content.

**PENINSULA** An area of land protruding into a lake or sea.

**PERIGLACIATION** The processes that take place as a result of the regular freeze-thaw cycle in high-latitude or high-altitude regions where there is no permanent ice cover. Features include ice wedges and pingos. See also *ice wedge, permafrost, pingo, tundra*.

**PERMAFROST** Rock or soil that remains frozen for at least two years. Permafrost is characteristic of periglacial and glacial regions. See also *periglaciation, tundra*.

**PHREATIC ERUPTION** An explosive eruption that occurs when magma or hot rock come into contact with groundwater or surface water, turning it to steam.

**PHYTOPLANKTON** The tiny photosynthesizing life-forms that float in the upper, sunlit layer of a lake or ocean and form the foundation of most aquatic food chains. See also *zooplankton*.

**PIEDMONT GLACIER** An area of slowly flowing ice that forms from the convergence of a number of valley glaciers at the foot of a mountain range. See also *valley glacier*.

**PILLOW LAVA** Molten rock that is extruded onto the ocean floor and cools quickly to form pillow-shaped lumps, usually of basalt. See also *lava*.

**PINGO** A periglacial landform comprising a dome-shaped hillock with a core of ice. Pingos form where groundwater – often from an old lake – rises into an area where it freezes and expands, pushing up the sediments above it. See also *periglaciation*.

**PLATE BOUNDARY** See *boundary* (panel).

**PLATEAU ▼** A large area of flat or gently undulating land whose surface is significantly higher than the surrounding landscape.

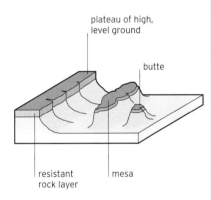

plateau of high, level ground

butte

resistant rock layer     mesa

**PLAYA** A flat-bottomed depression in a desert basin. Seasonally, a playa may be inundated to a shallow depth, but with no natural outlet, all the water evaporates.

**PLINIAN ERUPTION** The largest and most violent type of volcanic eruption, involving the release of enormous quantities of ash, lava, and gas. These eruptions are so violent that they can destroy large sections of the volcano itself. See also *eruption* (panel).

**PLUG** A mass of igneous rock filling the vent of an active volcano. If gas or magma is trapped under the plug, pressure will build beneath it, increasing the

likelihood of the next eruption being particularly violent. See also *magma*.

**PLUTON** A large body of magma that solidified slowly underground to form a coarse-grained igneous rock, such as granite. See also *batholith*.

**PLUTONIC** Relating to igneous processes taking place, or igneous rocks formed, deep below Earth's surface. See also *batholith, igneous intrusion* (panel), *pluton*.

**POINT BAR** The area of sand, silt, or gravel on the inside of a river meander, where water flow is slowest. See also *bar, meander*.

**POLDER** An area of low-lying land reclaimed from the sea and enclosed within dikes.

**POLYNA** A stretch of open water in an expanse of sea-ice. See also *pack ice, sea-ice*.

**POLYTHERMAL GLACIER** A glacier whose base is partly "warm ice" (at 0°C, 32°F) and partly cold ice (below 0°C, 32°F). Typically, the central areas are warmer than the margins.

**PRECIPITATE** A substance that has come out of solution to form a solid deposit.

**PRECIPITATION** Water originating from the atmosphere that reaches Earth's surface, including rain, snow, hail, and drizzle.

**PRESSURE GRADIENT** The rate of change of atmospheric pressure with distance.

**PRESSURE SYSTEM** A weather pattern in which air circulates around an area of low pressure (a depression, or cyclone) or high pressure (an anticyclone). See also *anticyclone, cyclone, isobar*.

**PREVAILING WIND** The most common direction of airflow at a particular location.

**PUMICE** A light, porous volcanic rock, which solidified rapidly from gas-rich frothy magma. The numerous pores, or vesicles, were gas bubbles.

**PYROCLASTIC DEPOSIT** An accumulation of fragments of rock, including ash and volcanic bombs, erupted from a volcano.

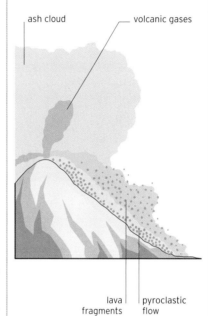

ash cloud     volcanic gases

lava fragments | pyroclastic flow

**PYROCLASTIC FLOW ▲** A dense, fast-moving, and destructive cloud of hot ash, fragments of lava, and gas ejected violently from a volcano. See also *eruption* (panel).

R

**RADIATION FOG** The fog created by the night-time cooling of air close to the ground until its content of water vapour condenses. See also *fog*.

**RAIN SHADOW** The area of low rainfall downwind of a mountain range. This is produced

as a result of the air shedding its moisture as it passes over the mountains.

**RAINBOW** A prism-like light effect caused by the Sun's rays being refracted into their constituent colours as they pass through raindrops.

**RAINFOREST** A forest with high rainfall and humidity all year round. Most rainforests are in the tropics or subtropics, with some in temperate regions.

**RAPIDS** Sections of a watercourse with a steep riverbed gradient, causing an increase in water velocity and turbulence.

**RECUMBENT FOLD** A fold of stratified rocks where the two limbs are almost parallel and close to the horizontal. See also *fold* (panel).

**REGIONAL METAMORPHISM** The processes of extreme heating and pressure by which the mineralogy and texture of rocks are transformed over a large area. See also *contact metamorphism, dynamic metamorphism, thermal metamorphism*.

**REGOLITH** The layer of soil and broken rock covering solid rock at Earth's surface.

**RELATIVE HUMIDITY** The amount of water vapour in a "parcel" of air, relative to the total amount that parcel can hold at that particular temperature. Warm air can hold more water vapour than cold air. See also *condensation, dew point*.

**REMNANT ARC** A formerly active volcanic island arc that has been moved away from where it was created by tectonic movement. See also *island arc, subduction*.

**RESURGENCE** The place where an underground watercourse emerges to become a surface stream or river. See also *karst, limestone*.

**REVERSE FAULT** An angled fault where one side has been pushed up relative to the other side. Reverse faults occur where two blocks of rock have been forced together by compression. See also *fault* (panel).

**RIA** A former river valley that has been drowned by a rise in sea level to form an inlet of the sea.

**RIDGE OF HIGH PRESSURE** A wedge-shaped area of relatively high atmospheric pressure, an extension of an anticyclone. See also *anticyclone*.

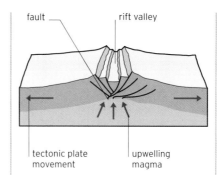

**RIFT VALLEY** ▲ A large block of land that has dropped vertically relative to the regions on either side. A rift valley, or graben, forms as the result of horizontal extension of Earth's crust and normal faulting. See also *normal fault*.

**RIVER BASIN** A region drained by a river and its tributaries where sediment accumulates.

**RIVER VALLEY** See panel, left.

**ROCK** A natural material made up of one or more minerals. See also *mineral*.

## RIVER VALLEY

A river valley is a depressed area, longer than it is wide, that has been carved by a river. A valley's shape depends on terrain, geology, and whether it has been modified by ice. Since a river has an upper, middle, and lower course, the shape of its valley reflects this. The upper section is V-shaped unless glacial ice has carved it into a U shape. Slopes are usually less steep in the middle course, and often very gentle in a valley's lower course, where the river may meander and deposit sediment on a floodplain, often dividing into channels called distributaries to form a delta. A river flows out to sea through its mouth and deposits more sediment on the seafloor.

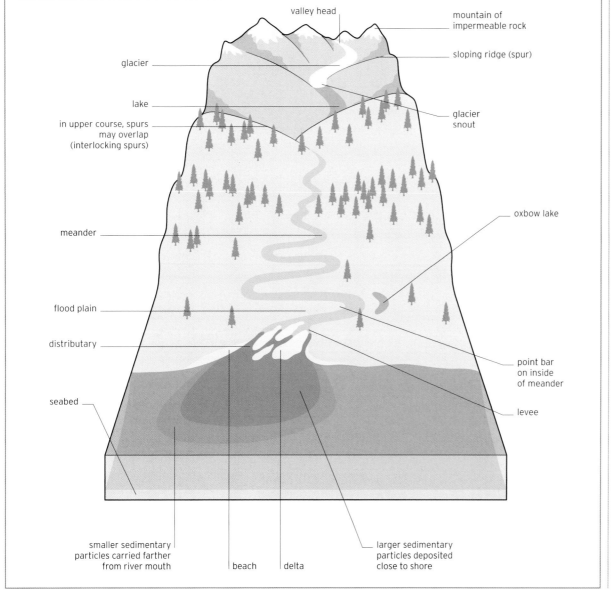

# S

**SALINITY** The concentration of dissolved salts in water, usually expressed in parts per thousand.

**SALT MARSH** See panel, right.

**SANDSTONE** A sedimentary rock composed mostly of sand-sized grains of quartz or feldspar that have been cemented together.

**SASTRUGI** Sharp, irregular grooves or ridges on an ice or snow surface. Sastrugi, or zastrugi, patterns are formed by wind erosion, particularly the blasting of the surface by ice crystals.

**SAVANNA** A mixed grassland-woodland ecosystem of the tropics where the trees are not close enough to form a closed canopy. Often found between deserts and tropical forests, savannas are characterized by seasonal rainfall and regular wildfires. See also *cerrado*.

**SEA ARCH** A natural arch formed in a sea cliff by marine erosion.

**SEA CAVE** A cave eroded into a sea cliff by marine erosion.

## SALT MARSH

Coastal wetlands that are flooded and drained by salt water brought in by tides are called salt marshes. Salt-tolerant vegetation grows in mud and peat, which may be deep and characterized by very low oxygen levels. Many areas of salt marsh are bordered by tidal flats. See also *tidal flat*.

upper high marsh · pool · lower high marsh · highest spring tide · mean sea level · mean high tide · upland · high marsh · low marsh · mudflat

**SEA STACK ▼** An offshore column of rock remaining after the erosion of adjacent sea cliffs, which are often of less-resistant rock.

straightened coast · sea stack · wave action · remains of collapsed arch

**SEA-ICE** Frozen ocean water. The sea freezes in several stages, first grease ice, then pancake ice, before forming continuous sheets of sea-ice. See also *grease ice, pancake ice*.

**SEAFLOOR SPREADING** A process that occurs at mid-ocean ridges where new oceanic crust is formed and gradually moves away from the ridge. See *mid-ocean ridge* (panel), *spreading ridge, tectonic plate*.

**SEAMOUNT** An undersea mountain, usually volcanic in origin, whose peak does not reach the surface. See also *guyot*.

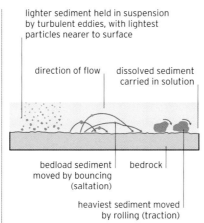

lighter sediment held in suspension by turbulent eddies, with lightest particles nearer to surface

direction of flow · dissolved sediment carried in solution · bedload sediment moved by bouncing (saltation) · bedrock · heaviest sediment moved by rolling (traction)

**SEDIMENT ▲** Mud, sand, gravel, and other particles that have been transported by water, wind, gravity, or volcanic processes and later deposited. Moving water can transport its sediment load by several different processes.

**SEDIMENTARY ROCK** Material formed when mud, sand, gravel, or other particles are deposited in an ocean or lake, or on land, and are then hardened by the processes of compaction and cementation.

**SEIF DUNE** See *linear dune*.

**SEISMIC WAVE** A shock wave generated by an earthquake. There are three types of seismic wave: P-waves, S-waves, and surface waves.

**SEMI-DESERT** An arid region that receives enough rainfall to support some plant life.

**SERAC** A pinnacle, or ridge, of glacial ice.

**SHEET LIGHTNING** An electrical discharge between two charged regions of one storm cloud. All or part of the actual discharge is obscured by cloud. It is also called intracloud lightning. See also *forked lightning, thunderstorm*.

**SHIELD** A large area of continental crust with ancient metamorphic rocks at the surface.

**SHIELD VOLCANO** The largest volcano type in terms of area, but with gently sloping sides. Shield volcanoes are the least violent, built mostly from liquid lava flows. See also *volcano* (panel).

**SILL** A thin, sheet-like igneous intrusion that forms when igneous rock forces its way between layers of sedimentary rocks. Most sills are roughly horizontal. See also *batholith, dike, igneous intrusion* (panel).

**SINKHOLE** A depression that forms where water has dissolved limestone bedrock. Sinkholes are characteristic of karst landscapes. See also *karst, limestone*.

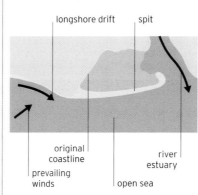

longshore drift · spit · original coastline · prevailing winds · river estuary · open sea

**SPIT ▲** A peninsula of sand or shingle connected to the shore at one end. Spits are usually created by longshore drift, often where the coastline changes direction abruptly. See also *bar, longshore drift, tombolo*.

**SPREADING RIDGE** The raised area on the ocean floor where two tectonic plates are slowly moving apart. Molten rock from the mantle wells up along a spreading ridge to solidify into new oceanic crust. Situated between the Pacific and Nazca plates, the East Pacific Rise contains one of Earth's fastest spreading ridge systems. See also *boundary* (panel), *mid-ocean ridge* (panel), *seafloor spreading, tectonic plate*.

**SPRING ▼** A surface outlet for water that has flowed from porous rocks underground. See also *groundwater*.

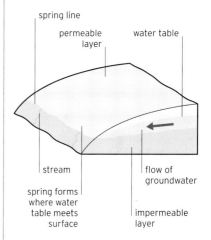

spring line · permeable layer · water table · stream · flow of groundwater · spring forms where water table meets surface · impermeable layer

**SPRING TIDE** The tide with the greatest difference between low and high water. Spring tides occur twice a month when the gravitational pull on Earth's oceans of the Moon and Sun are working with each other. It falls much lower than a neap tide, exposing more of the shore, as well as rising much higher a few hours later. See also *neap tide*.

**SPUR** A sloping ridge projecting from the side of a mountain or valley. A truncated spur has a blunt end as a result of erosion by a glacier or faulting.

**STALACTITE ▼** A deposit, usually of calcite, hanging from the roof of a cave. Stalactites are produced when minerals precipitate from dripping water. See also *calcite, karst, precipitate, stalagmite*.

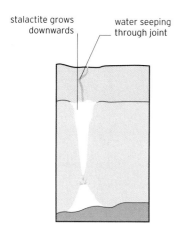

stalactite grows downwards

water seeping through joint

**STALAGMITE ▼** A deposit, usually of calcite, rising up from the floor of a cave. Like stalactites, stalagmites are produced when minerals precipitate from dripping water. See also *calcite, karst, precipitate, stalactite*.

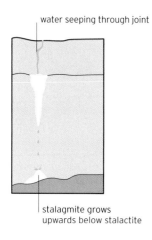

water seeping through joint

stalagmite grows upwards below stalactite

**STAR DUNE** A pyramid-shaped mound of sand with three or more "arms" radiating from the centre. Star dunes usually form where the wind blows from several directions, and they tend to grow up rather than spread out. See also *dune* (panel).

**STEPPE** A temperate grassland plain with very few trees, especially in regions with hot, dry summers and cold winters.

**STORM SURGE ▼** An abnormal rise of sea level, higher than the predicted tide, generated by a storm. The surge is caused by the storm's winds pushing water onshore.

surge    storm tide

mean sea level    high tide

**STRATOSPHERE** The layer of the atmosphere between the top of the troposphere, at 8–16km (5–10 miles), and the bottom of the mesosphere, at about 50km (30 miles). See also *mesosphere, troposphere*.

**STRATOVOLCANO** A conical volcano composed of many layers of ash, pumice, and solidified lava. Unlike a shield volcano, its eruptions are usually violent. Stratovolcanoes are common near subduction zones. See also *pumice, subduction, volcano* (panel).

**STRATUM (PL. STRATA)** A layer of sedimentary rock. See also *bedding plane, sedimentary rock*.

**STRIKE-SLIP FAULT** A near-vertical fault with horizontal displacement of the rocks on either side. The San Andreas Fault in California is a strike-slip fault. See also *fault* (panel).

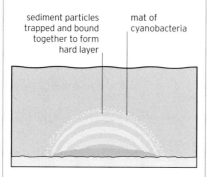

sediment particles trapped and bound together to form hard layer

mat of cyanobacteria

**STROMATOLITE ▲** A layered mound, column, or sheet formed by layers of single-celled cyanobacteria in ancient seas. When cyanobacteria became common about 2.5 billion years ago, the oxygen they produced through photosynthesis began to change the atmosphere in a way that enabled other life-forms to develop.

**STROMBOLIAN ERUPTION** A short-lived explosive volcanic eruption characterized by the ejection of ash, lava, and volcanic bombs. See also *phreatic eruption*.

**SUBDUCTION** The convergence of two tectonic plates, as a result of which an oceanic plate descends beneath another plate. Subduction zones can be classified as either ocean–ocean or ocean–continent, depending upon the nature of the two converging plates. See also *boundary* (panel), *deep-sea trench, tectonic plate*.

**SUBLIMATION** The change in a substance, such as glacier ice, from a solid to a gas without passing through the liquid phase.

**SUCCESSION** The gradual change from one community of plants to another over time in a particular area. For example, grassland may change to forest if grazing animals are removed from it.

**SUPERCELL CLOUD** A large cloud characterized by the presence of a deep, persistently rotating updraft (a mesocyclone), lightning, torrential rain and hail, strong winds, and sometimes tornadoes. See also *thunderstorm, tornado*.

**SURGE-TYPE GLACIER** A glacier whose movement features periods of normal flow alternating with episodes of much more rapid flow, up to 100 times faster. The fast–slow cycle varies greatly between different surge-type glaciers. See *glacier*.

**SURTSEYAN ERUPTION** A volcanic eruption that takes place in relatively shallow water, with magma or lava interacting explosively with the water. Surtseyan eruptions usually take place when an undersea volcano has finally grown large enough to break the water's surface. See also *phreatic eruption*.

**SWAMP** A freshwater or saltwater wetland that is dominated by trees. See also *mangrove swamp, marsh*.

**SWASH** The rush of seawater up a beach when a wave breaks.

**SYMMETRICAL FOLD** A geological structure in which once-horizontal layers of rock have been folded in such a way that the fold's axis is vertical. See also *asymmetrical fold, fold* (panel).

**SYNCLINE** A fold of stratified rocks in which the newest strata (layers) are at the core. Synclines result from massive horizontal compression. See also *anticline, fold* (panel).

# T

**TAIGA** The moist coniferous forest (boreal forest) biome covering much of northern Europe, Asia, and North America, south of the tundra. See also *biome, coniferous, tundra*.

**TECTONIC PLATE** One of the large, rigid sections into which Earth's lithosphere is divided. Most earthquakes and volcanic activity are driven by the relative motion of plates. See also *lithosphere* (panel).

**TEKTITE** A glassy particle sometimes formed when a large meteorite strikes the Earth. Tektites are thought to have been created when a meteorite impact melted or vaporized surface rocks upon impact.

**TEMPERATE** Relating to those parts of Earth between the tropics and the polar regions. See also *tropical*.

**TERMINAL MORAINE** The ridge of silt, sand, or gravel that builds up at the snout (front end) of a glacier. It marks the maximum extent of the glacier's advance. See also *lateral moraine, medial moraine, moraine* (panel).

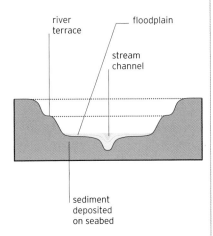

river terrace
floodplain
stream channel
sediment deposited on seabed

**TERRACE ▲** A flat area in a river valley that is raised above the present floodplain. Terraces are remnants of former floodplains created when rivers flowed at a higher level. See also *floodplain*.

**THERMAL METAMORPHISM** The transformation of rocks brought about mainly by intense heat, rather than pressure. See also *contact metamorphism, dynamic metamorphism*.

**THERMOCLINE** A region in an ocean, a lake, or the atmosphere where temperature changes more rapidly with depth or height.

**THERMOHALINE** A term describing the temperature and salt content of water. Together, these properties determine water's density and influence ocean currents.

**THERMOPAUSE** The upper edge of the thermosphere, about 640 km (400 miles) above Earth's surface and below the exosphere. See also *thermosphere*.

**THERMOSPHERE** The layer of the atmosphere between the mesosphere and the exosphere (the layer separating the atmosphere from space). It extends between about 80–640 km (50–400 miles) above Earth's surface.

**THRUST FAULT ▼** A type of reverse fault, which is an angled fault where one side (the hanging wall) has been pushed up relative to the other side (the footwall). In a thrust fault, the angle of the dip is less than 45°. See also *fault* (panel), *reverse fault*.

stress    thrust fault
older rock layers | movement along faults | younger rock layers

**THUNDERSTORM** A storm with discharges of lightning and its acoustic effect, thunder. Thunderstorms are often associated with massive cumulonimbus clouds and characterized by heavy rain, hail, strong but localized winds, and occasionally tornadoes. See also *lightning, supercell cloud*.

**TIDAL BORE** A single large wave sometimes created when a rising tide enters a narrowing channel such as a river estuary. See also *estuary*.

**TIDAL FLAT** An almost horizontal muddy or sandy area exposed at low tide and covered at high tide. Tidal flats are characteristic of sheltered areas such as estuaries. See also *estuary*.

**TIDAL RACE** A strong current created when a tide-generated water flow moves through a narrow channel.

**TIDAL RANGE** The vertical difference between a high tide and the succeeding low tide. At any location, the tidal range varies according to the relative positions of the Sun and Moon.

**TIDE ▼** The rise and fall (usually twice each day) of seawater on the shore caused by the gravitational pull of the Moon and the Sun and Earth's rotation. See also *neap tide, spring tide, tidal range*.

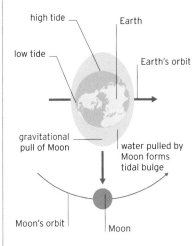

high tide    Earth
low tide    Earth's orbit
gravitational pull of Moon    water pulled by Moon forms tidal bulge
Moon's orbit    Moon

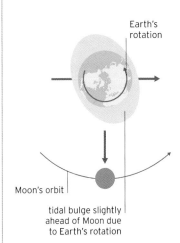

Earth's rotation
Moon's orbit
tidal bulge slightly ahead of Moon due to Earth's rotation

**TILL** A mixture of clay, sand, and rock fragments deposited by a glacier. Till that was left behind by glaciers in the last ice age covers many regions today. Also called boulder clay. See also *moraine* (panel).

**TOMBOLO** A spit that connects an island to the mainland. See also *spit*.

**TORNADO** A localized, violently destructive windstorm characterized by a tall funnel connecting a cloud with the ground. See *supercell cloud*.

**TRANSFORM FAULT** A type of fault in which two tectonic plates slide past each other. See also *boundary* (panel), *fault* (panel), *tectonic plate*.

**TRANSPORT** The movement of eroded and weathered material by water, ice, or wind. See also *weathering, erosion*.

**TRAVERTINE** A sedimentary rock, mostly calcium carbonate, that is precipitated around the edges of a hot spring. See also *precipitate, sedimentary rock*.

**TREE LINE ▼** The altitude or latitude above which conditions are too harsh for trees to grow in a particular location.

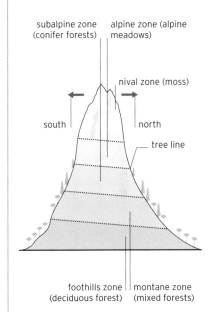

subalpine zone (conifer forests)    alpine zone (alpine meadows)
nival zone (moss)
south    north
tree line
foothills zone (deciduous forest) | montane zone (mixed forests)

**TRENCH** See *ocean trench*.

**TRIBUTARY** A river or stream that flows into a larger river.

**TROPICAL** Relating to those regions to the north and south of the equator, between the Tropic of Cancer (latitude 23.26°N) and the Tropic of Capricorn (latitude 23.26°S). See also *temperate*.

**TROPICAL CYCLONE** See *hurricane*.

## TSUNAMI

Sometimes mistakenly called a tidal wave, a tsunami is not created by tidal action. It is a fast-moving ocean wave generated by an earthquake, volcanic eruption, or landslide above or below water. If a tsunami moves into shallow water it gains height rapidly, in extreme cases reaching a height of tens of metres and causing coastal destruction on a vast scale. The 2004 Indian Ocean tsunami killed at least 230,000 people in 14 countries.

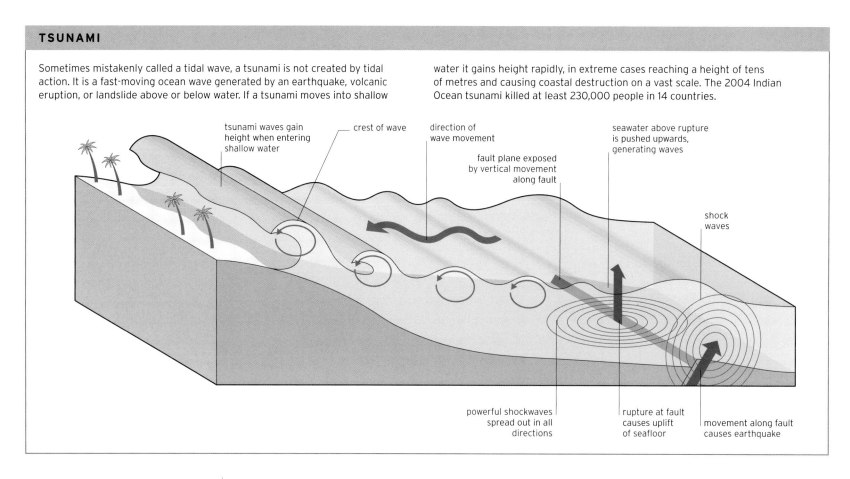

tsunami waves gain height when entering shallow water

crest of wave

direction of wave movement

fault plane exposed by vertical movement along fault

seawater above rupture is pushed upwards, generating waves

shock waves

powerful shockwaves spread out in all directions

rupture at fault causes uplift of seafloor

movement along fault causes earthquake

---

**TROPOPAUSE** The boundary between the troposphere and the stratosphere, at altitudes of about 16 km (10 miles) at the equator and 8 km (5 miles) over the poles. Above the tropopause, the air gets warmer with height. See also *stratosphere, troposphere*.

**TROPOSPHERE** The lowest, most oxygen-rich, layer of the atmosphere, where most weather events take place. The upper limit of the troposphere is lower at the poles than at the equator. See also *tropopause*.

**TROUGH** In meteorology, a long, relatively narrow area of low atmospheric pressure. See also *front*.

**TSUNAMI** See panel, above.

**TUFF** An igneous rock made of fine-grained pyroclastic material, such as ash, produced by a violent volcanic eruption. See also *pyroclastic deposit*.

**TUNDRA** The treeless biome of low-growing, cold-tolerant plants lying north of the taiga in northern Europe, Asia, and North America. See also *biome, taiga*.

**TURBIDITE** The sediment deposited by a turbidity current.

**TURBIDITY CURRENT** A rapidly moving flow of sediment-laden water, such as a flow down a continental slope onto the ocean floor.

**TYPHOON** See *hurricane*.

**ULTRAVIOLET** Invisible radiation with a short wavelength. Ultraviolet is a major component of sunlight. See also *infrared*.

**UNDERSTOREY** The layer of smaller trees and shrubs growing mainly in the shade beneath a forest canopy. See also *canopy*.

**UPWELLING** The process whereby cold, and usually nutrient-rich, water from ocean depths rises towards the surface.

## V

**VALLEY FOG** A type of fog that can form in low-lying areas, on cold, still nights. Heavy, chilled air settles in such places, and the water vapour it contains condenses when the temperature falls below the dew point. See also *radiation fog*.

**VALLEY GLACIER** A stream of flowing ice that is confined within the sides of a valley, often a V-shaped pre-glacial valley that is modified by the erosive action of the ice. See also *glacier*.

**VEIN** A fracture in rock containing mineral deposits. Veins are usually formed when minerals are precipitated from a hot liquid in which they are dissolved. See also *hydrothermal vein*.

**VENTIFACT** A desert rock or pebble that has been polished by the action of wind-blown sand.

**VERTICAL TRANSPORT** The ascent (upwelling) or descent of nutrient-rich water in the oceans.

**VOLATILE** A term used to describe water, carbon dioxide, and some other gases dissolved in magma that form bubbles at low pressure near the surface. When the bubbles combine, they can make an eruption more explosive, showering sprays of lava. See also *eruption* (panel).

**VOLCANIC ARC** A chain of volcanoes formed above a subduction zone and arranged in an arc shape. The volcanoes may form a chain of islands in the ocean. See also *island arc, subduction, tectonic plate*.

## VOLCANO

A volcano is a mountain with an opening, or vent, through which molten lava, volcanic ash, and gases are erupted. Eruptions range from violent andesitic eruptions to the gentle lava flows. Structurally, there are many types of volcanoes, their shapes being largely dependent on the type of eruptions and the quality of the magma. A dome volcano forms steep sides because the viscous lava it erupts cannot flow very far. Stratovolcanoes are formed of many layers of ash and solidified lava.

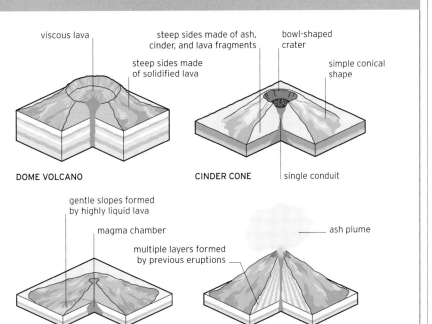

viscous lava — steep sides made of ash, cinder, and lava fragments — bowl-shaped crater

steep sides made of solidified lava

simple conical shape

**DOME VOLCANO**

**CINDER CONE** — single conduit

gentle slopes formed by highly liquid lava

magma chamber

multiple layers formed by previous eruptions

ash plume

**SHIELD VOLCANO**

**STRATOVOLCANO**

**VOLCANIC PLUG** A mass of igneous rock filling the vent of an active volcano. If gas or magma is trapped beneath the plug, pressure will build up, increasing the likelihood of the next eruption being violent. See also *magma*.

**VOLCANO** See panel, above.

**VORTEX** A rapidly circulating mass of air or water. See *eddy, tornado, whirlpool*.

**VULCANIAN ERUPTION** An explosive type of volcanic eruption. Vulcanian eruptions occur when the pressure of trapped gases in magma becomes strong enough to blow off an overlying crust of solidified magma. See also *eruption* (panel), *magma*.

# W

**WADI** A desert riverbed or valley that contains water only at times of heavy rainfall.

**WATER TABLE** ▼ The upper surface of the groundwater zone. Below the water table, the soil or rocks are permanently saturated, but its position fluctuates, often seasonally. See also *groundwater*.

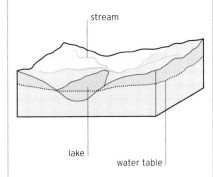

stream

lake — water table

**WATERFALL** ▼ A cascade of water formed when a stream or river flows over a cliff or steep incline.

hard rock

fallen rocks

plunge pool

soft rock eroded by water and small swirling rocks

**WATERSHED** An imaginary line that separates neighbouring drainage basins. See also *catchment area, drainage basin*.

**WAVE** A disturbance in the ocean, usually generated by wind. For a wave crossing the sea, the water itself does not move very much, except up and down as the wave passes. The high point of a wave is its crest, and the low point its trough.

**WAVELENGTH** The distance between successive crests (or troughs) of a wave.

**WEATHERING** The gradual breakdown of rocks on Earth's surface by chemicals in rainwater, temperature changes, and biological activity such as the growth of lichens and plants. Rocks are broken into small fragments, which are then transported away. See also *erosion*.

**WHIRLPOOL** A rapidly rotating body of water in a river or the ocean. See also *eddy, vortex*.

**WHITE SMOKER** A hydrothermal vent on the ocean floor emitting light-coloured minerals such as those containing barium, calcium, and silicon. White smokers tend to be cooler than black smokers. See also *black smoker, hydrothermal vent*.

# Y

**YARDANG** A wind-sculpted desert rock that has been eroded by wind-borne particles of sand and dust.

# Z

**ZOOPLANKTON** Animal life, mostly microscopic, that drifts in the water column of oceans, lakes, and rivers. See also *phytoplankton*.

**ZOOXANTHELLAE** Single-celled algae that live in corals and supply them with vital nutrients to grow. See also *coral bleaching*.

# INDEX

Page numbers in *italics* refer to photographs and illustrations.

## A

Abraham, Lake 49
acacia trees 203, 206, *206–207*
Acadia Coastline 65
Acadia National Park 384
Aconcagua, Mount *92–93*
Adirondack Mountains 363
Adrar Plateau 210
Adriatic Sea 171
Afar Depression 183, 184
Africa 178–79, 180–81
   *see also individual countries*
African Great Lakes 193, 196
African Plate 133, 183, 186, 199
Agassiz, Lake 23
Ahaggar Mountains 349–50
Akiyoshi-do 380
Alaska Range 24
Albania
   Dinaric Alps 348, *348*
Aldabra Atoll 314, *314–15*
Aletsch Glacier 138, 150–51
algae *97*, *192*, 368, *368–69*, *375*
Algarve Coast 170
Algeria
   Atlas Mountains 178, *181*, 182–83
   Hoggar Mountains 349–50
   Mediterranean Evergreen Forest 394–95
   Sahara Desert *181*, 208–209, 211, 349, 401
Almirante, Mount 99, *99*
Alpine Fault 273, 278, 279
alpine lakes 155
alpine tundras 25, 72
Alps 136, 138–39, 359, 372
   *see also* Southern Alps
Altai Mountains 175, 224, 237, 266, *266–67*
Altiplano 92, 94, *95*, 96–97, 105
Amalfi Coast 388
Amazon (river) 84, 87, *87*, 106–107
   basin 108, *108–109*
Amazon Rainforest *82–83*, 106, 120–21, 190
America *see* United States of America
American Falls *see* Niagara Falls

ammonites 168, *168–69*
Amur 377
Andaman Sea Reefs 253
Andean Yungas 119
Andes 92–93, 94–95, 105, 125
Angel Falls *90–91*, 91
Angola
   Central and Eastern Miombo Woodlands 398
   Congo (river) 106, 192, 200
   Namib Desert *176–77*, 212–13
   Okavango Delta 196–97
animals 16
   *see also* wildlife
Annamite Mountains 245
Antarctic Ice-sheet 39, 306–307, 308, 309
Antarctic Plate 94, 305
Antarctica 273, *300–301*, 302–303
   Antarctic Ice-sheet 39, 306–307, 308, 309
   Antarctic Tundra 309
   McMurdo Dry Valleys 309
   Mount Erebus 304–305
   Ross Ice Shelf *306–307*, 308–309
   Transantarctic Mountains 304, *307*
Antelope Canyon 80
Anti-Atlas mountains 182
Aoraki (Mount Cook) *278*, 279, 281
Apennines 347
Appalachians 22, 35, 70
aquatic plant life 105, 112, *160*
   *see also* marine life
Arabian Desert 262–63
Arabian Highlands 352
Aral Sea 377, 403
Aran Islands 166
archaea 16, *16*, *268–69*
Archean Era 15, 180, 272
archipelagos
   *see also* islands
   Dalmatian Coast 171
   Galápagos Islands 320–21
   Hawaiian Islands 13, *13*, 316–17, 318–19
   Lofoten Islands 386
   Nansei Shoto Archipelago Forest 397
   Santorini 348
   Yaeyama Islands 389
Arctic tundra 72, 175
Ardennes 394
Arenal 344–45, *344*

Argentina
   Altiplano 92, 94, *95*, 96–97, 105
   Andean Yungas 119
   Andes 92–93, 94–95, 105, 125
   General Carrera Lake 113
   Hornocal Mountains 98
   Iguaçu Falls 110–11
   Ojos del Salado 345
   Pampas 122–23
   Paraguay River 368–69
   Patagonian Desert 127
   Perito Moreno Glacier 102–103
   Salar de Arizano 400
   Salinas Grandes 400
   South American Dry Forest 392
   Southern Patagonian Ice-field 101, 103
   Talampaya National Park 400
   Valdivian Temperate Forest 119
   Valle de la Luna 127
Arizona Desert 79, 80
Arkansas River 364, *364*
Armenia
   Caucasus Mountains 221
Ar Rub 'al Khali *262–63*
ash clouds *141*, *143*, 345
Asia 216–17, 218–19
   *see also individual countries*
Assiniboine, Mount *24–25*
asthenosphere *312*
Aswan High Dam 191
Atacama Desert 92, *94–95*, 97, 124–25, 392
Atlantic Coast Restinga Forest 392, *392*
Atlantic Forest *110*
Atlas Mountains 178, *181*, 182–83
atmosphere, Earth's *10*, 11, *11*
atolls *see* lagoons; reefs
aurorae 338–39
Australia 272
   Australian Alps 274, 275
   Bungle Bungles 357, *357*
   Daintree Rainforest 292–93
   East Australia Temperate Forest 292
   Gibson Desert 298
   Great Barrier Reef 284–85, 286–87
   Great Dividing Range 274–75
   Great Sandy Desert 295, 298
   Great Victoria Desert 295, 298, 299

Australia *continued*
   Horizontal Falls 390
   Hyden Rock 278
   Jenolan Caves 283
   Kata Tjuta 358
   Kati Thanda-Lake Eyre *268–69*, 271, *299*
   Koonalda Caves 383
   Little Sandy Desert 403
   Murray–Darling 274, 282–83
   North Australian Savannas 294
   Pinnacles 298
   Shark Bay 288–89
   Simpson Desert 295, 296–97
   Twelve Apostles 290
   weather 334–35
   Whitehaven Beach 390, *390*
Australia and New Zealand (continent) 270–71, 272–73
Australian Plate 270, 273
Australian Shield 272
Austria
   Alps 136, 138–39, 359, 372
   Danube 156–57, 161, 173
   Lake Constance 373, *373*
   Kalkalpen National Park 394, *394*
   Pasterze Glacier 359, *359*
Auyan Tepui 91
avalanches *89*, 226, *232–33*
Avalonia 132
Ayers Rock (Uluru) *296–97*
Azerbaijan 221, 352

## B

Badwater Basin 75
Bagley Ice-field *40–41*, 41
Bahia Coastline 386
Baikal, Lake 237, 238–39
Baishuitai Hot Springs and Terraces 355
Bale Mountains 204
Balkhash, Lake 377, *377*
Baltoro Glacier 225, 232–33
Bavarian Forest 172–73
Banc d'Arguin 388
Bangladesh 332
   Ganges 106, 240, 260
   Sundarbans 241
   tornadoes in 332

baobab trees 202, *202–203*, 203
Baradla-Domica cave complex 159
Bardarbunga volcano 147
Barrancas del Cobre (Copper
    Canyon) 37, *37*
barrier reefs
    *see also* reefs
    Great Barrier Reef 284–85,
        286–87
basalt rock 165, *165*, 189, 198
    flood basalt 110, 318
base camps *226*
basins, river 48, 87, 197, 238, *238*
    Amazon 106, 108–109
    Lake Ladoga 160
Bassenthwaite Lake 153
Bay of Fundy 64–65
beaches *see* coastlines
Belarus
    Białowieża Forest 174
Belgium
    Ardennes 394
    Hallerbos 172
Belize
    Great Blue Hole 114–15, 384
    Lighthouse Reef 384, *384–85*
Beluka, Mount 224
Benagil Beach *170*
Bengal, Bay of 241
Ben Nevis 134
Beppu Hot Springs 354
Bering Glacier 40–41
Bhutan
    Eastern Himalayan Forest 256
    Himalayas 94, 219, 222–23, 240
    Terai-Duar Savannas 260–61
Biafo Glacier 234
Białowieża Forest 174
Biebrza Marshes 158–59
Bigar Waterfall 374–75
Big Muddy (Mississippi–Missouri)
    53, 106, 190, 364
Big Room, Carlsbad Caverns 58
Big Sur 63
Binggou Danxia *228*
biomes 66, 392
bird life
    in coastal and island regions *117,*
        *166, 321*
    in desert habitats *210, 265, 296*
    in forest habitats *68, 120, 200,*
        *255, 256*
    in mountainous regions *189, 279*
    in river and lake habitats *52, 110,*
        *157, 191, 196, 246, 282–83*
    in wetland habitats *61, 154, 241*
Bird's Foot Delta *53*
Blackburn, Mount 44, *45*
Black Forest 393
Black Rapids Glacier 24, 43

Black Sea 375
black smokers (hydrothermal vents)
    322, *322–23*
"blackwater" swamps 59
bluebells 172, *172*
Blue Forest (Hallerbos) 172, *172*
blue holes 114–15, 251, 384
Blue Mountains 274, *274–75*, 292
Blue Nile 191, *191*
Blue Ridge Mountains 35
Blyde River Canyon *180*, 376, *376*
Bogoria, Lake 351
bogs 59, 158, 174, *174*
Bohemian Forest 173
Bolivia
    Altiplano 92, 94, *95*, 96–97, 105
    Amazon Rainforest 82–83, 106,
        120–21, 190
    Andean Yungas 119
    Andes 92–93, 94–95, 105, 125
    Laguna Colorada 368, *368–69*
    Lake Titicaca 105
    Pantanal 112–13
    Salar de Uyuni 96, 126
    South American Dry Forest 392
Boma National Park *204*
boreal forest 21, 66, 174
    North American Boreal Forest
        56
    Siberian Taiga 254–55
Borneo
    Bornean Rainforest 258–59
    Gunung Mulu 382–83, *382*
Borra Caves 378
Botswana
    Kalahari Desert 197, 211
    Makgadikgadi Pan *402*, 403
    Okavango Delta 196–97
    Zambezi 195, 376
Brahmaputra river 240
Brandberg Massif 188
Brazil
    Amazon (river) 84, 87, *87,*
        106–107
    Amazon Rainforest *82–83*, 106,
        120–21, 190
    Atlantic Coast Restinga Forest
        392, *392*
    Bahia Coastline 386
    Caatinga 400
    Cerrado 398, *398*
    Gruta do Janelão 369
    Guiana Highlands 90–91, 366
    Iguaçu Falls 110–11
    Lençóis Maranhenses 116–17
    Pampas 122–23
    Pantanal 112–13
    Paraguay River 368–69
    South American Dry Forest 392
    Sugarloaf Mountain 98

Brazo Rico Ice Dam 103, *103*
Bridal Veil Falls (Niagara) 52
Briksdalsbreen *149*
Bromo, Mount 230, *230*
Brunei 258
Bryce Canyon 78
Buenos Aires, Lake 113
Bulgaria
    Black Sea 375
    Danube 156–57, 161, 173
    Stone Desert 401
Bungle Bungles 357, *357*
Burma (Myanmar) 243, 253, 380
Burntcoat Head 64
Burundi
    Congo (river) 106, 192, 200
    Nile 106, 190–91, 196
buttes 35, *76*, 77
Byrd Glacier *306*

# C

Caatinga 400
cacti 81, *81, 126*
Cairngorms 134
calcite deposits *58, 59, 245, 283*
    stalactites and stalagmites 114,
        *252–53, 283*, 369
calcium carbonate 220, 221, 285
caldera volcanoes 32–33, 89, 230,
    *231, 321*
    Crater Lake 28–29
Caledonian Forest 392
Caledonian Mountains 134
Californian Chaparral and
    Woodland 391
Camargue 154
Cambodia
    Mekong 243
Cambrian period 16
Cameroon
    Cameroonian Highlands Forest
        395
    Congo (river) 106, 192, 200
    Congo Rainforest 200–201
    Kapsiki Peak 350
    Lake Chad 375
    Lake Nyos 350
Canada
    Appalachians 22, 35, 70
    Bay of Fundy 64–65
    Coast Ranges 343
    Columbia River 362–63
    Great Bear Lake 362
    Great Lakes 23, 50–51, 238, *336*
    Great Plains 73
    Kaskawulsh Glacier 42

Canada *continued*
    Lake Abraham 49
    Mackenzie Dikes 342
    Manicouagan Lake 363
    Niagara Falls 51, 52
    North American Boreal Forest
        66
    North American Tundra 72
    Pacific Northwest Rainforest 67
    Rocky Mountains (Rockies) *22,*
        *23, 24–25*
    Spotted Lake 362, *362*
    weather 336–37, 338
    Yukon 48–49
Canadian Shield 21, *22*
Cango Caves 376–77
Caño Cristales 105
canopies (trees) *67, 120, 121, 255*
Cantabrian Mountains 346, 373
Canyonlands National Park 79, *79*
canyons 31, 37, 62
    Antelope Canyon 80
    Blyde River Canyon *180*, 376, *376*
    Bryce Canyon 78
    Mesa Arch 79
    Grand Canyon 54–55, 56–57
    Valle de la Luna 127
    Verdon Gorge 155
    Yarlung Tsangpo Grand Canyon
        379
Cap Blanc Nez 167
Cape de Creus 387
Cape Floral Region 205, 389
Cape Horn 117
Cape of Good Hope 188
Cape Perpetua 62
Cape Point 389
Cap Ferret 387
Cape York Peninsula 294
Cappadocia 348–49, *348–49*
Caribbean Islands
    Les Pitons 385
    Mount Pelée, Martinique 344
    Soufrière Hills 89
    Windsor Caves 366
Carrizo Plains *30*
Carlsbad Caverns 58
Carpathian Mountain Forests 394
Cascade Range 27
Caspian Sea 221
Catania 140
catchment area (river basins) 48, 87,
    197, 238, *238*
    Amazon 106, 108–109
    Lake Ladoga 160
Cathedral Beach 387
Cathedral Chamber, Jenolan Caves
    283
Caucasus Mountains 221
Cave of Crystals 36

caves and cave systems 248, 251
    Akiyoshi-do 380
    Asia 246
    Borra Caves 378
    Cango Caves 376–77
    Carlsbad Caverns 58
    Cave of Crystals 36
    Central and South America 114
    decorations 58, 249
    Dinaric Alps 348
    Grotte Casteret 372
    Gruta do Janelão 369
    Gunung Mulu 382, 382–83
    Ha Long Bay 250–51
    Hang Son Doong 244–45
    Jeita Grotto 378, 378
    Jenolan Caves 283
    Koonalda Caves 383
    Krubera Caves 221, 378
    Lascaux 372
    Mammoth Cave 364
    Optimisticheskaya 375
    Ox Bel Ha 364
    Sac Actun 365
    sea caves 64–65, 166, 170
    Sistema Huautla 364–65, 365
    Slovak Karst Caves 159
    stalactites 114, 252–53, 283
    stalagmites 245, 283, 369
    Vercors Caves 372, 372
    Waitomo Caves 283
    Windsor Caves 366
cedar tress 68, 395, 395
Cenozoic Era 14
Central African Republic
    Congo (river) 106, 192, 200
    Congo Rainforest 200–201
Central America 84–85, 86–87
    see also individual countries
Central and Eastern Miombo
    Woodlands 398
Cerrado 398, 398
Cerro Azul 345
Cerro Paine Grande 99
Cerulean Lake 24–25
Chad
    Lake Chad 375
    Sahara Desert 181, 208–209, 211,
    349, 401
Chaîne des Puys 346, 346
chalk cliffs 167, 167, 387
chalk deserts 210
Challenger Deep 324
Chamonix Valley 150
Chapada Diamantina National Park
    86–87
Cherokee National Forest 70
Chesil Beach 169
Chignecto Bay 64
Chihuahuan Desert 81

Chile
    Altiplano 92, 94, 95, 96–97, 105
    Andes 92–93, 94–95, 105, 125
    Atacama Desert 92, 94–95, 97,
    124–25, 392
    Cape Horn 117
    Cerro Azul 345
    Chilean Coast Range 125
    Chilean Fiordland 92, 116
    Chilean Matorral 392
    El Tatio Geysers 97
    General Carrera Lake 113
    Ojos del Salado 345
    Patagonian Desert 127
    Southern Patagonian Ice-field
    101, 103
    Torres del Paine 99
    Valdivian Temperate Forest 119
Chimborazo volcano 87
China
    Altai Mountains 175, 224, 237,
    266–67, 266
    Amur 377
    Dragon Hole 251
    Eastern Steppe 260
    Gobi Desert 224, 266–67, 398
    Himalayas 94, 219, 222–23, 240
    Inylchek Glacier 360
    Karakoram Range 225, 234
    Mekong 243
    Mongolian–Manchurian Steppe
    398
    Mount Everest 219, 226–27, 235,
    361
    mountains and volcanoes 214–15,
    354–55, 355
    Ob–Irtysh 237
    rivers and lakes 379–82, 380–81
    Rongbuk Glacier 361
    South China karst 246–47,
    248–49
    Taklamakan Desert 264–65
    Tian Shan Mountains 224
    Upper Yangtze Forests 257
    Xishuangbanna Rainforest 397
    Yangtze 190, 242
    Yarlung Tsangpo Grand Canyon
    379
    Yellow River 242–43
    Yulong Glacier 232
    Zhangye Danxia 228
Chocoan Rainforest 392
Chugach Mountains 40, 41
Churun, River 91
Chukchi Peninsula 261
cinder cones 28, 135, 135
    see also pyroclastic flows
Cirque de Gavarnie 137
cirque glacier 100, 232
cirques 138

Claron Formation 78
cliffs
    Algarve Coast 170
    Cape de Creus 387
    Cliffs of Dover 133, 167
    Cliffs of Moher 166
    Étretat Cliffs 387, 387
    Los Gigantes Cliffs 198
    Palisades 342, 342–43
    Turkish Steps 388, 388
    Twelve Apostles 290
climates 11, 15
    see also rainfall
    in Africa 178, 205
    in Antarctica 303
    in Asia 217, 227, 255, 257, 261
    in Australia and New Zealand
    271, 294
    in Central and South America 84,
    113, 119, 122
    in Europe 130
    microclimates 119, 155, 245, 292
    in North America 20, 71
clonal groves 70
cloud forests 91, 118–19
clouds 330, 331
Clutha River 279
coastal deserts 212, 298
coastal mountain ranges 39, 63
coastal temperate rainforests 67,
    119
coastal wetlands 154
coastlines
    see also fiords
    Acadia Coastline 65
    Algarve Coast 170
    Bahia Coastline 386
    Banc d'Arguin 388
    Bay of Fundy 64–65
    Big Sur 63
    Cape Horn 117
    Cape Point 389
    Cliffs of Dover 133, 167
    Coffee Bay 388
    Dalmatian Coast 171
    Dune du Pilat 167
    El Nido 390
    Europe 386–88
    Giant's Causeway 152, 164–65
    Jeffreys Bay 389
    Jurassic Coast 168–69
    Krabi Coastline 252–53
    Lençóis Maranhenses 116–17
    Moeraki Beach 290
    Ninety Mile Beach 390
    Paracas Peninsula 385
    Puget Sound 384
    Shark Bay 288–89
    Twelve Apostles 290
    Whitehaven Beach 390, 390

Coast Ranges 343
Coffee Bay 388
Colca River 367
cold deserts 74, 264, 309
Colombia
    Amazon (river) 84, 87, 87,
    106–107
    Amazon Rainforest 82–83, 106,
    120–21, 190
    Andes 92–93, 94–95, 105,
    125
    Caño Cristales 105
    Galeras 345
    La Guajira Desert 399
    Llanos 104
    Orinoco 366
Colorado Plateau 399, 399
Colorado River 54, 55, 56, 57
Columbia Glacier 40
Columbia River 362–63
columnar rock formations 35, 152,
    165
complex volcanoes 89
Cone Peak, Big Sur 63
confluence, river 56, 109, 240
Congo, Democratic Republic of the
    Congo (river) 106, 192, 200
    Congo Rainforest 200–201
    Nile 106, 190–91, 196
    Sudanian Savannas 204
Congo, Republic of the
    Congo (river) 106, 192, 200
    Congo Rainforest 200–201
Congo Brazzaville see Congo,
    Republic of the
Congo-Kinshasa see Congo,
    Democratic Republic of the
conifer trees 68, 69, 255, 255, 293,
    293
    in boreal forests 66
    in rainforests 67
    in temperate forests 119
Constance, Lake 371, 373, 373
Continental Divide 25
continental crust 10, 10, 132
continental rifting 12, 186–87, 322,
    346
continental shelf 109, 286
continental shift 12
convergent boundaries 13, 13
Cook, Mount (Aoraki) 278, 279,
    281
Coorong 283
Copper Canyon 37, 37
coral bleaching 286, 286
coral reefs
    Andaman Sea Reefs 253
    Dragon Hole 251
    Great Barrier Reef 284–85,
    286–87

coral reefs *continued*
    Great Blue Hole 114–15, 384
    Great Chagos Bank 315
    Lighthouse Reef 384, *384–85*
    Maldives 315
    Nusa Tenggara 253
    Red Sea Coast 198–99
    Shiraho Reef 251
Cordillera Cantábrica 346
Coriolis force 328
Costa Rica
    Arenal *344*, 344–45
    Monteverde Cloud Forest 118–19
Cotahuasi River 367
Cotopaxi 87, 93
courses, river 156, 243
craters
    Crater Lake 28–29
    man-made *264*
    Meteor Crater 79
    volcanic 89, *143*, 275, 351
cratons 132, 180, 181, 272
    *see also* continental crust
crevasses, glacial *227*, 233, 235, 281,
    *361*
Croatia
    Danube 156–57, 161, 173
    Plitvice Lakes 374, *374*
    Dalmatian Coast 171
Crooked Forest 393, *393*
crust, Earth's *10*, 10–11
Cuernos del Paine 99, *99*
cyanobacteria 288, 375, *375*
cyclones 328–29, *330*
cypress swamps *60–61*

**D**

Daba Mountains 257
Daintree Rainforest 292–93
Dalmatian Coast 171
Damietta river 191
Danube 156–57, 161, 173
Darling River 274, 282–83
Dasht-e-Kavir Desert 403, *403*
Dasht-e-Lut 263
Dead Sea 192, 236
deadwood 174
Dean's Blue Hole 251
Death Valley 75, *75*
Deccan Traps 218, *218*, 353
deep sea life 322, *322*, *323*, *325*
deforestation *109*, 120, 202, 258
deltas, river *15*, *340–41*
    Danube 156, *157*
    Lena *239*
    Mekong *243*

deltas, river *continued*
    Mississippi–Missouri 53, *53*
    Nile Delta 191
    Okavango Delta 196–97
    Rhône Delta 154, 155
    Sundarbans 240–41
    Volga Delta 161, *161*
Democratic Republic of the Congo
    *see* Congo, Democratic Republic
    of the
Denali 24, *24*
Denmark
    North Jutland Dunes 386
depressions (weather) 328
Derwentwater *152*
Desert Island, Mount 65, 384
desert pavement formations 266
deserts
    Arabian Desert 262–63
    Atacama Desert 92, *94–95*, 97,
      124–25, 392
    Caatinga 400
    Chihuahuan Desert 81
    Colorado Plateau 399, *399*
    Dasht-e-Kavir Desert 403, *403*
    Dasht-e-Lut 263
    Gibson Desert 298
    Gobi Desert 224, 266–67, 398
    Great Basin Desert 74
    Great Sandy Desert 295, 298
    Great Victoria Desert 295, 298,
      299
    Indus Valley Desert 403
    Kalahari Desert 197, 211
    Karakum Desert 264
    Karoo 211
    Kyzylkum Desert 403
    La Guajira Desert 399
    Little Sandy Desert 403
    McMurdo Dry Valleys 309
    Meteor Crater 79
    Mojave Desert 74–75
    Monument Valley 76–77
    Namib Desert *176–77*, 212–13
    Navajo 34
    Nyiri Desert 401
    Patagonian Desert 127
    Sahara Desert *181*, 208–209, 211,
      349, 401
    Sechura Desert 399
    Simpson Desert 295, 296–97
    Sonoran Desert 76
    Stone Desert 401
    Syrian Desert 403
    Tabernas Desert 400–401
    Taklamakan Desert 264–65
    Talampaya National Park
      400
    Thar Desert 265
    White Desert 210

Dettifoss 369
Devil's Cataract, Victoria Falls 195,
    *195*
Devil's Golf Course, Death Valley
    *75*
Devil's Postpile 342–43
Devil's Tower 35
Devin Gate 156
Devon Ria Coast 386–87
dike swarms 342
Dinaric Alps 348
dipterocarps trees 258
divergent boundaries *12*, 13,
    *312–13*
Djibouti
    Afar Depression 183, 184
Dobruja Hills 156
Dobšinská Ice Cave 159
dolines *244*, 245, 248, *249*
    *see also* sinkholes
Dolomites 136
Doñana 373
Doubtful Sound 291
Dover, Cliffs of 133, 167
Dragon Hole 251
drainage systems (rivers)
    Amazon 106, 108–109
    basins 48, 87, 197, 238, *238*
    Danube 156, *157*
    deltas *15*, *340–41*
    Lake Ladoga 160
    Lena *239*
    Mekong *243*
    Mississippi–Missouri 53, *53*
    Nile Delta 191
    Okavango Delta 196–97
    Rhône Delta 154, 155
    Sundarbans 240–41
    Volga Delta 161, *161*
Drakensberg *180*, 189, 376
drought regions *see* rainfall
Dry Falls 23
dry forests 202–203, 392
Dudhsagar Falls 379
Dugi Otok 171
Dune du Pilat 167
dunes, sand 81, 265, 297, 299
    Arabian Desert 263
    Atacama Desert *124–25*
    Dune du Pilat 167
    Erg Chebbi Sand Dunes 401,
      *400–401,*
    Great Sandy Desert 295
    Lençóis Maranhenses 116–17
    Namib Desert *176–77*, 212,
      *212–13*
    North Jutland Dunes 386
    Pinnacles, the 298
    Taklamakan Desert 264–65
Dungeness Spit 384

Durdle Door *169*
Durung Drung 361
duststorms *326–27*, 334–35

**E**

Earth
    *see also* plate tectonics
    crust 10, *10*, 12
    life on 16–17
    mantle 10, *10*
    mantle convection 10, *10*
    mantle plumes 33, 186, 187, *319*
    origins of 14–15
    structure of 10–11, *10*
earthquakes 10, 30, 92
East African Coastal Forests 395
East African Rift System 184,
    186–87
East Australian Temperate Forest
    292
Eastern Cataract, Victoria Falls *195*
Eastern Steppe 260
East European Craton 132
East Himalayan Forest 256
East Pacific Rise 322–23
ecosystems
    Africa 179
    Antarctica 302
    Asia 217
    Australia and New Zealand 270
    Central and South America 85
    Europe 131
    North America 21
Ecuador
    Amazon Rainforest 82–83, 106,
      120–21, 190
    Andes 92–93, 94–95, 105, 106
    Chocoan Rainforest 392
    Cotopaxi 87, 93
    Galápagos Islands 320–21
    Tungurahua volcano *87*
Egypt
    Arabian Desert 262–63
    Nile 106, 190–91, 196
    Red Sea Coast 198–99
    Sahara Desert *181*, 208–209, 211,
      349, 401
    White Desert 210
Eiger 138
Elbert, Mount 25
Elbras, Mount 221
El Capitan ("the Chief") 26
El Chichón 344
*El Hongo 127*
El Nido 390
El Niño 294, 399

El Popo (Popocatépetl) 37
El Tatio Geysers 97
emergents 120, *259*, *293*
endemic species
    animals 53, *184*, *238*, *283*
    fish *51*, 192–93
    in forests *172*, 201, *201*, 202, *293*
    plants 68, *91*, *106*, 294, *314*
endorheic basins 197
England
    Cliffs of Dover 133, 167
    Devon Ria Coast 386–87
    Jurassic Coast 168–69
    Lake District 152–53
    Severn 370
    Thames 370
Erebus, Mount 304–305
Erg Chebbi Sand Dunes *400–401*, 401
ergs 209, 401
Erie, Lake *50–51*, 51
Eritrea 204
    Afar Depression 183, 184
    Nile 106, 190–91, 196
    Red Sea Coast 198–99
Erta Tale volcano 181
eruptions, volcanic 29, 142, *143*
    *see also* lava
    in Asia 229, 230, 231
    in Central and South America 88, *88*, 89, 344–45, *344*
    in Europe 140, 144, 145, 345
    in North America 37, 316, 342, 343, *343*, 344
escarpments 37, 186, 188, *295*, *350–51*
Estonia
    Jagala Waterfall 370, *370*
estuaries 60, *60*, *109*
Ethiopia
    Afar Depression 183, 184
    Ethiopian Highlands 350–51, *350–51*
    Ethiopian Montane Grasslands 204–205
    Nile 106, 190–91, 196
    Sudanian Savannas 204
Etna, Mount 140–41, 142–43
Étretat Cliffs 387, *387*
eucalypt trees 292
eukaryotes 15, 16
Eurasian Plate 133, 218, 273
Europe 130–31, 132–33
    *see also individual countries*
Everest, Mount 219, 226–27, 235, 361
    *see also* Khumbu Glacier
Everglades 60–61
"exploding lakes" 350

extinction, of life 16
Eyjafjallajökull 369
Eyre, Kati Thanda-Lake *268–69*, 271, *299*

F

faults *13*, 24
    continental rifting 12, 186–87, 322, 346
    San Andreas Fault 23, *23*, 30
    thrust faults 24, 94, *95*
Fedchenko Glacier 232
fenglin towers *248*
Fingal's Cave 152, 166
Finke River *272*
fiords
    Chilean Fiordland 92, 116
    Ilulissat Icefiord 384, *384*
    New Zealand Fiords 291
    Norwegian Fiords 162–63
fish life
    in lagoons *51*, 116, *116*
    in lakes and lagoons *192–93*, 193
    in river habitats *106*
fissures 142, *313*, 322
Five Finger Rapids *49*
flank craters *142*
floods 104, 108
    flash floods 147
    flood basalts *110*, 318
    floodplains *48–49*, 108, *109*, *112–13*, *240*, 243
fog
    in deserts *124*, 213
    in forests *70*, 118, *172–73*
forests 71, 120
    Andean Yungas 119
    Ardennes 394
    Atlantic Coast Restinga Forest 392, *392*
    Bavarian Forest *172–73*
    Białowieża Forest 174
    Black Forest 393
    Caledonian Forest 392
    Cameroonian Highlands Forest 395
    Carpathian Mountain Forests 394
    Cherokee National Forest 70
    Chilean Matorral 392
    Crooked Forest 393, *393*
    East African Coastal Forests 395
    East Australia Temperate Forest 292
    East Himalayan Forest 256
    Forest of the Cedars of God 395, *395*

forests *continued*
    Green Mountain National Forest 71
    Hainich National Park 393
    Hallerbos 172
    Jalisco Dry Forest 392
    Kalkalpen National Park 394, *394*
    Madagascan Dry Forest 202–203
    Mediterranean Evergreen Forest 394–95
    Mudumalai Forest 396
    Nansei Shoto Archipelago Forest 397
    Redwood National Forest 391
    Russian Far East Temperate Forest 395
    Sagano Bamboo Forest 396, *397*
    Siberian Taiga 254–55
    Sierra Madre Pine-Oak Forest 391
    South American Dry Forest 392
    Southwestern Ghats Moist Forest 396, *396–97*
    Sri Lankan Moist Forest 396
    Taiheiyo Montane Forest 256
    Tongass National Forest 391, *391*
    Upper Yangtze Forests 257
    Waipoua Forest 293
    Western Himalayan Temperate Forest 396
fossils 14, 64, 127, 167
    Jurassic Coast *168–69*, 169
Fox Glacier 361, *361*
France
    Alps 136, 138–39, 359, 372
    Ardennes 394
    Camargue 154
    Cap Ferret 387
    Chaîne des Puys *135*, 346, *346*
    Dune du Pilat 167
    Étretat Cliffs 387, *387*
    Jura 346
    Lake Geneva 155, 359
    Lascaux 372
    Loire 371
    Massif Central *135*, 346
    Mediterranean Evergreen Forest 394–95
    Mer de Glace 150
    Pyrenees 137
    Rhine Rift 346
    Rhône Glacier 372
    Seine 371
    Vercors Caves 372, *372*
    Verdon Gorge 155
Franz Josef Glacier 280
French Guiana
    Amazon Rainforest 82–83, 106, 120–21, 190
frost-wedging 78
Fuji, Mount 229

fumaroles 33, 97, *276*, 305
fungi 16, 174
Furnace Creek 75
fynbos vegetation 205, *205*

G

Gabon
    Congo (river) 106, 192, 200
    Congo Rainforest 200–201
Galápagos Islands 320–21
Galeras 345
Galician Coast 387
Ganges 106, 240, 260
Garonne river 137
Gasherbrum Group 225, *232*, 233
General Carrera Lake 113
General Sherman, Giant Forest 68
Geneva, Lake 155, 359
geology
    Africa 179
    Antarctica 302
    Asia 216
    Australia and New Zealand 270
    Central and South America 85
    Europe 131
    North America 21
Georgia
    Black Sea 375
    Caucasus Mountains 221
    Krubera Caves 221, 378
geothermal activity 10, *18–19*, *276–77*
    Afar Depression 183, 184
    Baishuitai Hot Springs and Terraces 355
    Beppu Hot Springs 354
    El Tatio Geysers 97, *97*
    fumaroles 33, 97, *276*, 305
    geysers 277
    hydrothermal vents 322, *322–23*
    Pamukkale Hot Springs 220–21
    Rotorua *276–77*
    Strokkur Geyser 135
    Terme di Saturnia 347, *347*
    Valley of Geysers 352, *352*
    Yellowstone *18–19*, *30–31*, 31, 32
Germany
    Alps 136, 138–39, 359, 372
    Ardennes 394
    Bavarian Forest *172–73*
    Black Forest 393
    Danube 156–57, 161, 173
    Hainich National Park 393
    Lake Constance 373, *373*
    Rhine *370–71*, 371
    Rhine Rift 346

geysers 277
   *see also* hot springs; hydrothermal
      vents
   El Tatio Geysers 97
   Rotorua 276–77
   Strokkur Geyser 135
   Valley of Geysers 352, *352*
   Yellowstone 31
Ghash Mastan 220
Giant Forest 68–69
Giant's Causeway 152, 164–65
Gibson Desert 298
Giles Corridor, Great Victoria Desert
   299
glacial lakes *101*, 279, *306*
   General Carrera Lake 113
   Lake Abraham 49
Glacier Bay 47, *47*
glaciers
   *see also* fiords; ice-sheets;
      moraines
   Aletsch Glacier 138, 150–51
   Antarctic Ice-sheet 39, 306–307,
      308, 309
   Antarctica *300–301*
   Asia 360–61, *361*
   Baltoro Glacier 225, 232–33
   Bering Glacier 40–41
   Biafo Glacier 234
   Black Rapids Glacier 24, 43
   Columbia Glacier 40
   Fedchenko Glacier 232
   Fox Glacier 361, *361*
   Franz Josef Glacier 280
   Greenland Ice-sheet 38–39
   Jostedalsbreen Ice-field *133*, 149
   Kaskawulsh Glacier 42
   Kennicott Glacier 44–45
   Khumbu Glacier *227*, 234–35
   Kongsvegen Glacier 359
   Malaspina Glacier 42
   Margerie Glacier 47
   Mendenhall Glacier 46–47
   Mer de Glace 150
   Monaco Glacier 148
   Pasterze Glacier 359, *359*
   Pastoruri Glacier 100
   Perito Moreno Glacier 102–103
   retreat, glacial 40, 151, 281
   Rhône Glacier 359
   Ross Ice Shelf *306–307*, 308–309
   Tasman Glacier 280–81
   Yulong Glacier 232
glaze ice 337
global warming 15, 39, 100, 285
Glossglockner 138
Goat Island, Niagara Falls 52
Gobi Desert 224, 266–67, 398
Godwin-Austen Glacier 233
Godwin-Austen, Mount *225*, 233

Gondwana 86, 132, 180, 181, 272,
   303
gorges 195, 246
   *see also* canyons
   Grand Canyon 54–55, 56–57
   Verdon Gorge 155
gours *248*
Graah Glacier *38–39*
graben 160, 236, 238
Grampians 134
Grand Canal 242
Grand Canyon 54–55, 56–57
Grand Prismatic Spring, Yellowstone
   *30–31*
granite formations 26, 98, 99, 188
grasslands
   Central and Eastern Miombo
      Woodlands 398
   Cerrado 398, *398*
   Eastern Steppe 260
   Ethiopian Montane Grasslands
      204–205
   Great Plains 73
   Hortobágy Wetlands 158
   Kalahari Desert 197, 211
   Llanos 104
   Mongolian–Manchurian Steppe
      398
   North Australia Savannas 294
   Pampas 122–23
   Serengeti 206–207
   Sudanian Savannas 204
   Terai-Duar Savannas 260–61
   Western Steppe 175, 260
grass species 73, *122–23*, 123, *260*,
   260
Great Australian Bight 282
Great Barrier Reef 284–85, 286–87
Great Basin Desert 74
Great Bear Lake 362
Great Blue Hole 114–15, 384
Great Britain 133, *133*
   *see also* Ireland
   Caledonian Forest 392
   Cliffs of Dover 133, 167
   Devon Ria Coast 386–87
   Fingal's Cave 152, 166
   Giant's Causeway 152, 164–65
   Jurassic Coast 168–69
   Lake District 152–53
   Loch Ness 370
   Scottish Highlands 132, *132*, 134
   Severn 370
   Thames 370
Great Chagos Bank 315
Great Dike 351
Great Dismal Swamp 364
Great Dividing Range 274–75
Great Glen Fault 134, 370
Great Hungarian Plain 156, 158

Great Lakes 23, 50–51, 238, *336*
Great Plains 73
Great Rift Valley 13, 184–85,
   186–87, 193, 196
Great Sandy Desert 295, 298
Great Smokey Mountains National
   Park *70*
Great Victoria Desert 295, 298, 299
Greece
   Mediterranean Evergreen Forest
      394–95
   Pindus Mountains 348, *348*
   Santorini 348
Greenland
   Ilulissat Icefiord 384, *384*
   North American Tundra 72
   Greenland Ice-sheet 38–39
Green Mountain National Forest 71
Grey Glacier *101*
Grímsvötn 147, 345
Grose Valley *274–75*
Grossglockner 359
Grotte Casteret 372
Gruta do Janelão 369
Guadalupe Mountains 58
Guam Bay 325
Guanabara Bay 98, *98*
Guatemala 88
Guiana Highlands 90–91, 366
Guiana Shield 86, *86*
Guinea
   Sudanian Savannas 204
Gunung Mulu *382*, 382–83
Gurudongmar Lake 380, *380*
Guyana
   Amazon Rainforest 82–83, 106,
      120–121, 190
   Guiana Highlands 90–91, 366
   Kaieteur Falls 366–67, *366*

Haakon VII Land 148
haboob *326–27*, 334
Hadean Era 15
hailstones 330
Hainich National Park 393
Half Dome 26
half-graben basins 238
Hallerbos 172
Hall of Giants, Carlsbad Caverns 58
Ha Long Bay 250–51
hamadas 209
Hamelin Pool 288
Hammersley Range *272*
hanging glacier 232, *232*
Hang Son Doong 244–45

Hawaiian Islands 13, *13*, 316–17,
   318–19
headlands *see* coastlines
Hengduan Mountains 257
Hengifoss 152
High Atlas mountains *181*, 182
high-plateau regions 92, *95*, 96,
   189, 224
Hillman Peak, Crater Lake *28*
Himalayas 94, 219, 222–23, 240
Hispar Glacier 234
Hjeltefiord *162*
Hogenakkal Falls 379, *379*
Hoggar Mountains 349–50
hoodoos 78, *78*, 348, *348–49*
Horizontal Falls 390
Hornocal Mountains 98
Horseshoe Falls (Niagara) 52, *52*
horsetail formation (waterfalls)
   *152*
Hortobágy Wetlands 158
hot springs
   *see also* geysers
   Afar Depression 183, 184
   Baishuitai Hot Springs and
      Terraces 355
   Beppu Hot Springs 354
   El Tatio Geysers 97, *97*
   Pamukkale Hot Springs
      220–21
   Rotorua 276–77
   Terme di Saturnia 347, *347*
   Yellowstone *18–19*, *30–31*, 31,
      *32*
hotspots, volcanic 10, 13, 318
Huangshan Mountains 355
Hudson River 363
Huka Falls 383
Humboldt Current *124*
Hungary
   Danube 156–57, 161, 173
   Hortobágy Wetlands 158
   Slovak Karst Caves 159
Huron, Lake *50*, 51
hurricanes *11*, 328
Hyden Rock 278
hydroelectric energy 242, 363,
   373
hydrothermal vents 322, *322–23*

Iberia 137
ice ages 15, 23, 114, 132–33
icebergs *148*, *307*, 309
   Greenland Ice-sheet *38–39*, 39
   calving of 47, 103

ice-caps
    Kilimanjaro Ice-cap 360
    Quelccaya Ice-cap 100–101
    Vatnajökull 146–47, 369
ice caves *100, 146–47, 149, 151*
    Mount Erebus 305, *305*
    Mendenhall Glacier 46, *46–47*
ice dams 103, *103*
ice divides 39, 307
icefalls 235, *235*
ice-fields
    Jostedalsbreen Ice-field *133,* 149
    Southern Patagonian Ice-field
      101, 103
ice fins *43*
ice floes *161, 237, 306*
Iceland
    Dettifoss 369
    glaciers and ice-sheets *128–29*
    Grímsvötn 345
    Litlanesfoss 152
    Seljalandsfoss 369
    Skógafoss 369, *369*
    Strokkur Geyser 135
    Surtsey 345, *345*
    Vatnajökull 146–47, 369
    weather 338–39
ice-sheets 23, 162
    *see also* glaciers
    Antarctic Ice-Sheet 39, 306–307,
      308, 309
    Greenland Ice-sheet 38–39
ice shelves *306–307,* 308–309
ice storms 336–37
igneous rocks 10
Iguaçu Falls 110–11
Iguaçu River *110*
Ilulissat Icefiord, Greenland 384,
    *384*
India
    Deccan Traps 218, *218,* 353
    Durung Drung 361
    East Himalayan Forest 256
    Ganges 106, 240, 260
    Himalayas 94, 219, 222–23, 240
    Indus 240
    Karakoram Range 225, 234
    Kolahoi Glacier 361
    Mudumalai Forest 396
    rivers and lakes 378–80
    Siachen Glacier 360–61, *360–61*
    Southwestern Ghats Moist Forest
      396, *396–97*
    Sundarbans 240–41
    Terai-Duar Savannas 260–61
    Thar Desert 265
    Tibetan Plateau 355
    Western Himalayan Temperate
      Forest 396
Indo-Gangetic Plain 240

Indonesia
    Bornean Rainforest
      258–59
    Kelimutu 356
    Lake Toba 356
    Mount Merapi 356
    Nusa Tenggara 253
    Tambora 356
    Tengger Volcanic Complex
      230
Indus 240
Indus Valley Desert 403
Inner Hebrides 166
inselbergs 188
interglacial periods 15
Inylchek Glacier 360
Iran
    Dasht-e-Kavir Desert
      403, *403*
    Dasht-e-Lut 263
    Zagros Mountains 220
Iraq
    Arabian Desert 262–63
    Syrian Desert 403
    Zagros Mountains 220
Ireland, Republic of
    *see also* Northern Ireland
    Fingal's Cave 152, 166
    Lough Derg 370
Irrawaddy, Myanmar 380
Irtysh River 237
Ischigualasto 127
Isla Incahuasi 126, *126*
islands
    Dalmatian Coast 171
    El Nido 390
    Galápagos Islands 320–21
    Guam 325
    Ha Long Bay 250–51
    Hawaiian Islands 13, *13,* 316–17,
      318–19
    Krabi Coastline 252–53
    lakebound *28, 51, 239*
    Les Pitons 385
    Lofoten Islands 386
    Long Island 384
    Maldives 315
    Mount Desert Island 384
    Nansei Shoto Archipelago
      Forest 397
    Santorini 348
    Seychelles 314–15
    Surtsey 345
    volcanic island chains 312, *313,*
      315, 318–19
    Yaeyama Islands 389
Israel
    Dead Sea 192, 236
    Red Sea Coast 198–99
Isthmus of Panama 84

Italy
    Alps 136, 138–39, 359, 372
    Amalfi Coast 388
    Dolomites 136
    Mediterranean Evergreen Forest
      394–95
    Mount Etna 140–41
    mountains and volcanoes 347–48
    Po 374
    Stromboli 144
    Turkish Steps 388, *388*
    Vesuvius 145
Itasca, Lake 53

## J

Jagala Waterfall 370, *370*
Jakobshavn Glacier 39
Jalisco Dry Forest 392
Japan
    Akiyoshi-do 380
    Beppu Hot Springs 354
    Mount Fuji 229
    Mount Unzen 354
    Nansei Shoto Archipelago Forest
      397
    Sagano Bamboo Forest 396, *397*
    Sakurajima 354–55, *355*
    Shiraho Reef 251
    Taiheiyo Montane Forest 256
    Yaeyama Islands 389
Jasovská Cave *159*
Jeffreys Bay 389
Jeita Grotto 378, *378*
Jenolan Caves 283
Jiuzhaigou Valley National Park
    *380–81,* 381
Jordan
    Arabian Desert 262–63
    Dead Sea 192, 236
    Red Sea Coast 198–99
    Syrian Desert 403
Joshua trees *74–75,* 75
Jostedalsbreen Ice-field *133,* 149
Juneau Ice-field 46
Jungfraujoch 150
Jura 346
Jurassic Coast 168–69

## K

K2 (mountain) 225, *225,* 233
Kagera river 196
Kaibab Plateau *57*
Kaieteur Falls *366,* 366–67

Kakahi Falls 276
Kalahari Desert 197, 211
Kalkalpen National Park 394,
    *394*
Kamchatka Peninsula 229
Kaokoveld Desert 212
Kapsiki Peak 350
Karakoram Range 225, 234
    *see also* Biafo Glacier
Karakum Desert 264
Karelian Isthmus 160
Karoo 211
karst landscapes 248–49, 252,
    348
    Bungle Bungles 357, *357*
    Ha Long Bay 250–51
    South China karst 246–47,
      *248–49*
Kaskawulsh Glacier 42
Kata Tjuta 358
Kati Thanda-Lake Eyre *268–69,* 271,
    *299*
Katmai, Mount 342
kauri trees 293, *293*
Kazakhstan
    Altai Mountains 175, 224, 237,
      266, *266–67*
    Aral Sea 377
    Inylchek Glacier 360
    Kyzylkum Desert 403
    Lake Balkhash 377, *377*
    Ob–Irtysh 237
    Tian Shan Mountains 224
    Urals 130, 144–45, *218*
    Western Steppe 175, 260
Kelimutu 356
Kennicott Glacier 44–45
Kenya
    East African Coastal Forests
      395
    Lake Bogoria 351
    Lake Magadi *8–9*
    Lake Nakuru 375
    Lake Victoria 196
    Nile 106, 190–91, 196
    Nyiri Desert 401
    Serengeti 206–207
Khumbu Glacier *227,*
    234–35
Kierkfjord *162–63*
Kilauea *316–17, 319*
Kilimanjaro, Mount *187*
Kilimanjaro Ice-cap 360
Kliuchevskoi 229
Klondike Gold Rush 48
Kluane Lake 42
Kolahoi Glacier 361
Kola Peninsula Tundra 175
Kongsvegen Glacier 359
Konkordiaplatz 151

Koonalda Caves 383
kopjes 206
Kornati Archipelago *171*
Kosciuszko, Mount 274
Krabi Coastline 252–53
Krubera Cave 221, 378
Kukenan Tepui 91
Kuwait
    Arabian Desert 262–63
Kyrgystan
    Inylchek Glacier 360
    Tian Shan Mountains 224
Kyzylkum Desert 403

# L

Ladoga, Lake 160–61
Lago Argentino 103
lagoons *100*, 116, *283, 315–16,*
    383
La Guajira Desert 399
Laguna 69 367, *367*
Laguna Roja *97*
Laguna Colorada 368, *368–69*
Lake District 152–53
lakes
    Asia 377–81, *380–81*
    Crater Lake 28–29
    Dead Sea 192, 236
    General Carrera Lake 113
    Great Bear Lake 362
    Great Lakes 23, 50–51, 238, *336*
    in Great Rift Valley 184, *184*
    Kati Thanda-Lake Eyre *268–69,*
        271, *299*
    Laguna 69, 367, *367*
    Laguna Colorada 368, *368–69*
    Laguna Miscanti *96–97*
    Lake Abraham 49
    Lake Baikal 238–39
    Lake Chad 375
    Lake Constance 373, *373*
    Lake District 152–53
    Lake Geneva 155, 359
    Lake Ladoga 160–61
    Lake Nakuru 375
    Lake Natron *340–41*, 375, *375*
    Lake Nicaragua 366
    Lake Nyos 350
    Lake Retba 192
    Lake Seneca 363
    Lake Titicaca 105
    Lakes of Covaonga 373
    Loch Ness 370
    Lough Derg 370
    Manicouagan Lake 363
    Mono Lake *362–63*, 363

lakes *continued*
    Plitvice Lakes 374, *374*
    Spotted Lakes 362, *362*
    Waituna Lagoon 383
Laos
    Mekong 243
Lascaux 372
lateral moraines 42, *44*, 150
Laurentia 22, 132
Lautaro volcano 101
lava
    domes 88, 89, *89*
    flows 140, *140*, 152, *229*
    Hawaiian Islands *316–17, 319*
    lakes 89, *89*, 305, *305*
    pillow lavas *312*
    tube *140*
    in stratovolcanoes 142,
        *142–43*
Lebanon 378, 395
    Forest of the Cedars of God 395,
        *395*
    Great Rift Valley 13, 184–85,
        186–87, 193, 196
    Jeita Grotto 378, *378*
Legend People (hoodoos) 78
Legzira 388
Lena 237, *239*
Lençóis Maranhenses 116–17
Les Pitons 385
Lhotse peak *226, 227*
Libya
    Sahara Desert *181*, 208–209, 211,
        349, 401
Liefdefiord 148
Lighthouse Reef 384, *384–85*
lightning storms 330, *330, 331*
Lisan Peninsula 236
lithosphere 12
Litlanesfoss 152
Little Sandy Desert 403
Llanos 104
Llao Rock *28*
lochs *see* lakes
Lofoten Islands 386, *386–87*
Loire 371
Loktar Lake 380
Lonar Lake 378
Long Island 384
loppoliths 351
Los Gigantes Cliffs 198
Los Glaciares National Park
    103
loughs *see* lakes
lunette dunes *299*
Luxembourg
    Ardennes 394

# M

Machu Picchu *119*
Mackenzie Dikes 342
Madagascar 181
    Madagascan Dry Forest 202–203
    Madagascan Rainforest 201
Magadi, Lake *8–9*
magma 10
    chambers 32–33, *143*, 319
    plugs 353, *353*
magnetic poles 312, 338
magnetite *221*, 263
Makgadikgadi Pan *402*, 403
Malaspina Glacier 42
Malawi
    Lake Malawi 192–93
Malaysia
    Bornean Rainforest 258–59
Maldives 315
Mali
    Sahara Desert *181*, 208–209, 211,
        349, 401
Mammoth Cave 364
mangroves 240–41, *288–89*
Manicouagan Lake 363
Manpupuner Rock Formations 349
mantle 10, *10*
    convection 10, *10*
    plume 33, 186, 187, *319*
maple trees *71*
Marble Cathedral 113, *113*
Margerie Glacier 47
Mariana Trench 324–25
marine life
    *see also* fish life
    aquatic plant life 105, 112, *160*
    in coastal regions *171*, 288, *288*
    deep sea life 322, *322, 323*, 325
    in fiord regions 291
    Great Barrier Reef 284–85, 285,
        *286, 287*
    Red Sea Coast *198–99*, 199
    in reef habitats 251, 253, 285, 315,
        *315*
marine terraces 63
Marmolada 136
Masaya Volcano 89
Massif Central 135, 346
Matterhorn 138, *139*
Matusevich Glacier *300–301*
Maumturk mountains 166
Mauritania
    Adrar Plateau 210
    Banc d'Arguin 388
    Sahara Desert *181*, 208–209, 211,
        349, 401
Mayon Volcano 355

Mazama, Mount 29
    *see also* Crater Lake
McMurdo Dry Valleys 309
meanders, river 48, *53, 57, 65*, 108
medial moraines *42, 44, 45*, 150,
    *150–51*
Mediterranean Evergreen Forest
    394–95
Mediterranean Sea 130, 133, *133*,
    171
Meghna river 240
Mekong 243
Melanesia 272
meltwater 46, *128–29*
    lakes *45*, 280, 281, *281*
    streams *38, 234, 280*
    tunnels carved from *146, 151*
Mendenhall Glacier 46–47
Merapi, Mount 356
Mer de Glace 150
Mesa Arch 79
mesas 77, 79
Mesozoic Era 14
metamorphic rocks 10
Meteor Crater 79
methane gas 49, *49*
Mexico
    Californian Chaparral and
        Woodland 391
    Cave of Crystals 36
    Chihuahuan Desert 81
    El Chichón 344
    Jalisco Dry Forest 392
    Ox Bel Ha 364
    Parícutin 343
    Popocatépetl 37
    Rio Grande 363
    Sac Actun 365
    Sierra Madre 37
    Sierra Madre Pine-Oak Forest
        391
    Sistema Huautla 364–65, *365*
    Sonoran Desert 76
    Volcán de Colima 343, *343*
Michigan, Lake 51
microclimates *119, 155, 245*, 292
Mid-Atlantic Ridge 312–13
Milford Sound 291, *291*
Minas Basin 64
minerals 14–15
mires 59
Mississippi–Missouri 53, 106, 190,
    364
Moeraki Beach 290
Moher, Cliffs of 166
Mojave Desert 74–75
Moldova
    Danube 156–57, 161, 173
Monaco Glacier 148
Monch mountain 150

Mongolia
  Altai Mountains 175, 224, 237, 266, *266–67*
  Eastern Steppe 260
  Gobi Desert 224, 266–67, 398
  Mongolian–Manchurian Steppe 398
  Ob–Irtysh 237
monkey puzzle trees *119*
Mono Lake *362–63, 363*
monoliths *113, 127*
  Devil's Tower 35
  Monument Valley 77
  Sugarloaf Mountain 98
  White Desert 210
Mont Blanc 138, 150, *150*
Monterey Bay 62
Monteverde Cloud Forest 118–19
Montserrat 89
monuments 34, *34*, 76–77, 342–43
Monument Valley 76–77
moraines *42*, 361
  lateral moraines 42, *44*, 150
  medial moraines *42, 44, 45*, 150, *150–51*
Morocco
  Atlas Mountains 178, *181*, 182–83
  Erg Chebbi Sand Dunes *400–401*, 401
  Legzira 388
  Mediterranean Evergreen Forest 394–95
  Sahara Desert *181*, 208–209, 211, 349, 401
mountain climbing 226, *234*, 235, *235*
mountain ranges 94–95
  *see also* mountains
  Africa 178
  Alaska Range 24
  Alps 136, 138–39, 359, 372
  Altai Mountains 175, 224, 237, *266–67*, 266
  Andes 92–93, 94–95, 105, 125
  Appalachians 22, 35, 70
  Asia 352, 355
  Atlas Mountains 178, *181*, 182–83
  Cascade Range 27
  Caucasus Mountains 221
  Coast Ranges 343
  coastal ranges 39, 63
  Dolomites 136
  Drakensberg *180*, 189, 376
  Ethiopian Highlands 350–51, *350–51*
  Europe 346–47
  Great Dividing Range 274–75
  Guiana Highlands 90–91, 366
  Himalayas 94, 219, 222–23, 240
  Hoggar Mountains 349–50

mountain ranges *continued*
  Hornocal Mountains 98
  Karakoram Range 225, 234
  Lake District 152–53
  Pyrenees 137
  Rocky Mountains (Rockies) *22, 23*, 24–25
  Scottish Highlands 132, *132*, 134
  Sierra Madre 37
  Sierra Nevada 342
  Southern Alps 273, *273*, 278–79
  Tian Shan Mountains 224
  Torres del Paine 99
  Transantarctic Mountains 304, *307*
  Urals 130, 144–45, *218*
  Zagros Mountains 220
mountains
  Brandberg Massif 188
  El Capitan and Half Dome 26
  Mount Bromo 230, *230*
  Mount Everest 219, 226–27, 235, 361
  Mount Fuji 229
  Sugarloaf Mountain 98
  Table Mountain 188
Mozambique
  Central and Eastern Miombo Woodlands 398
  Zambezi 195, 376
  East African Coastal Forests 395
  Lake Malawi 193
  Great Rift Valley 13, 184–85, 186–87, 193, 196
mud pools 31, 32, 276, *276*
Mudumalai Forest 396
Mull, Isle of 166
Murchison Falls 191
Murray–Darling 274, 282–83
Myanmar 243, 253, 380
mycorrhiza 174

Narodnaya, Mount 144
native species
  animals 53, *184, 238, 283*
  fish *51*, 192–93
  in forests *172*, 201, *201*, 202, *293*
  plants 68, *91, 106*, 294, *314*
Natron, Lake *340–41*, 375, *375*
Navajo Nation 34, 77, 80
Nazca Plate 86–87, 94, 320
Nepal
  East Himalayan Forest 256
  Khumbu Glacier *227*, 234–35
  Himalayas 94, 219, 222–23, 240
  Mount Everest 219, 226–27, 235, 361
  Terai-Duar Savannas 260–61
  Western Himalayan Temperate Forest 396
Neva river 160
New Zealand 273
  *see also* Australia and New Zealand (continent)
  Fox Glacier 361, *361*
  Franz Josef Glacier 280
  Huka Falls 383
  Moeraki Beach 290
  New Zealand Fiords 291
  Ninety Mile Beach 390
  Rotorua 276–77
  Ruapehu 358, *358*
  Southern Alps 273, *273*, 278–79
  Tasman Glacier 280–81
  Tongariro Volcanic Park 275, 276
  Waimangu 358
  Waipoua Forest 293
  Waitomo Caves 283
  Waituna Lagoon 383
Ngauruhoe 275
Ngami, Lake 197
Ngorongoro Crater *187*
Niagara Falls 51, 52
Nicaragua
  Masaya Volcano 89
  Lake Nicaragua 366
Nicobar Islands 253
Nigardsbreen 149
Niger
  Lake Chad 375
  Sahara Desert *181*, 208–209, 211, 349, 401
Nigeria
  Cameroonian Highlands Forest 395
  Lake Chad 375
Nile 106, 190–91, 196
Ninety Mile Beach 390
North America 20–21, 22–23
  *see also individual countries*

North American Boreal Forest 66
North American Plate 22, 30, 33, 229
North American Tundra 72
North Australian Savannas 294
Northern Ireland 164–65
  *see also* Ireland
Northern Lights 338–39
North Jutland Dune 386
Norway
  Jostedalsbreen Ice-field *133*, 149
  Kongsvegen Glacier 359
  Lofoten Islands 386, *386–87*
  Monaco Glacier 148
  Norwegian Fiords 162–63
nunatak 44
Nuptse peak *226, 227*
Nusa Tenggara 253
Nyiri Desert 401
Nyos, Lake 350

Oakover River *295*
Ob–Irtysh 237
oceanic crust 10, *10*
oceans 14, *94*, 310–11, 324
  East Pacific Rise 322–23
  Mariana Trench 324–25
  Mid-Atlantic Ridge 312–13
ogives 150
Ohio River 53
Ojos del Salado 345
Okanagan, Lake 362, *362*
Okavango Delta 196–97
Okefenokee Swamp 59
Old Faithful geyser, Yellowstone 31
Ol Doinyo Lengai 184, *185*
Oman 262
Ontario, Lake 51
Optimisticheskaya 375
Öræfajökull volcano 147
orchids *121, 188, 259*
Orinoco 366
overthrusts 24
Ox Bel Ha 364
oxbow lakes *48–49*, 49

Pacific Northwest Rainforest 67
Pacific Ocean 87, *124*, 125
Pacific Plate 20, 22, 270, 324, *324*

Pakistan
  Baltoro Glacier 225, 232–33
  Biafo Glacier 234
  Himalayas 94, 219, 222–23, 240
  Indus 240
  Indus Valley Desert 403
  Karakoram Range 225, 234
  Thar Desert 265
  Western Himalayan Temperate
    Forest 396
Painted Grotto, Carlsbad Caverns
  58
Paleozoic Era 14
Palisades 342, 342–43
Palo Duro Canyon 364
Palouse River 23
Pampas 122–23
Pamukkale Hot Springs
  220–21
Panama
  Chocoan Rainforest 392
Panamint Range 75
Pando Aspen Grove 70
Pangaea 22, 132, 180–81, 218
Pantanal 112–13
Papua New Guinea 272, 272
  Rabaul 357
Paracas Peninsula 385
Paraguay
  Pantanal 112–13
  Paraguay River 368–69
  South American Dry Forest
    392
Parícutin 343
Pasterze Glacier 359, 359
Pastoruri Glacier 100
Patagonian Desert 127
Patagonian Shield 86
Patagonian steppe 99
Pearl River 382
peat bogs see bogs
Pelée, Mount 344
peninsulas see coastlines
penitentes 124, 125
Perito Moreno Glacier
  102–103
permafrost 72, 175
Peru
  Altiplano 92, 94, 95, 96–97,
    105
  Amazon Rainforest 82–83, 106,
    120–21, 190
  Amazon (river) 84, 87, 87,
    106–107
  Andean Yungas 119
  Andes 92–93, 94–95, 105,
    125
  Colca River 367
  Cotahuasi River 367
  Laguna 69 367, 367

Peru continued
  Lake Titicaca 105
  Paracas Peninsula 385
  Pastoruri Glacier 100
  Quelccaya Ice-cap
    100–101
  Sechura Desert 399
  Tres Hermanas Falls 367
Phantom Ship, Crater Lake 29
Phi Phi Le 252
Philippine Plate 324, 325
Philippines
  El Nido 390
  Mayon Volcano 355
  Mount Pinatubo 231
  Puerto Princesa Subterranean
    River National Park 382
  Taal Volcano 356, 356
Pico Cão Grande 350
Pico de Teide 349
piedmont glaciers 42
Pieter Botte, Mount 293
pillar formations 58, 145, 214–15,
    237
  Pinnacles, the 298
  sea stacks 170, 170, 290, 290
pillow lavas 312
Pinatubo, Mount 231
Pindus Mountains 348, 348
Pinnacles, the 298
pit craters (volcanoes) 89
Piton de la Fournaise 351
phytoplankton 310–11, 315
plains see plateaus; savannas
plankton 310–11, 315
plant life
  see also trees
  in African deserts 211, 211
  in African forests 200, 200–201,
    202, 202–203
  aquatic plant life 105, 112,
    160
  in Asian deserts 264, 264–65
  in Asian forests 255, 256, 257,
    257, 258, 259
  in Australian deserts 296, 297,
    298, 299
  in European forests 173, 174
  in forest habitats 120, 121
  forests of Australia and New
    Zealand 292, 292–93, 293
  in glacial conditions 41
  in grasslands 122–23, 123, 207
  in mountainous regions 188, 189,
    189, 205, 205, 229
  in North American deserts 74,
    74–75, 76, 81
  in North American forests 67, 67,
    70
  in river habitats 156–57, 190

plant life continued
  in swamplands 59, 59, 241
  in tundra regions 175, 175, 309, 309
  in wetland habitats 60, 112,
    158–59
plateaus 109
  Adrar Plateau 210
  Altiplano 92, 94, 95, 96–97, 105
  Atacama Desert 92, 94–95, 97,
    124–25, 392
  in Sahara Desert 209, 209
  Salar de Uyuni 96, 126
  Tibetan Plateau 355
plate tectonics 12–13, 134, 223
  African Plate 133, 183, 186, 199
  continental rifting 12, 186–87,
    322, 346
  Eurasian Plate 133, 218, 273
  formation of mountain chains 94,
    94–95
  formation of volcanic island
    chains 318–19
  Nazca Plate 86–87, 94, 320
  North American Plate 22, 30, 33,
    229
  Pacific Plate 20, 22, 270, 324, 324
  San Andreas Fault 23, 23, 30
  thrust faults 24, 94, 95
Plitvice Lakes 374, 374
plug, volcanic 34
plutons 26
Po 374
Pohutu Geyser 277
Poland
  Białowieża Forest 174
  Biebrza Marshes 158–59
  Crooked Forest 393, 393
Popocatépetl 37
Port Campbell National Park 290
Portugal
  Algarve Coast 170
  Tagus 373
pothole arches 79, 79
Praia da Roche 170
prairies 73
  see also grasslands
President, The, Giant Forest 68–69
Prince William Sound 40
prokaryotes 15, 16
Proterozoic Era 15
Puerto Princesa Subterranean River
    National Park 382
Puget Sound 384
Pukaki, Lake 279
Pulpit Rock 162
Puy de Côme 135
Pyrenees 137
pyroclastic flows 89, 89, 231, 344
  see also cinder cones

Q

Qatar 262
Qinghai Hu 380–81
Qinling Mountain 257, 257
Quelccaya Ice-cap 100–101
Queen Maud Mountains 304

R

Rabaul 357
Rainbow Falls, Victoria Falls 195
rainfall
  in desert regions 76, 125, 211,
    401
  Asian 263, 264, 265, 266, 403
  Australian 295, 296, 299
  in forests 118, 120, 201, 392, 396
  in mountainous regions 96, 279
rainforests
  Amazon Rainforest 82–83, 106,
    120–21, 190
  Bornean Rainforest 258–59
  Chocoan Rainforest 392
  Congo Rainforest 200–201
  Daintree Rainforest 292–93
  Madagascan Rainforest 201
  Monteverde Cloud Forest 118–19
  Pacific Northwest Rainforest 67
  Valdivian Temperate Forest 119
  Xishuangbanna Rainforest 397
rain shadows 93
Ranier, Mount 27, 27
Rao Thuong River 245
Red Beach 380
Red River of the South 364
Red Sea Coast 198–99
Redwood National Forest 391
redwood trees 68, 391
reedbeds 154, 159
reefs
  Andaman Sea Reefs 253
  Dragon Hole 251
  Great Barrier Reef 284–85,
    286–87
  Great Blue Hole 114–15, 384
  Great Chagos Bank 315
  Lighthouse Reef 384, 384–85
  Maldives 315
  Nusa Tenggara 253
  Red Sea Coast 198–99
  Shiraho Reef 251
Republic of the Congo
  Congo Rainforest 200–201
Retba, Lake 192
retreat, glacial 40, 151, 281

Rhine 370–71, 371, 373
Rhine Rift 346
Rhône (river) 371, 372
Rhône Delta 154, 155
Rhône Glacier 359, 372
ria landscapes 386–87
Richat Structure 209, 210
ridge systems (oceanic)
    East Pacific Rise 322–23
    Mid-Atlantic Ridge 312–13
rifting, continental 12, 186–87, 322,
    346
    see also Great Rift Valley
"Ring of Fire" volcanoes 27
Rio Grande 363
Rio Negro 116
river canyons 56–57
rivers
    Amazon 84, 87, 87, 106–107
    Arkansas River 364, 364
    Asia 377, 380, 382
    basins 48, 87, 197, 238, 238
    Blyde River Canyon 180, 376, 376
    Caño Cristales 105
    Colca River 367
    Columbia River 362–63
    Congo 192
    Cotahuasi River 367
    Danube 156–57, 161, 173
    deltas, river 15, 340–41
    Europe 374, 373
    Ganges 106, 240, 260
    Hudson River 363
    Lena 237, 239
    Loire 371
    Mekong 243
    Mississippi–Missouri 53, 106,
        190, 364
    Murray–Darling 274, 282–83
    Nile 106, 190–91, 196
    Ob–Irtysh 237
    Orinoco 366
    Paraguay River 368–69
    Red River of the South 364
    Rhine 370–71, 371
    Rhône 372
    Rio Grande 363
    Seine 371
    Severn 370
    systems 108
    Thames 370
    Volga 161
    Yangtze 190, 242
    Yellow River 242–43
    Yukon 48–49
    Zambezi 195, 376
rock and stone formations
    Bungle Bungles 357, 357
    Devil's Tower 35
    Ha Long Bay 250–51

rock and stone formations continued
    hoodoos 78, 78, 348,
        348–49
    Hyden Rock 278
    karst landscapes 248–49, 252,
        348
    Moeraki Beach 290
    Monument Valley 76–77
    pillar formations 58, 145, 214–15,
        237
    Pinnacles, the 298
    sea stacks 170, 170, 290, 290
    South China karst 246–47,
        248–49
    Valle de la Luna 127
rock cycle 11, 11
rock layers (strata) 14, 56, 57,
    228
Rocky Mountains (Rockies) 22, 23,
    24–25
Rodinia 22
Romania
    Białowieża Forest 174
    Bigar Waterfall 374–75
    Black Sea 375
    Danube 156–57, 161, 173
Rongbuk Glacier 361
Root Glacier 44, 45
root systems 70, 258
Roraima, Mount 91, 91
Rosetta river 191
Ross Ice Shelf 306–307, 308–309
Ross Sea 304, 304
Rotorua 276–77
Royal Chitwan National Park
    260–61
Ruapehu 275, 358, 358
Ruhuhu River 193
Russell Glacier 38
Russia
    Altai Mountains 175, 224, 237,
        266, 266–67
    Amur 377
    Black Sea 375
    Caucasus Mountains 221
    Eastern Steppe 260
    Kliuchevskoi 229
    Kola Peninsula Tundra 175
    Lake Baikal 238–39
    Lake Ladoga 160–61
    Lena 237, 239
    Manpupuner Rock Formations
        349
    Mongolian–Manchurian Steppe
        398
    Ob–Irtysh 237
    Russian Far East Temperate Forest
        395
    Siberian Taiga 254–55
    Siberian Traps 352

Russia continued
    Siberian Tundra 261
    Urals 130, 144–45, 218
    Valley of Geysers 352, 352
    Volga 161
    Western Steppe 175, 260
Rwanda
    Congo (river) 106, 192,
        200
    Nile 106, 190–91,
        196

# S

Sac Actun 365
Sagano Bamboo Forest 396,
    397
Sahara Desert 181, 208–209, 211,
    349, 401
"sailing stones" 74, 75
Sainte-Croix, Lake 155
Sakurajima 354–55, 355
Salar de Arizaro 400
Salar de Uyuni 96, 126
Salinas Grandes 400
saline lakes see salt lakes
salt crusts 126
salt deserts 263, 400
salt flats
    Makgadikgadi Pan 402, 403
    Great Basin Desert 74
    Salar de Arizano 400
    Salar de Uyuni 96, 126
    Salinas Grandes 400
salt lakes 95, 183, 184, 299
    Dead Sea 192, 236
    Lake Natron 340–41, 375, 375
    Lake Retba 192
    Sambhar Salt Lake 378
    Spotted Lake 362, 362
salt pans see salt flats
Sambhar Salt Lake 378
San Andreas Fault 23, 23,
    30
sand dunes 81, 265, 297, 299
    Arabian Desert 263
    Atacama Desert 124–25
    Dune du Pilat 167
    Erg Chebbi Sand Dunes 400–401,
        401
    Great Sandy Desert 295
    Lençóis Maranhenses 116–17
    Namib Desert 176–77, 212,
        212–13
    North Jutland Dunes 386
    Pinnacles 298
    Taklamakan Desert 264–65

sand seas 212–13
sandstorms 264, 326–27,
    334–35
Santa María 88
Santiago Crater 89
Santiaguito Complex 88
Santorini 348
São Tomé and Príncipe 350
Saudi Arabia
    Arabian Desert 262–63
    Arabian Highlands 352
    Red Sea Coast 198–99
    Syrian Desert 403
savannas
    see also grasslands; steppes
    Kalahari Desert 197, 211
    Llanos 104
    North Australia Savannas 294
    Serengeti 206–207
    Sudanian Savannas 204
    Terai-Duar Savannas 260–61
Scafell Pike, England 153
Schreckhorn 138
Scotland
    Caledonian Forest 392
    Fingal's Cave 152, 166
    Loch Ness 370
    Scottish Highlands 132, 132, 134
sea arches 169, 169, 170
sea caves 64–65, 166, 170
sea levels 15, 133
sea life
    see also fish life
    aquatic plant life 105, 112, 160
    in coastal regions 171, 288, 288
    deep sea life 322, 322, 323, 325
    in fiord regions 291
    Great Barrier Reef 284–85, 285,
        286, 287
    in reef habitats 251, 253, 285, 315,
        315
    Red Sea Coast 198–99, 199
sea stacks 170, 170, 290, 290
Sechura Desert 399
sedimentary rocks 10, 14
    strata 14, 56, 57, 228
seiches 155
seif dunes 265
Seine 371
Selenga River 239
Seljalandsfoss 369
Semeru, Mount 230, 230
Seneca, Lake 363
Senegal 192
Seongsan Ilchulbong 353
Sequoia National Park 68
sequoia trees 68
seracs 233, 233, 235
Serbia
    Danube 156–57, 161, 173

Serengeti 206–207
Serrania de la Macarena 105
Severn 370
Seychelles 314–15
shamals (desert winds) 262–63
Shark Bay 288–89
shield volcanoes 142, 318
shingle banks 169, 169
Ship Rock 34
Shiraho Reef 251
Shire River 193
shrinking glaciers 40, 151, 281
Siachen Glacier 360–61, 360–61
Siberia see Russia
Siberian Platform 218
Siberian Taiga 254–55, 261
Siberian Traps 352
Siberian Tundra 261
Sichuan Basin 257
Sierra Madre 37, 81
Sierra Madre Pine-Oak Forest 391
Sierra Negra volcano 321
Sierra Nevada (Spain) 347
Sierra Nevada (USA) 342
Sigiriya 353, 353
Simien Mountains 204, 204–205
Simpson Desert 295, 296–97
sinkholes 29, 246, 248, 249, 251
    see also dolines
    Dragon Hole 251
    Great Blue Hole 114–15, 384
    Lighthouse Reef 384, 385
    Xiaozhai Tiankeng 381
Sistema Huautla 364–65, 365
Skeleton Coast 213
Skógafoss 369, 369
Skye, Isle of 134
Slovakia
    Carpathian Mountain Forests 394
    Danube 156–57, 161, 173
    Slovak Karst Caves 159
Slovak Karst Caves 159
Slovenia
    Dinaric Alps 348
Smólikas, Mount 348, 348
snow forests 66, 254
snow formations see penitentes
Snow Lake 234
Socotra 388, 388–89
Sognefiord 162
Solfatara 348
Somalia
    East African Coastal Forests 395
Somma, Mount 145
Sonoran Desert 76
Soufrière Hills 89
South Africa
    Blyde River Canyon 180, 376, 376
    Cango Caves 376–77
    Cape Floral Region 205, 389

South Africa continued
    Cape Point 389
    Coffee Bay 388
    Drakensberg 180, 189, 376
    Jeffreys Bay 389
    Kalahari Desert 197, 211
    Karoo 211
    Namib Desert 176–77, 212–13
    Table Mountain 188
South America 84–85, 86–87
    see also individual countries
South American Dry Forest 392
South American Plate 95
South China karst 246–47, 248–49
South Sudan
    see also Sudan
    Nile 106, 190–91, 196
Southern Alps 273, 273, 278–79
Southern Lights 338–39
Southern Patagonian Ice-field 101, 103
South Korea
    Seongsan Ilchulbong 353
Southwestern Ghats Moist Forest 396, 396–97
Spain
    Cape de Creus 387
    Cathedral Beach 387
    Cordillera Cantábrica 346
    Doñana 373
    Galician Coast 387
    Grotte Casteret 372
    Lakes of Covadonga 373
    Los Gigantes Cliffs 198
    Mediterranean Evergreen Forest 394–95
    Pyrenees 137
    Sierra Nevada 347
    Tabernas Desert 400–401
    Tagus 373
spits see coastlines
Spotted Lake 362
spruce trees 66, 172, 173, 174
Sri Lanka 353, 353, 396
Sri Lankan Moist Forest 396
stalactites 58, 58, 114, 252–53, 283
stalagmites 58, 245, 283, 369
Stara Baska 171
St Elias Mountains 42, 47
steppes
    Eastern Steppe 260
    Mongolian–Manchurian Steppe 398
    Western Steppe 175
Stirling Falls 291
St Helens, Mount 27, 27
St Lawrence River 51
St Martins, Bay of Fundy 64–65

St Marys river 59
Stone Desert 401
strangler fig 258
strata 14, 56, 57, 228
stratovolcanoes 29, 142–43
Strokkur Geyser 135
stromatolites 16, 288, 288
Stromboli 144
subducting plates 324, 324–25
Sudan
    Nile 106, 190–191, 196
    Red Sea Coast 198–99
    Sahara Desert 181, 208–209, 211, 349, 401
    Sudanian Savannas 204
Sudanian Savannas 204
Sugarloaf Mountain 98
Sunburst Peak 24–25
Sundarbans 240–41
supercells (thunderstorms) 330, 330–31, 331
supercontinents 12
Superior, Lake 51, 51
supervolcanoes 32–33
Suriname 391
    Amazon Rainforest 82–83, 106, 120–21, 190
Surtsey 345, 345
Suwannee river 59
Stanley, Mount 187
Svalbard see Norway
swamps
    Everglades 60–61
    Great Dismal Swamp 364
    Okefenokee Swamp 59
    Sundarbans 241
Switzerland
    Aletsch Glacier 138, 150–51
    Alps 136, 138–39, 359, 372
    Lake Constance 373, 373
    Lake Geneva 155, 359
    Jura 346
    Rhine 370–71, 371
    Rhône (river) 372
    Rhône Glacier 359
Syria
    Syrian Desert 403

T

Taal Volcano 356, 356
Tabernas Desert 400–401
Table Mountain 188
Tagus 373
taiga 255
    see also boreal forest
Taiheiyo Montane Forest 256

Tajikistan
    Fedchenko Glacier 232
    Karakoram Range 225, 234
Taklamakan Desert 264–65
Talampaya National Park 400
Tambora 356
Tana, Lake 191
Tanami Desert 295
Tanganyika, Lake 186
Tanggula Mountains 242
Tanzania
    Central and Eastern Miombo Woodlands 398
    Congo (river) 106, 192, 200
    East African Coastal Forests 395
    Great Rift Valley 13, 184–85, 186–87, 193, 196
    Kilimanjaro Ice-cap 360
    Lake Malawi 193
    Lake Natron 340–41, 375, 375
    Nile 106, 190–91, 196
    Serengeti 206–207
    Lake Victoria 196
Tarim Basin 264
Tarn Gorge 135
tarns 153
Tasman Glacier 280–81
Tassili du Hoggar 209
Taupo, Lake 276
Tavurvur 272
tectonic plates 12–13, 134, 223
    African Plate 133, 183, 186, 199
    continental rifting 12, 186–87, 322, 346
    Eurasian Plate 133, 218, 273
    faults 13, 24
    formation of mountain chains 94, 94–95
    formation of volcanic island chains 318–19
    Nazca Plate 86–87, 94, 320
    North American Plate 22, 30, 33, 229
    Pacific Plate 20, 22, 270, 324, 324
    San Andreas Fault 23, 23, 30
    thrust faults 24, 94, 95
Tekapo, Lake 279
Telescope Peak 75
Tell Atlas mountains 182
temperate climates see climates
Tenerife
    Los Gigantes Cliffs 198
    Pico de Teide 349
Tengger Volcanic Complex 230
Teno Mountains 198
tepuis 91
Terai-Duar Savannas 260–61

Terme di Saturnia 347, *347*
terraces *see* hot springs
Teton Range *25*
Thailand
    Andaman Sea Reefs 253
    Krabi Coastline 252–53
    Mekong 243
Thames 370
Thar Desert 265
Thor's Well 62
Three Gorges Dam 242
thrust faults 24, 94, *95*
thunderstorms *73*, *93*, 330–31
Tian Shan Mountains 224
Tianzi Mountains *214–15*, *354–55*,
    355
Tibetan Plateau 216, *219*, 222, 242,
    355
tidal ranges 64
tidewater glaciers 47
Tien Shan Mountains *see* Tian Shan
    Mountains
Tierra del Fuego 116, 117
Titicaca, Lake 105
Toba, Lake 356
Tongariro Volcanic Park 275, 276
Tongass National Forest 391, *391*
Tornado Alley 332
tornadoes *73*, 330, *330*, 332–33
Torres del Paine 99
Toukbai, Mount *182–83*
trade winds 119
Trango Towers 225, *225*
Transantarctic Mountains 304,
    *307*
transform boundaries 13, *13*
travertine 220, 221
trees 68, 120–21, 255
    acacia trees 203, 206, *206–207*
    conifer trees 66, 67, 68, *69*, 119,
        255, *255*, 293, *293*
    eucalypt trees 292
    Joshua trees *74–75*, 75
    kauri trees 293, *293*
    maple trees *71*
    Pando Aspen Grove 70
    redwood trees 68, 391
    spruce trees *66*, *172*, 173,
        174
Tres Hermanas Falls 367
tributaries, river 108
tropical climates *see* climates
tsunamis 155
tufa *362–63*, 363
Tularosa Basin *81*
tundra
    Antarctic Tundra 309
    Kola Peninsula Tundra 175
    North American Tundra 72
    Siberian Tundra 261

Tunisia
    Atlas Mountains 178, *181*, 182–83
    Mediterranean Evergreen Forest
        394–95
    Sahara Desert *181*, 208–209, 211,
        349, 401
Turkey
    Black Sea 375
    Cappadocia 348–49, *348–49*
    Pamukkale Hot Springs 220–21
    Zagros Mountains 220
Turkish Steps 388, *388*
Turkmenistan 264, 403
Twelve Apostles 290
typhoons 328


# U

UAE *see* United Arab Emirates
Uganda
    Lake Victoria 196
    Nile 106, 190–91, 196
Ukraine
    Black Sea 375
    Carpathian Mountain Forests
        394
    Danube 156–57, 161, 173
    Optimisticheskaya 375
    Western Steppe 175, 260
Uluru *296–97*
United Arab Emirates
    Arabian Desert 262–63
United States of America
    Acadia Coastline 65
    Alaska Range 24
    Antelope Canyon 80
    Appalachians 22, *35*, 70
    Bering Glacier 40–41
    Big Sur 63
    Black Rapids Glacier 24, 43
    Bryce Canyon 78
    Carlsbad Caverns 58
    Cherokee National Forest 70
    Chihuahuan Desert 81
    Columbia Glacier 40
    Crater Lake 28–29
    Devil's Tower 35
    Dungeness Spit 384
    El Capitan and Half Dome
        26
    Everglades 60–61
    forests 391
    Giant Forest 68–69
    Grand Canyon 54–55, 56–57
    Great Basin Desert 74
    Great Lakes 23, 50–51, 238, 336
    Great Plains 73

United States of America *continued*
    Green Mountain National Forest
        71
    Hawaiian Islands 13, *13*, 316–17,
        318–19
    Kennicott Glacier 44–45
    Long Island 384
    Malaspina Glacier 42
    Margerie Glacier 47
    Mendenhall Glacier 46–47
    Mesa Arch 79
    Meteor Crater 79
    Mississippi–Missouri 53, 106,
        190, 364
    Mojave Desert 74–75
    Monterey Bay 62
    Monument Valley 76–77
    Mount Desert Island 384
    mountains and volcanoes
        342–43
    Niagara Falls 51, 52
    North American Boreal Forest
        66
    North American Tundra 72
    Okefenokee Swamp 59
    Pacific Northwest Rainforest
        67
    Pando Aspen Grove 70
    Puget Sound 384
    rivers and lakes 362–64
    Rocky Mountains (Rockies) 22,
        23, 24–25
    San Andreas Fault 23, *23*, 30
    Ship Rock 34
    Sonoran Desert 76
    Thor's Well 62
    weather 332–33, 334–35, 336–37
    Yellowstone National Park *18–19*,
        25, 30–31, 32–33
    Yukon 48–49
Unzen, Mount 354
Upper Yangtze Forests 257
Urals 130, 144–45, *218*
Urique Canyon 37
USA *see* United States of America
Uzbekistan
    Aral Sea 377, 403
    Kyzylkum Desert 403


# V

Valaam Island, Lake Ladoga 160
Valdivian Temperate Forest 119
Valdai Hills 161
Valle de la Luna 127
Valley of Geysers 352, *352*
Vatnajökull 146–47, 369

vegetation
    *see also* trees
    aquatic plant life 105, 112, *160*
    in African deserts 211, *211*
    in African forests 200, *200–201*,
        202, *202–203*
    in Asian deserts 264, *264–65*
    in Asian forests 255, 256, 257, *257*,
        258, *259*
    in Australian deserts 296, *297*,
        *298*, 299
    forests of Australia and New
        Zealand 292, *292–93*, 293
    in European forests 173, 174
    in glacial conditions *41*
    in grasslands *122–23*, 123, 207
    in mountainous regions *188*, 189,
        *189*, 205, *205*, 229
    in North American deserts 74,
        *74–75*, *76*, *81*
    in North American forests 67, *67*,
        70
    in river habitats *156–57*, *190*
    in swamplands 59, *59*, 241
    in tundra regions 175, *175*, *309*,
        309
    in wetland habitats 60, 112,
        *158–59*
Venezuela
    Amazon Rainforest 82–83,
        106, 120–21, 190
    Andes 92–93, 94–95, 105,
        125
    Guiana Highlands 90–91,
        366
    Llanos 104
    Orinoco 366
Vercors Caves 372, *372*
Verdon Gorge 155
Vesuvius 145
Victoria, Lake 186, 191,
    196
Victoria Falls 194–95
Victory Peak 224
Vietnam
    Dragon Hole 251
    Ha Long Bay 250–51
    Hang Son Doong 244–45
    Mekong 243
Vitus Lake *40*, 41
Volcán de Colima 343, *343*
volcanoes
    *see also* volcanic activity
    Afar Depression 183, 184
    Asia 354–57, 355, 356
    Cascade Range 27
    Central and South America
        344, 344–45
    Chaîne des Puys 346, *346*
    Devil's Tower 35

volcanoes *continued*
  Europe 348–49, *348–49*
  extinct 92, 95, 134, 135
  Galápagos Islands 320–21
  Great Rift Valley 184
  Grímsvötn 345
  Hawaiian Islands 13, 13, 316–17, 318–19
  island chains 312, 313, 315, 318–19
  Kapsiki Peak 350
  Kliuchevskoi 229
  Masaya Volcano 89
  Mount Erebus 304–305
  Mount Etna *140–41*
  Mount Fuji 229
  Mount Pinatubo *231*
  mud volcanoes 352
  North America 342–44, *343*
  Pico Cão Grande 350
  Pico de Teide 349
  Piton de la Fournaise 351
  Popocatépetl 37
  Ruapehu 358, *358*
  Santa María 88
  Ship Rock 34
  Soufrière Hills 89
  Stromboli *144*
  supervolcanoes 32–33
  Surtsey 345, *345*
  Tengger Volcanic Complex *230*
  Tongariro Volcanic Park 275, 276
  Vatnajökull 146–47, 369
  Vesuvius 145
  Waimangu 358
volcanic activity 10, 13
  *see also* volcanoes
  ash clouds *141*, *143*, 345
  in Asia 229, 230, 231
  in Central and South America 88, *88*, 89, 344–45, *344*
  eruptions 29, 142, *143*
  in Europe 140, 144, 145, 345
  lakes 356, 385
  lava domes 88, 89, *89*
  lava flows 140, *140*, 152, *229*
  lava lakes 89, *89*, 305, *305*
  lava tubes *140*
  lava in stratovolcanoes 142, *142–43*
  in North America 37, 316, 342, 343, *343*, 344
  pillow lavas *312*
  pyroclastic flows 89, *89*, 231, 344

volcanic activity *continued*
  in rift faults 186, *186*, 187
  undersea *324*
  Volcanic Explosivity Index 32
Volga 161
Vulcano 348

# W

Wadi Rum (Valley of the Moon) *263*
Waimangu 358
Waipoua Forest 293
Waitomo Caves 283
Waituna Lagoon 383
Wales
  Severn 370
Warrumbungles 274, *274*
Wastwater 153
waterfalls 111
  Bigar Waterfall 374–75
  Dettifoss 369
  Drakensberg *189*
  Dudhsagar Falls 379
  Guiana Highlands 90–91, 366
  Hogenakkal Falls 379, *379*
  Huka Falls 383
  Iguaçu Falls 110–11
  Jagala Waterfall 370, *370*
  Kaieteur Falls 366–67, *366*
  Litlanesfoss 152
  New Zealand Fiords 291, *291*
  Niagara Falls 51, 52
  Plitvice Lakes 374, *374*
  Pyrenees 137
  Seljalandsfoss 369
  Skógafoss 369, *369*
  Tres Hermanas Falls 367
  Victoria Falls 194–95
watershed 108, *108*
waterspouts *333*
weather 11
  aurorae 338–39
  cyclones 328–29, *330*
  ice storms 336–37
  sandstorms and duststorms 264, *326–27*, 334–35
  tornadoes 73, 330, *330*, 332–33
  thunderstorms 73, *93*, 330–31
wells, natural 62
Western Himalayan Temperate Forest 396
Western Interior Seaway 23

Western Steppe 175, 260
wetlands
  Biebrza Marshes 158–59
  Camargue 154
  Everglades 60–61
  Great Dismal Swamp 364
  Hortobágy Wetlands 158
  Okefenokee Swamp 59
  Pantanal 112–13
Wet Tropics World Heritage Site 292
White Desert 210
Whitehaven Beach 390, *390*
White Mountains *35*
White Nile 191, 196
Whitney, Mount 342
wildlife
  *see also* bird life; fish life; marine life
  in African deserts *209*, *211*, *212*, 213
  in African forests 200–201, 200
  in African grasslands 204, *204*, 206, *206*
  in Asian deserts 262, *262*, 264, 265, *266*
  in Asian forests *202*, *203*, *257*, 257, *258*, *259*
  in Asian grasslands 260, *260–61*
  in Australian deserts 295, 298, *298*, 299, *299*
  in deserts of the Americas 77, 125
  in forests of the Americas 66, *66*, 70, *70*, 71, *118* 125, 127
  in glacial regions 46, *46*, *147*, 148, *148*, 234
  in grasslands of the Americas 73, *123*, 123
  in mountainous regions *54*, *92*, *183*, *222*, 224
  in polar regions *39*, 304, *308*
  in river habitats 116, *156*, *194*, *195*, *282*
  in tundra *72*, *72*, 261, *261*
  in wetland habitats 60, *61*, 154, 158, 241
Windermere 153
winds, desert 209, *262–63*, 263
Windsor Caves 366
Wizard Island, Crater Lake *28*
woodland *156–57*, 174, 204
  Californian Chaparral and Woodland 391
  Central and Eastern Miombo Woodlands 398
  Hallerbos 172
Wrangell-St Elias National Park 44

# X

Xiaozhai Tiankeng 381
Xishuangbanna Rainforest 397

# Y

Yacoraite formation 98
Yaeyama Islands 389
Yangtze 190, 242
  *see also* Upper Yangtze Forests
Yarlung Tsangpo Grand Canyon 379
Yellow River 242–43
Yellowstone National Park *18–19*, 25, 30–31, 32–33
Yemen
  Arabian Desert 262–63
  Arabian Highlands 352
  Red Sea Coast 198–99
  Socotra 388, *388–89*
Yosemite National Park 26, *26*
Yukon Flats 48
Yukon 48–49
Yulong Glacier 232

# Z

Zagros Mountains 220
Zambezi River 195, 376
Zambia
  Zambezi 195, 376
  Central and Eastern Miombo Woodlands 398
  Congo (river) 106, 192, 200
  Victoria Falls 194–95
Zanskar river *240*
Zealandia 273, 275
Zhangye Danxia 228
Zimbabwe
  Great Dike 351
  Victoria Falls 194–95
  Zambezi 195, 376

# ACKNOWLEDGMENTS

**Dorling Kindersley** would like to thank the following people for their work on this book: Professor Simon Lamb, Dr Tony Waltham, and Tim Harris for fact-checking; Katie John for proof-reading; Rob Houston for editorial help; Gregory McCarthy, Duncan Turner, and Adam Spratley for design assistance; Steve Crozier for Photoshop retouching; and Simon Mumford for cartographic work and advice.

**DK India** would like to thank Aisvarya Misra and Nisha Shaw for editorial assistance; Himshikha, Konica Juneja, Avinash Kumar, and Anjali Sachar for design assistance; and Deepak Negi for picture research assistance.

Data for the geological maps on the introductions to the continents is taken from the Generalized Geological Map of the World, published by the Geological Survey of Canada, and contains information licensed under the Open Government Licence – Canada.

Data and imagery for artworks has been used from the following sources: **pp.28-29, Crater Lake** USGS/USDA: NAIP Digital Ortho Photo Image and USGS NED; **pp.32–33, Yellowstone** Landsat 8 and USGS SRTM; **pp.44-45, the Kennicott Glacier** modified Copernicus Sentinel data (2016) and USGS NED; **pp.94-95, the Andes** Landsat 8 and USGS SRTM; **pp.108-109, the Amazon Basin** Blue Marble/NASA Earth Observatory and ETOPO1–NOAA; **pp.186-87, the Great Rift Valley** Landsat 8 and USGS SRTM; **pp.226–27, Mount Everest** NASA Earth Observatory image by Jesse Allen and Robert Simmon, using EO-1 ALI data from the NASA EO-1 team, archived on the USGS Earth Explorer and ASTER GDEM (a product of NASA and METI); **pp.238–39 Lake Baikal** Landsat 8 and USGS SRTM; **pp.286-87, the Great Barrier Reef** Landsat 8 and Deepreef Explorer, Beaman, R.J., 2010. Project 3DGBR: a high-resolution depth model for the Great Barrier Reef and Coral Sea. Marine and Tropical Sciences Research Facility (MTSRF) Project 2.5i.1a Final Report, MTSRF, Cairns, Australia, pp.13 plus Appendix 1; **pp.306-307, the Antarctic Ice-sheet** British Antarctic Survey BedMap2 (www.bas.ac.uk/project/bedmap-2/); **pp.312–13, the Mid-Atlantic Ridge** Blue Marble/NASA's Earth Observatory and ETOPO1– NOAA; **pp.318-319, the Hawaiian Islands** Landsat 8 and Hawaii Mapping Research Group/School of Ocean and Earth Science and Technology - Main Hawaiian Islands Multibeam Bathymetry and Backscatter Synthesis: University of Hawai'i at Manoa; **pp.324–25, the Mariana Trench** Landsat 8 and Bathymetric Digital Elevation Model of the Mariana Trench - NOAA: National Geophysical Data Center (NGDC).

The publisher would like to thank the following for their kind permission to reproduce their photographs:

(Key: a-above; b-below/bottom; c-centre; f-far; l-left; r-right; t-top)

**Endpapers:** AirPano Images
**1 Getty Images:** DigitalGlobe. **2-3 Peter Franc. 4 Alamy Stock Photo:** National Geographic Creative (fcla). **Getty Images:** Mike Lanzetta (ca); G & M Therin-Weise (fcra). **Imagelibrary India Pvt Ltd:** Jinhu Wang (cla). **Philip Klinger (philip-klinger. photography):** (cra). **5 Getty Images:** Sue Flood / Oxford Scientific (ca); Buena Vista Images / DigitalVision (fcla); Wil Meinderts / Buiten-beeld / Minden Pictures (cra); James D. Morgan (fcra). **Matt Hutton:** (cla). **6-7 Alamy Stock Photo:** Steven Sandner. **8-9 Getty Images:** Yann Arthus-Bertrand. **11 Getty Images:** Danita Delimont (t); NOAA (b). **12 naturepl.com:** Alex Mustard (br). **13 Getty Images:** Education Images (bc); Mark Hannaford (bl); Douglas Peebles (br). **14 NASA:** JPL–Caltech (cl). **J. W. Valley, University of Wisconsin-Madison:** (tr). **14-15 Getty Images:** John Lund / Tom Penpark. **15 NASA:** (cr). **16 Alamy Stock Photo:** Science History Images (cra). **Allen Nutman, University of Wollongong:** (cl). **Science Photo Library:** (ca). **17 Alamy Stock Photo:** Life On White. **18-19 Getty Images:** Peter Adams. **22 Getty Images:** Ron Garnett (tc); Mike Grandmaison (tl). **22-23 Zack Frank:** (t). **23 Alamy Stock Photo:** Kip Evans (cra). **Science Photo Library:** W.K. Fletcher (ca). **24-25 Imagelibrary India Pvt Ltd:** Victor Aerden (t). **24 Carl Battreall / photographalaska.com:** (bl). **25 Copyright Tom Lussier Photography 2017:** (br). **26 Dorling Kindersley:** National Birds of Prey Centre, Gloucestershire (bc). **Imagelibrary India Pvt Ltd:** Mike Wilson (t). **27 Alamy Stock Photo:** Image Source (clb). **Science Photo Library:** USDA / Science Source (cr). **28 Dreamstime.com:** Maria Luisa Lopez Estivill (tl). **Getty Images:** Thomas Winz / Lonely Planet Images (bl). **29 123RF.com:** William Perry (tl). **Alamy Stock Photo:** Gerhard Zwerger-Schoner / Imagebroker (bl). **30 Getty Images:** Kevin Schafer (clb). **30-31 Alamy Stock Photo:** Christian Handl / Imagebroker (t). **National Geographic Creative:** Michael Nichols (b). **31 Brett Lange:** (bc). **32 Alamy Stock Photo:** robertharding (cra). **Getty Images:** Babak Tafreshi / National Geographic (cla, bc). **33 Alamy Stock Photo:** Gaertner (cb). **34 Alamy Stock Photo:** Brad Mitchell (tr). **Getty Images:** Wild Horizon / Contributor (b). **35 Getty Images:** DenisTangneyJr (br). **Imagelibrary India Pvt Ltd:** Josh Baker (tc). **36 Getty Images:** Carsten Peter / Speleoresearch & Films / National Geographic (bl). **37 Alamy Stock Photo:** Leonardo Díaz Romero / age fotostock (br). **Getty Images:** Manfred Gottschalk

(cl). **38-39 Steve Morgan:** (t). **38 Getty Images:** Jason Edwards / National Geographic (br); Patrick Robert / Corbis Premium Historical (clb). **39 Dorling Kindersley:** Jerry Young (cb). **40-41 Alamy Stock Photo:** NASA / Dembinsky Photo Associates (t). **40 Alamy Stock Photo:** Frans Lanting Studio (br). **Getty Images:** DEA / M. Santini / De Agostini (bl). **National Geographic Creative:** Design Pics Inc (clb). **41 Jason Hollinger:** (cb). **42 Alamy Stock Photo:** Marion Bull (crb). **David P. Reilander:** (cr). **43 Carl Battreall / photographalaska.com:** (c). **Imagelibrary India Pvt Ltd:** Lee Petersen (cla). **44 Getty Images:** Daniel A. Leifheit (tr). **45 123RF. com:** Galyna Andrushko (br). **Alamy Stock Photo:** John Schwieder (tr); Zoonar GmbH (crb). **46-47 Getty Images:** John Hyde. **46 Getty Images:** Sergey Gorshkov / Minden Pictures (clb). **47 Larry McCloskey:** (cb). **48-49 Getty Images:** Kevin Smith / Design Pics. **48 Alamy Stock Photo:** Fred Lord (clb). **49 Ardea:** Steffen & Alexandra Sailer (tc). **Getty Images:** Emmanuel Coupe / Photographer's Choice (br). **50 Imagelibrary India Pvt Ltd:** Jeff Moreau (br). **50-51 Dave Sandford:** (t). **51 Getty Images:** David Doubilet / National Geographic (crb); Rolf Hicker / All Canada Photos (bl). **52 Dorling Kindersley:** Neil Fletcher (br). **Imagelibrary India Pvt Ltd:** Jin Kim (t). **53 Alamy Stock Photo:** NASA / Landsat / Phil Degginger (tl). **Getty Images:** Cameron Davidson / Photographer's Choice (tr); Joel Sartore / National Geographic (crb). **54-55 Getty Images:** Paul Rojas / Moment Select. **54 123RF.com:** Tom Tietz (c). **56 Alamy Stock Photo:** Inge Johnsson (cb). **57 Alamy Stock Photo:** B.A.E. Inc. (bc); John Barger (tr). **Getty Images:** Pete Mcbride / National Geographic (ca). **58 Getty Images:** Joel Sartore / National Geographic (cla). **Imagelibrary India Pvt Ltd:** Cynthia Spence (tr). **SuperStock:** Keith Kapple (br). **59 Getty Images:** David Sieren / Visuals Unlimited (bc). **Diane Kirkland / dianekirklandphoto.com:** (t). **60 Getty Images:** Jupiterimages / Photolibrary (c). **iStockphoto.com:** Donyanedomam (clb). **Robert Harding Picture Library:** David Fleetham / Okapia (cra). **61 Courtesy of National Park Service, Lewis and Clark National Historic Trail:** G. Gardner. **62 Getty Images:** Nick Boren Photography / Moment (tr). **SeaPics.com:** Phillip Colla (bc). **63 Getty Images:** James P. Blair / Contributor / National Geographic (cb). **Imagelibrary India Pvt Ltd:** Clemens Ruehl (t). **64-65 Khanh Ngo, landscape photographer in Atlantic, Canada:** (b). **65 Getty Images:** Dale Wilson / Photographer's Choice (c). **Imagelibrary India Pvt Ltd:** Arun Sundar (br). **66 Getty Images:** Thomas Kitchin & Victoria Hurst (cr). **Imagelibrary India Pvt Ltd:** 500px / Jakub Sisak (bl). **67 Alamy Stock Photo:** All Canada Photos (bc); Spring Images (bl). **Getty Images:** Konrad Wothe (t). **68 Alamy Stock Photo:** age fotostock (bc); Tierfotoagentur (c). **68-69 National Geographic Creative:** Micheal Nichols. **70 Dreamstime.com:** Betty4240 (bc). **Getty**

**Images:** Danita Delimont (tl). **Stacey Putman Photography:** (br). **71 iStockphoto.com:** Gary R Benson (t). **William Neill Photography:** (cb). **72 Alamy Stock Photo:** FLPA (br). **Getty Images:** Steven Kazlowski / Science Faction (t). **73 Alamy Stock Photo:** Tom Bean (br). **Getty Images:** Antonyspencer / E+ (cr). **74 Getty Images:** Witold Skrypczak / Lonely Planet Images (cb). **iStockphoto.com:** Avatar Knowmad (bl). **74-75 iStockphoto.com:** Gary Kavanagh. **75 FLPA:** Mark Newman (bc). **Getty Images:** Andrew Kennelly / Moment Open (crb). **76 Getty Images:** Danita Delimont / Gallo Images (bl). **Imagelibrary India Pvt Ltd:** Jinhu Wang (tr). **76-77 Alamy Stock Photo:** Phil Degginger (b). **78 Getty Images:** Jeremy Duguid Photography / Moment (bc). **Katharina Winklbauer / www.winka-photography.de:** (t). **79 Getty Images:** Chris Saulit / Moment (cra). **Kris Walkowski / kriswalkowski.com:** (clb). **80 Imagelibrary India Pvt Ltd:** Michael T. Lim (r). **Alex E. Proimos:** (bc). **81 Dreamstime.com:** Eutoch (b). **FLPA:** Yva Momatiuk &, John Eastcott / Minden Pictures (cra). **82-83 Getty Images:** Mint Images - Frans Lanting. **86 Getty Images:** Apomares (tl). **86-87 Alamy Stock Photo:** Pulsar Images (t). **87 Getty Images:** EyeEm / André Reis (tc); Mint Images / Frans Lanting (cra). **iStockphoto.com:** Pxhidalgo (cla). **88 Getty Images:** traumlichtfabrik / Moment (c). **Rex Shutterstock:** WestEnd61 / REX (t). **89 Getty Images:** Andoni Canela / age fotostock (br). **Martin Rietze:** (tr). **90-91 AirPano images. 91 Ardea:** Adrian Warren (clb). **naturepl.com:** Luiz Claudio Marigo (c). **92 123RF.com:** Eric Isselee (br). **92-93 Ricardo La Piettra. 93 Alamy Stock Photo:** Francisco Negroni / Biosphoto (bl). **94 Alamy Stock Photo:** Kseniya Ragozina (tr). **95 Alamy Stock Photo:** Hemis (cra). **Getty Images:** Mint Images / Art Wolfe (br/flying flamingo ). **iStockphoto.com:** DC_Colombia (br); Elisalocci (tl). **96-97 Getty Images:** Hans Neleman / Stone. **96 123RF.com:** Jarous (clb). **97 Getty Images:** Richard I'Anson / Lonely Planet Images (br). **Science Photo Library:** Bernhard Edmaier (bl). **98 Thanat Charoenpol:** (ca). **Getty Images:** Marisa L?pez Estivill / Moment Open (bl). **99 Getty Images:** Gina Pricope / Moment (c). **Imagelibrary India Pvt Ltd:** Craig Holden (b). **100 Alamy Stock Photo:** Minden Pictures (br). **Getty Images:** Bobby Haas / National Geographic (cra). **100-101 ESA:** KARI (t). **101 Alamy Stock Photo:** Tino Soriano / National Geographic Creative (bc). **Lucas Cometto / www.lucascometto.com:** (cra). **102-103 Imagelibrary India Pvt Ltd:** Alejandro Ferrand. **103 Getty Images:** Walter Diaz / Stringer / Afp (br); Travel Images / UIG (c). **104 Thomas Marent:** (t). **105 Getty Images:** Veronique Durruty (cb). **Claudio Sieber:** (tl). **106-107 Getty Images:** Mint Images / Frans Lanting. **106 Alamy Stock Photo:** Photoconix (crb). **108 James Contos:** (tr). **FLPA:** Kevin Schafer / Minden Pictures (crb). **109 Alamy Stock Photo:** blickwinkel (cra); Jacques Jangoux (bc). **NASA:** GSFC / JPL, MISR Team (crb). **110 Alamy Stock Photo:** Junior Braz (cra);

Chris Howarth / Argentina (cl); Michele Burgess (clb). **111 Getty Images:** Norberto Duarte / AFP (cb); Javier Larrea (cla). **112 123RF.com:** Ana Vasileva / ABV (br). **112-113 Getty Images:** Minden Pictures / Luciano Candisani (t). **113 Alamy Stock Photo:** imagebroker (cr). **Getty Images:** Barcroft Media / Linde Waidehofer (br). **114 Getty Images:** Tim Rock / Lonely Planet (tc). **Ramon F. Llaneza:** (cb, crb). **115 Getty Images:** David Doubilet (br); Brian J. Skerry (tr). **116-117 Alamy Stock Photo:** Giovanni Monterzino (t). **116 Getty Images:** Holger Leue / Lonely Planet Images (br). **Johnny Jensen:** (c). **117 Alamy Stock Photo:** Stichelbaut Benoit / Hemis (bc). **Dorling Kindersley:** Blackpool Zoo, Lancashire, UK (crb). **118-119 Getty Images:** Mike Lanzetta (t). **118 Getty Images:** Javier Fernández Sánchez (br); Panoramic Images (bc). **119 Alamy Stock Photo:** Ian Watt (br). **naturepl.com:** Luiz Claudio Marigo (cra). **120 Alamy Stock Photo:** Nature Picture Library (tr). **FLPA:** Minden Pictures / Chris van Rijswijk (tc). **121 Ardea:** Nick Gordon (br). **Getty Images:** Hemis.fr / Aurélien Brusini (tr). **naturepl.com:** Luiz Claudio Marigo (cla). **122-123 iStockphoto.com:** NormaZaro (t). **122 Ardea:** Yves Bilat (br). **123 Alamy Stock Photo:** Robert Eastman (cr). **iStockphoto.com:** Stephen Meese (bc). **124-125 Imagelibrary India Pvt Ltd:** Goetze. **124 Laurent Abad:** (br). **ESO:** (clb). **126 Dreamstime.com:** Jorg Hackemann (cra). **Imagelibrary India Pvt Ltd:** Thomas Heinze (b). **127 Alamy Stock Photo:** Kevin Schafer (crb). **Imagelibrary India Pvt Ltd:** Martin Marilungo (tr). **Javier Etcheverry Photography:** (br). **128-129 Getty Images:** Werner Van Steen. **132 Alamy Stock Photo:** David Gowans (cla); Chris Warham (tr). **133 Alamy Stock Photo:** Hemis (tl); Worldspec / NASA (ca); Stefano Politi Markovina (tr). **134 Alamy Stock Photo:** Art Directors & TRIP (bc). **Getty Images:** Frank Krahmer / Corbis Documentary (crb). **Imagelibrary India Pvt Ltd:** George Turner (t). **135 Getty Images:** Picavet / Photodisc (bc). **Imagelibrary India Pvt Ltd:** Marc Brulard (tr); Hans-Peter Deutsch (t). **136 Getty Images:** Mariusz Kluzniak / Moment (l). **137 Imagelibrary India Pvt Ltd:** Denis Roschlau (t). **138-139 Getty Images:** Katarina Stefanovic / Moment Open. **138 Alamy Stock Photo:** Ashley Cooper pics (bc). **Dreamstime.com:** Isselee (c). **140-141 Imagelibrary India Pvt Ltd:** Fernando Famiani. **140 National Geographic Creative:** Robbie Shone (crb). **142 Alamy Stock Photo:** Richard Roscoe / Stocktrek Images, Inc. (bc); Tromp Willem van Urk (cra). **143 Alamy Stock Photo:** Wead (cr). **Getty Images:** Vittoriano Rastelli / Corbis Documentary (tc). **144-145 Sergei Proshchenko / https://500px.com/sergurai:** (b). **144 Dorling Kindersley:** Natural History Museum, London (cb). **Dreamstime.com:** Bierchen (cra). **145 Alamy Stock Photo:** Andrey Vishin (crb). **Getty Images:** Alberto Incrocci / The Image Bank (cra). **146 Getty Images:** Danita Delimont (bc). **146-147 Imagelibrary India Pvt Ltd:** Frank Kaiser (b). **iStockphoto.com:** golfer2015 (t). **148**

**Getty Images:** Olaf Kruger (clb); Kevin Schafer / Photolibrary (r). **149 Getty Images:** Kim Walker / robertharding (ca). **iStockphoto.com:** konstantin32 (b). **150 Getty Images:** Hagenmuller Jean-Francois / Hemis.fr (b). **150-151 Getty Images:** Achim Thomae / Moment Open. **151 Tobias Hunziker:** (b). **152-153 Jason Chambers Photography:** (t). **152 Alamy Stock Photo:** Danita Delimont (bl); Prisma by Dukas Presseagentur GmbH (ca). **154 Marc Lelievre / Paysages Corse Provence:** (b). **155 Getty Images:** Samuel Gachet (crb). **iStockphoto.com:** Marcobarone (ca). **156-157 Mark D. Babbidge:** (t). **156 123RF.com:** Rudmer Zwerver (cb). **157 Alamy Stock Photo:** Stelian Porojnicu (bl). **Science Photo Library:** Planet Observer (bc). **158 Alamy Stock Photo:** Nature Picture Library (br). **Getty Images:** Minden Pictures / NiS / Grzegorz Lesniewski (c). **158-159 © Manfred Bächler / Naturelodge.info:** (t). **159 Imagelibrary India Pvt Ltd:** Dmytro Gilitukha (br). **160-161 Fedor Lashkov:** (t). **161 Dorling Kindersley:** Robert Royse (ca). **Imagelibrary India Pvt Ltd:** Lilinum (cra). **Fedor Lashkov:** (bl). **NASA:** NASA image courtesy Jeff Schmaltz, MODIS Land Rapid Response Team at NASA GSFC (crb). **162-163 Imagelibrary India Pvt Ltd:** Adnan Bubalo. **162 Getty Images:** Copyright Morten Falch Sortland / Moment (crb). **Imagelibrary India Pvt Ltd:** Johannes Hulsch (cb). **164-165 Getty Images:** Mammuth / E+. **165 Alamy Stock Photo:** Siim Sepp (c). **Getty Images:** Chris Hill / National Geographic (br); Peter Langer / Perspectives (clb). **166 Alamy Stock Photo:** Chris Gomersall (crb). **Getty Images:** Jim Richardson / National Geographic (cra); Peter Unger / Lonely Planet Images (bc). **167 Alamy Stock Photo:** John Miller / The National Trust Photolibrary (cra). **iStockphoto.com:** Ian_Redding (ca). **Jean-Baptiste Meunier:** (bl). **169 Alamy Stock Photo:** Skyscan Photolibrary (crb). **Geoff Griffiths:** (bc). **170 Dorling Kindersley:** Ruth Jenkinson / Holts Gems (bc). **Imagelibrary India Pvt Ltd:** Radius Images (br). **Juan Pablo de Miguel:** (t). **171 123RF.com:** Andrei Pop (br). **Getty Images:** Romulic-Stojcic (clb). **172-173 Philip Klinger (philip-klinger.photography)**. **172 Alamy Stock Photo:** Blickwinkel (tr). **Kilian Schönberger:** (bl). **173 123RF.com:** Eric Isselee (cra). **174 123RF.com:** Aleksander Bolbot (b). **Alamy Stock Photo:** imagebroker (cla). **175 123RF.com:** Maxim Tatarinov (cl). **Alamy Stock Photo:** Iryna Rasko (bc). **Getty Images:** Yevgen Timashov (cra). **176-177 NASA:** USGS EROS Data Center Satellite Systems Branch. **180 Hougaard Malan:** (t). **181 123RF.com:** Vladislav Gajic (tr). **Alamy Stock Photo:** Prisma by Dukas Presseagentur GmbH (ca). **Getty Images:** Radius Images (cl); Visions of Our Land / The Image Bank (tl). **182-183 David Kiff**. **182 U.S. Geological Survey:** (bc). **183 Imagelibrary India Pvt Ltd:** Michael Wilhelmi (bl). **National Geographic Creative:** Carsten Peter (br). **184-185 Getty Images:** Nigel Pavitt / AWL Images. **184 Alamy Stock Photo:** John Downer /

Bluegreen Pictures (crb). **Getty Images:** Tom Brakefield / Corbis Documentary (c). **186 Getty Images:** Werner Van Steen (tr). **187 Getty Images:** Christopher Kidd (tl); Westend61 (cra); Pete Turner (br). **188 Alamy Stock Photo:** AfriPics.com (bc). **Getty Images:** Stocktrek Images (cra); 4FR / Vetta (crb). **189 AirPano images:** (t). **Alamy Stock Photo:** FLPA (bl); Ilko SouthAfrica (crb). **190 123RF.com:** Oleg Znamenskiy (br). **190-191 Johann Stritzinger, Wels, Austria:** (t). **191 Alamy Stock Photo:** Aleksandra Kossowska (bl). **Getty Images:** Morgan Trimble (cr). **192-193 Elena Kis:** (t). **192 Getty Images:** Yann Arthus-Bertrand (bc). **Science Photo Library:** Massimo Brega, The Lighthouse (c). **193 Alamy Stock Photo:** Graham Prentice (bc). **Imagelibrary India Pvt Ltd:** David Hobcote (crb). **194-195 Marsel van Oosten**. **196 Alamy Stock Photo:** Universal Images Group North America LLC (bl). **SuperStock:** Roger de la Harpe (tr). **196-197 SuperStock:** age fotostock / Roger de la Harpe (b). **197 Dorling Kindersley:** Peter Janzen (ca). **198 Getty Images:** Zu Sanchez Photography / Moment Open (bl). **198-199 Getty Images:** Georgette Douwma / Stockbyte. **199 Alamy Stock Photo:** Jeff Rotman (br). **200 Alamy Stock Photo:** Ivan Kuzmin (bc). **Dreamstime.com:** Ricardo De Paula Ferreira (clb). **200-201 FLPA:** Imagebroker / Robert Haasmann (t). **Dr Michele Menegon:** (b). **201 Getty Images:** Keren Su (br). **202-203 Getty Images:** G & M Therin-Weise. **203 naturepl.com:** Bernard Castelein (tc); Peter Oxford (tr). **204-205 Getty Images:** BremecR / E+ (t). **204 Alamy Stock Photo:** FLPA (ca). **National Geographic Creative:** George Steinmetz (bl). **205 Alamy Stock Photo:** AfriPics.com (br); GFC Collection (bc). **206-207 Getty Images:** Russell Burden / Photodisc. **206 123RF.com:** StarJumper (c). **Getty Images:** James Hager / robertharding (bl); Ariadne Van Zandbergen / Lonely Planet Images (bc). **208 Getty Images:** Yann Arthus-Bertrand. **209 Getty Images:** George Steinmetz (bc). **NASA:** JPL / NIMA (clb). **210 123RF.com:** Michael Lane (cr). **Getty Images:** Eric Teissedre (tr). **Michal Huniewicz / m1key. me:** (b). **211 Alamy Stock Photo:** AfriPics. com (br); Friedrichsmeier (cl). **212-213 Getty Images:** Mariusz Kluzniak (t). **Katja Schilling:** (b). **213 Alamy Stock Photo:** Martin Harvey (cb). **214-215 AirPano images**. **218 123RF.com:** Anton Starikov (ca). **Alamy Stock Photo:** Dinodia Photos (tr). **Imagelibrary India Pvt Ltd:** 500px / Vadim Balakin (tl). **219 123RF.com:** Raimond Klavinsh (cr). **Science Photo Library:** NASA (b). **220-221 Getty Images:** Mauro Cociglio - Turin - Italy / Moment Open (t). **220 Dreamstime.com:** Isselee (br). **Marek Zgorzelski, Poland:** (bl). **221 Dorling Kindersley:** Colin Keates / Natural History Museum, London (bc). **iStockphoto.com:** Andrew_Mayovskyy (br). **222 Dorling Kindersley:** Wildlife Heritage Foundation, Kent, UK (br). **222-223 Imagelibrary India Pvt Ltd:** Bibi Bielekova. **223 Alamy Stock Photo:** Daniel J. Rao (cb). **224 123RF.com:** Dmytro Pylypenko (cra). **Getty Images:** Feng Wei

Photography / Moment Open (br). **Imagelibrary India Pvt Ltd:** George Balyasov (c). **225 Syed Sadaqat Ali:** (l). **Getty Images:** Colin Monteath / Hedgehog House / Minden Pictures (bc). **226 Alamy Stock Photo:** Vick Fisher (tr); National Geographic Creative (br). **227 Robert Downie:** (tr). **Getty Images:** Barry C. Bishop / Contributor / National Geographic (crb). **228 Getty Images:** MelindaChan / Moment (t); Hou / Moment (crb). **229 AirPano images:** (br). **Dreamstime.com:** Suriyaphoto (cla). **Imagelibrary India Pvt Ltd:** Hidetoshi Kikuchi (tr). **230 Imagelibrary India Pvt Ltd:** EC_Tong (cra); Puripat Lertpunyaroj (b). **231 Alamy Stock Photo:** Josef Beck / Imagebroker (t). **Getty Images:** InterNetwork Media / Photodisc (crb). **232-233 Getty Images:** Colin Monteath / Hedgehog House / Minden Pictures. **232 Ranjit Doroszkiewicz:** (bc). **NASA:** JSC Gateway to Astronaut Photography of Earth (cra). **233 Oleg Bartunov / Sternberg Astronomical Instite, Moscow, Russia:** (cb). **234-235 Getty Images:** Bardon, Andrew / National Geographic. **234 Getty Images:** Feng Wei Photography / Moment (bl). **235 Alamy Stock Photo:** Iuliia Kryzhevska (cra). **Getty Images:** Lowe, Max / National Geographic (crb). **236 Alamy Stock Photo:** Albatross / Duby Tal (t). **Dreamstime.com:** Vvoevale (br). **237 123RF.com:** Victoria Ivanova (b); Vladimir Melnikov (cra). **238 AirPano images:** (cra). **Alamy Stock Photo:** Remo Savisaar (cb). **239 AirPano images:** (cb). **U.S. Geological Survey:** National Center for Earth Resources Observation and Science (EROS) and the National Aeronautics and Space Administration (NASA) (bl). **240 Imagelibrary India Pvt Ltd:** Issy3 (crb); Ajeet Maurya (cra). **240-241 Alamy Stock Photo:** B.A.E. Inc. (t). **241 123RF.com:** Thawat Tanhai (br). **Getty Images:** Md. Akhlas Uddin (clb). **242 Alamy Stock Photo:** Chris Mattison (c). **Getty Images:** Chen Hanquan (cra); Suttipong Sutiratanachai (br). **242-243 Dreamstime.com:** Qin0377 (t). **243 Getty Images:** Istvan Kadar Photography (crb). **Courtesy of Zeb Hogan:** (br). **244-245 Samuel Lainez. 245 Alamy Stock Photo:** Aurora Photos (bc). **Getty Images:** Carsten Peter (crb). **246 Getty Images:** John Roberts (br). **246-247 Karl Willson:** (t). **247 Getty Images:** National Geographic / Carsten Peter (bl). **248 naturepl.com:** Dong Lei (tr). **249 123RF.com:** millions27 (tl). **Alamy Stock Photo:** Geir Olaf Gjerden (bc). **Imagelibrary India Pvt Ltd:** Yulong (cra). **250-251 Getty Images:** SimonDannhauer / iStock / Getty Images Plus (b). **250 Getty Images:** Ly Hoang Long Photography / Moment Open (tr); Jed Weingarten / National Geographic My Shot (ca). **251 Getty Images:** The Asahi Shimbun Premium / Contributor (ca); VCG / Contributor (br). **252 Getty Images:** Pete Atkinson / Photographer's Choice (bl). **252-253 Imagelibrary India Pvt Ltd:** Alex Saluk. **253 Alamy Stock Photo:** Jose B. Ruiz / Nature Picture Library (br); WaterFrame (cra). **254-255 naturepl.com:** Bryan and Cherry Alexander. **255 FLPA:** (cra). **Getty Images:** Elena Liseykina (cb).

256 123RF.com: Yokokenchan (bc). **Alamy Stock Photo:** Ger Bosma (c). **Getty Images:** Feng Wei (cr). **257** 123RF.com: Okjyt (ca). **FLPA:** Minden Pictures / Mitsuaki Iwago (b). **258-259** Imagelibrary India Pvt Ltd: Kim Briers (t). **259** Getty Images: Peter Langer / Design Pics (bl). **naturepl.com:** Tim Laman (bc). **260-261** Getty Images: Jacek Kadaj / Moment Open (t). **260** Getty Images: Tuul and Bruno Morandi / The Image Bank (bc). **iStockphoto.com:** Roongzaa (ca). **261** 123RF.com: Iakov Filimonov (bc). **Imagelibrary India Pvt Ltd:** Robertharding (br). **262-263** Getty Images: Buena Vista Images / DigitalVision (b). **262** 123RF.com: Robhillphoto (br). **263** Getty Images: George Steinmetz / Corbis Documentary (br); Peter Unger / Lonely Planet Images (clb). **264** Rex Shutterstock: Amos Chapple / REX (cra). **264-265** Alamy Stock Photo: Liu Kuanxin / TAO Images Limited. **265** Dreamstime.com: Joan Egert (tr). **Priyankar K. Datta / www. priyankarkdatta.space:** (cra). **266-267** Getty Images: Timothy Allen / Photonica World. **266** Dreamstime.com: Aleksandr Frolov (crb); Matthijs Kuijpers (c). **268-269** Peter Elfes Photography. **272** Alamy Stock Photo: Redbrickstock.com (ca). **Getty Images:** Australian Scenics (tl); Ted Mead (tr). **273** Alamy Stock Photo: Ingo Oeland (cla). **Serge Horta:** (tr). **274-275** Gary P Hayes Photography. **274** Alamy Stock Photo: Christian Kapteyn (clb). **275** Getty Images: DEA / C. Dani I. Jeske / Contributor / De Agostini (br). **276** Alamy Stock Photo: Zoonar GmbH (cb). **Getty Images:** Frank Krahmer (ca). **277** Imagelibrary India Pvt Ltd: Jeremy Larrumbe (tc). **iStockphoto.com:** rusm (cb). **278-279** Adrian Liedtke / www. facebook.com/Captured.AdrianLiedtke. **278** Dreamstime.com: Joanne Harris (clb). **Getty Images:** Artie Photography (Artie Ng) / Moment (bl). **National Geographic Creative:** Ned Norton / Hedgehog House / Mind (crb). **279** 123RF.com: Eric Isselee (br). **280** Getty Images: Laurenepbath / Room (br). **Tim Hayes / www. thetravelyear.com:** (bl). **280-281** Getty Images: Colin Monteath / Hedgehog House / Minden Pictures. **281** Getty Images: Southern Lightscapes-Australia / Moment (bl). **282-283** Grant Schwartzkopff: (t). **282** Alamy Stock Photo: Juniors Bildarchiv / F259 (br). **283** Discover Waitomo: (br). **Getty Images:** Auscape (bl); EyeEm / Amrish Aroonda Manikoth (cra). **284-285** Jennifer Watson www.500px.com/jensphotos1: (t). **284** AirPano images: (br). **NASA:** (bc). **285** Alamy Stock Photo: Reinhard Dirscherl (c). **286** Alamy Stock Photo: Global Warming Images (bc). **Getty Images:** Oliver Lucanus / NiS / Minden Pictures (cra). **287** Getty Images: Auscape / UIG (cra); Len Zell / Lonely Planet Images (crb). **288-289** Matt Hutton. **288** Getty Images: Hiroya Minakuchi / Minden Pictures (c). **iStockphoto.com:** Totajla (clb). **290** AirPano images: (cra). **Pascal Kross:** (crb). **291** Imagelibrary India Pvt Ltd: Alok Gohil (b). **292-293** Marc Schmittbuhl www.marcsphoto.com: (t). **292** Alamy

Stock Photo: imagebroker (ca); Stephanie Jackson (br). **Fotolia:** Eric Isselée (bc). **293** Alamy Stock Photo: National Geographic Creative (tr). **Ardea:** Pat Morris (br). **Robert Menard:** (bl). **294** Getty Images: Auscape (tr); Minden Pictures / Ingo Arndt (b). **295** Alamy Stock Photo: Auscape International Pty Ltd (t). **Getty Images:** Robert McRobbie (bl); Ted Mead (br). **296** Getty Images: Minden Pictures / BIA / Simon Bennett (cra). **296-297** Getty Images: Ignacio Palacios (b). **297** Alamy Stock Photo: Ingo Oeland (cla). **Getty Images:** UIG / MyLoupe (ca). **298** Getty Images: Artie Photography (Artie Ng) (br). **Steve Strike:** (ca). **299** Ardea: Auscape (c). **Peter MacDonald / www. thesentimentalbloke.com:** (b). **300-301** NASA: NASA Earth Observatory image created by Jesse Allen and Robert Simmon, using EO-1 ALI data provided courtesy of the NASA EO-1 team. **304-305** Getty Images: George Steinmetz / Corbis Documentary. **304** Alamy Stock Photo: Minden Pictures (br). **NASA:** Michael Studinger (bl). **305** National Geographic Creative: Carsten Peter (bl). **306** Alamy Stock Photo: LWM / NASA / LANDSAT (clb). **Science Photo Library:** Maria-Jose Vinas, NASA (bc). **307** Alamy Stock Photo: Danita Delimont (cr); Kim Westerskov (cla); Tui De Roy / Minden Pictures (crb). **308-309** Getty Images: Sue Flood / Oxford Scientific. **308** Getty Images: Frank Krahmer / Photographer's Choice RF (bl). **309** Alamy Stock Photo: Colin Harris / Era-Images (bc). **Dr Roger S. Key:** (cra). **Getty Images:** NASA - digital version copyright Science Faction (bl). **Courtesy of the National Science Foundation:** Elizabeth Mockbee (crb). **310-311** NASA: Norman Kuring, NASA's Ocean Color Web. **312** NOAA: IFE, URI-IAO, UW, Lost City Science Party; NOAA / OAR / OER; The Lost City 2005 Expedition (cra). **Science Photo Library:** B. Murton / Southampton Oceanography Centre (cb). **313** Getty Images: Alex Mustard / Nature Picture Library (cra); Planet Observer (cr). **314-315** Getty Images: Rene van Bakel / Contributor / ASAblanca (b); Wil Meinderts / Buiten-beeld / Minden Pictures (t). **314** 123RF.com: Thomas Lenne (bc). **315** Alamy Stock Photo: Doug Perrine / Nature Picture Library (br). **Alasdair Harris:** (c). **316-317** naturepl.com: Doug Perrine. **316** Alamy Stock Photo: Tami Kauakea Winston / Photo Resource Hawaii (cb). **318** Getty Images: jimkruger (cb). **319** 123RF.com: Nikki Gensert (tc). **Alamy Stock Photo:** RGB Ventures / SuperStock (cra). **NOAA:** OAR / National Undersea Research Program (NURP); Univ. of Hawaii - Manoa (crb). **320** FLPA: David Hosking (cra). **320-321** Valeria Jaramillo Ramón / 500px.com/ valeriajaramillo. **321** Alamy Stock Photo: Michael Nolan / robertharding (ca). **Getty Images:** Mint Images - Frans Lanting (cla). **322** NOAA: Okeanos Explorer Program, Galapagos Rift Expedition 2011 (cra). **323** Ocean Networks Canada: (tl, tr). **Science Photo Library:** Dr Ken Macdonald (crb). **324** Alamy Stock Photo: Science History Images (tr). **325** 123RF.com: Jonghyun Kim

(cr). **NOAA:** NOAA Office of Ocean Exploration and Research, 2016 Deepwater Exploration of the Marianas. (bc). **326-327** NASA: Earth Science and Remote Sensing Unit, NASA JSC. **328** Alamy Stock Photo: World History Archive (cr). **Getty Images:** Mike Theiss (bc). **329** Getty Images: AFP / Stringer (crb). **NASA:** ESA / Terry Virts (ca). **330** ESA: NASA (cra). **331** Alamy Stock Photo: Cultura Creative (RF) (cr). **Getty Images:** Scott J McPartland - Footage (cra). **332-333** Brittany Dawson: (t). **333** Alamy Stock Photo: Luca Pescucci (bc). **Getty Images:** Jim Reed (bl). **334-335** Matthew McIver Photo & Design: (t). **334** Getty Images: James D. Morgan (br). **335** Alamy Stock Photo: CPRESS Photo Limited (bl). **336-337** Imagelibrary India Pvt Ltd: Michael Lanzetta. **336** NASA: Jeff Schmaltz, LANCE / EOSDIS MODIS Rapid Response Team (br). **337** Imagelibrary India Pvt Ltd: Elena Elisseeva (crb). **338-339** Getty Images: Sascha Kilmer / Moment (t). **338** NASA: ESA (br). **339** Getty Images: Babak Tafreshi / National Geographic (bl). **340-341** Alamy Stock Photo: National Geographic Creative. **342-343** Imagelibrary India Pvt Ltd: Stuart Gordon. **343** Getty Images: Hector Guerrero / Stringer / AFP (r). **344-345** Getty Images: Kevin Schafer / Corbis Documentary. **345** Science Photo Library: Omikron (tr). **346** Pierre Tichadou. **347** Alamy Stock Photo: Agenzia Sintesi. **348-349** Imagelibrary India Pvt Ltd: Miguel Angel Martín Campos. **348** Getty Images: Alexandros Maragos / Moment Open (bl). **350-351** Kjeld Friis. **352** Getty Images: AFP / Stringer. **353** Getty Images: Tuul and Bruno Morandi / The Image Bank. **354-355** Getty Images: Feng Wei Photography / Moment. **355** Getty Images: Moodboard / Cultura (tr). **356** Getty Images: Ted Spiegel / Contributor / National Geographic. **357** Francesco Riccardo Iacomino. **358** Getty Images: Piskunov / Vetta. **359** Robert Haasmann / www.roberthaasmann.com. **360-361** Alamy Stock Photo: Maxim Toporskiy. **361** Bernard Spragg: (br). **362** Alamy Stock Photo: Darrel Giesbrecht / All Canada Photos (r). **362-364** Joao Eduardo Figueiredo. **364** Mike Payne: (b). **365** Getty Images: Danita Delimont. **366** Alamy Stock Photo: Tom Till. **367** Getty Images: Westend61. **368-369** National Geographic Creative: Dordo Brnobic / National Geographic My Shot. **369** Imagelibrary India Pvt Ltd: Dirk Vonten (br). **370** Dreamstime.com: Gemini808. **370-371** age-m-fotoart / Martin Herrsche. **372** Imagelibrary India Pvt Ltd: Jean-Michel Deborde. **373** Alamy Stock Photo: Stefan Arendt / Imagebroker. **374-375** Imagelibrary India Pvt Ltd: Paniti Márta. **375** Getty Images: George Steinmetz / Corbis Documentary (tr). **376** iStockphoto. com: Freder. **377** Getty Images: Planet Observer / Universal Images Group. **378** Getty Images: Tim Gerard Barker / Lonely Planet Images. **379** Xsalto. **380** Alamy Stock Photo: Roop Dey (tl). **380-381** iStockphoto.com: Sahachat. **382** Getty Images: Alan Cressler / Moment. **383**

Getty Images: Ignacio Palacios / Lonely Planet Images. **384** Alamy Stock Photo: Zoonar GmbH. **385** NASA: Jesse Allen, using EO-1 ALI data provided courtesy of the NASA EO-1 Team. **386-387** Imagelibrary India Pvt Ltd: Nick Fox. **387** Gerry van Roosmalen / www.OneEyeArt. nl: (br). **388** iStockphoto.com: AlbertoLoyo (tl). **388-389** iStockphoto.com: javarman3. **390** Dreamstime.com: Tanya Puntti. **391** Getty Images: Carlos Rojas. **392** Renato Duarte da Cunha. **393** Radek Dranikowski. **394-395** Alamy Stock Photo: Yuriy Brykaylo. **395** Getty Images: jcarillet (br). **396-397** anilsphotography. **397** iStockphoto.com: AlxeyPnferov (tr). **398** Getty Images: Mark Jones Roving Tortoise Photos / Oxford Scientific. **399** Imagelibrary India Pvt Ltd: Gleb Tarro. **400** Alejandro Ferrand: (bl). **400-401** Getty Images: Cosmo Condina. **402** Alamy Stock Photo: Athol Pictures. **403** Alamy Stock Photo: Planet Observer / Universal Images Group North America LLC

All other images © Dorling Kindersley
For further information see:
www.dkimages.com